Preface

The tenth edition of *Cioffari's Experiments in College Physics* is the solution to the problem of laboratory texts that contain far more experiments than can be done in a one-year course and that include some that are too advanced for many courses. This text is the result of considerable study of an extensive questionnaire that was sent to many people. It consists of a "core volume" containing the experiments that seem to be most popular and a "pool" from which users can select any experiments that appeal to them, including advanced ones, which, while of great significance, may not be appropriate for the more general or elementary courses.

This book contains all the usual basic experiments, but several improvements have been made to make use of newly available equipment. Most notable among these is the use of Pasco's so-called "complete rotational system" (model no. ME-8950), which contains in one kit all that is needed for the three major experiments on rotation—Uniform Circular Motion, Rotational Motion, and Conservation of Angular Momentum. These experiments have been rewritten for the new equipment, although alternate instructions applicable to the apparatus described in earlier editions (which some users may still have) are included. The author cannot help noting that in former editions the Uniform Circular Motion and Conservation of Angular Momentum experiments called for much equipment that had to be made in-house, and some users may be relieved to find that such home-building is no longer necessary. The Pasco spring gun, or "projectile launcher," has also been specified because it provides for the attachment of a pair of light gates spaced exactly 10 cm apart and connected to a timer enabling direct measurement of the launcher's muzzle velocity. This removes the requirement that a projectile-motion experiment be done every time a value of this velocity is needed and thus makes it possible to have a separate and more detailed study of projectile motion, as the muzzle velocity for the ballistic pendulum and conservation of angular momentum experiments is easily determined in each case. In addition, the experiment on electromagnetic induction makes use of the disassemblable transformer provided by Pasco and omits consideration of motors and generators, partly because the Cenco "model generator" called for in earlier editions hasn't been available for years (it was replaced by a homemade version described in the Instructor's Guide) and partly because a new and improved experiment on motors and generators has been developed for inclusion in the "pool." Finally, the three experiments on radioactivity have been condensed to one (the experiment on the decay of a relatively short-lived isotope having been dropped), and an experiment on the photoelectric effect has been added. The latter may seem surprising in view of its omission from earlier editions because of its difficulty, but it has been reinstated in the core volume both by popular request and because new equipment makes obtaining good results possible without the difficulties previously encountered.

The creation of a pool of more advanced experiments allows this author to reintroduce some of his favorites that had been dropped from earlier editions because they were seen as too difficult, too costly, or both. These include the Millikan oil-drop experiment, the Franck-Hertz experiment, and the experiments on electronics that, in themselves, provide an introduction to electronics seldom included in freshman or sophomore texts but nonetheless considered by many to be essential preparation for various professions. Doctors, for example, will probably not have any opportunity to study electronics beyond the year of physics required in any premedical program but will be faced with a wall full of electronics the first time they walk into an operating room. Although the introduction to the cathode ray oscilloscope has been included (again by popular demand) in the core volume, the experiments on rectifiers, power supply circuits, amplification, and operational amplifiers, and a new experiment on digital electronics, now appear in the pool and can be ordered as a set. Finally, the existence of the pool has made possible the inclusion of an experiment on the velocity of light that employs a variation on the classic Michelson approach. This method won an award in the 1971 AAPT apparatus competition, and although it requires a high-speed rotating mirror, it has an advantage in that the actual value of this speed is not needed, Again, this experiment may be chosen alone or in combination with others from the pool for assembly into a supplementary volume designed specifically for a particular course or student.

ACKNOWLEDGMENTS

The tenth edition reflects ongoing feedback from many users of previous editions, whose reviews and evaluations were helpful in making improvements. In particular, and to a large extent, the contents of this edition are based on the results of a very comprehensive questionnaire. I would thus like to thank the following individuals for their valuable contributions:

Barbara E. Arens, Swarthmore College
Mary E. Barrows, Quinsigamond Community College
William DeBuvitz, Middlesex County College
Jeanne Calvert, Lutheran College
Richard Calhoun, Concordia University
Stanley Cloud, University of Nevada—Las Vegas
Leonard Cohen, Drexel University
Rory Coker, The University of Texas at Austin
Dave Cordes, Belleville Area College
James Coughlin, Northwestern Michigan College
Dan Durben, Black Hills State University
John Eries, Rancho Santiago College
Allan Fryant, Jamestown College
James Girard, Montana Technical College
John H. Grant, Jr., Pearl River Community College
M. Goldberg, Westchester Community College
Sam Guccione, Delaware Technical and Community College
Phillip Kelly, Belhaven College
Larry Kirkpatrick, Montana State University
Jim Knowles, North Lake College
Gary Kyle, New Mexico State University
William Lamb, Scott Community College
Sue Lawrence, Thomas Nelson Community College
Millard Lee, Paradise Valley Community College
Arlan W. Matnz, Connecticut College
Peter Mattocks, State University of New York—College at Fredonia
Michael N. Monce, Connecticut College
Steven Morris, Los Angeles Harbor College
R.J. Peterson, University of Colorado—Boulder
David A. Pierce, El Camino College
Steven Ratliff, Northwestern College—Minnesota
Dipankar Roy, Clarkson University
Thomas Smith, Mt. San Antonio College
Arnold D. Walker, Southeastern Oklahoma State University
Douglas B. Wood, North Hennepin Technical College

In addition, Professors George Zimmerman and Lawrence Sulak of the Boston University Physics Department deserve special mention for their continued suggestions and discussions. Thanks are also due to Jinara Reyes, who typed large portions of the manuscript. Erich Burton, the teaching laboratory curator, was invaluable in making equipment available and arranging for its use in testing the validity of several of the proposed experiments. Experiments were also conducted at the University of Western Ontario, where Professors Donald Moorcroft and Patrick Whippey of the Physics Department not only put laboratory space and materials at my disposal but also were always ready to help. In this connection, special credit should also go to Irold Schmidt, the department's instrument maker, who created some of the apparatus described in this text. And with regard to apparatus, I would like to extend most enthusiastic thanks to Paul Stokstad, president of the Pasco Scientific Company, not only for providing much of the equipment used in the development of these experiments but also for permitting extensive work with his staff members Robert Morrison and Mary-Ellen Niedzielski, who were particularly helpful in trying out some of my ideas and resolving questions I had about Pasco equipment.

Finally, a special thank-you goes to my wife, Wendy Edmonds, who proofread and critiqued much of the manuscript and made many excellent suggestions for its improvement, in addition to putting up with me while I suffered the usual traumas of book writing.

Dean S. Edmonds, Jr.

Introduction

TO THE STUDENT

The purpose of the physics laboratory is to supply the practical knowledge necessary for a well-rounded understanding of physics and the physicist's way of looking at the universe. A further aim is to develop familiarity with the experimental method of scientific investigation and to give you experience in the actual handling of laboratory apparatus. It is one thing to study a certain model of some physical phenomenon and deduce that certain results should be observed. It is quite another to set up an experiment in which these observations can be made and thus produce data on the basis of which the model's validity may be tested. In particular, obtaining experimental results depends on your ability to make accurate measurements of physical quantities in the real world. A major purpose of every experiment in this book is to provide practice in doing so. To this end, although modern equipment is provided, a lot of automatic data taking, computer interfacing, etc., has been left out so that instead of just pushing buttons, you must develop some dexterity in taking data and develop a feeling for just how precise you can be. Everyone, in every human endeavor, must walk before they can run, and an intimate sense of how the apparatus works will stand you in good stead even when automation later takes over much of the more boring and repetitive aspects of data taking.

INSTRUCTIONS

The instructions for each experiment include some basic theory on the phenomenon to be investigated and a description of the procedure to be used. You should study these carefully before coming to the laboratory to avoid waste of valuable laboratory time figuring out what should be done. You will be told well in advance which experiments are to be performed and the date for which each is scheduled so that there will be time for proper preparation. The necessary equipment will be laid out at each assigned place in the laboratory. Missing or defective apparatus should be reported to the laboratory instructor immediately. The instructor should also be consulted if you have any questions about the experiment.

Record all observations and data in the blank tables provided for this purpose in each experiment. Columns in these tables are already suitably labeled, but be careful to note the units in which each of the observed quantities is measured. Instruments should be read to the limit of their possibilities by estimating the last figure of the reading, that is, the fraction of the smallest scale division. Record each measurement directly on the data sheet exactly in the form in which it is made without any mental calculation. Do not use "scratch" data sheets from which data are to be transcribed onto the blank ones provided in this book. Very neat data sheets can be made out this way, but mistakes can also creep in. The instructor is interested in an *original* data sheet and is willing to put up with a certain amount of sloppy penmanship in order to see the direct recording of the actual data taken in the laboratory.

Whenever feasible, make calculations in the spaces provided in the manual, which have been made roughly proportional to the expected length of the calculation. If the required computations are too long, complete them on a separate sheet of paper, and include it in the report. Each set of calculations should be headed by the pertinent equation so that anyone reading the report can see what mathematical operations are being performed and why. An electronic calculator of your own may and indeed should be used, but be careful. Erroneous entries produce erroneous results, and all numbers should be looked at to be sure no careless mistake has crept in. A quick order-of-magnitude check by hand is sometimes useful, but bear in mind that practice in arithmetic is *not* an important goal of the physics laboratory.

The questions at the end of each experiment are to be answered in the spaces provided for this purpose. Use proper English so that communication between you and a reader will not be impeded. Complete all calculations and as much of the remainder of the report as possible during the laboratory period. However, unfinished reports may be completed outside the laboratory and should be handed in at the time and place specified by the instructor.

CARE OF APPARATUS

The apparatus provided with each experiment has been set up to work properly in the arrangement described for that experiment and is in some cases very delicate. Use extreme care in handling it. The instructions for each experiment include a list of the required equipment, and you should check this list against the items on the work-bench to make sure everything necessary is there and in good condition. Anything missing or broken should be reported to the instructor. At the end of the period, check the apparatus again and leave it neatly arranged.

Whenever an experimental setup has been assembled, it should be checked before being placed in operation so that any mistakes that might keep it from working properly or that might cause actual damage can be found and corrected. In particular, electrical circuits should be examined carefully for proper wiring. Application of power to a circuit containing wiring errors can cause serious damage. The source of power (battery or power supply) should always be connected last, and the circuit should be checked and approved by the instructor before this final connection is made. Special care should be exercised in setting meters to the proper range, as these items are expensive and easily destroyed if excessive current is allowed to pass through the movement. Whenever the range of a meter or any wiring in a circuit is to be changed, the source of power should always be disconnected first to eliminate the possibility of electrical shock or damage due to a temporary wrong connection.

THE REPORT

You must prepare a report of the work done in each experiment and hand it in at the beginning of the next laboratory period or at some other time designated by the instructor. The report will be graded and returned as soon as possible, after which it may be kept in a folder or binder for future reference. The report should include:

1. *A title page.* This should carry your name, the date, and the name and number of the experiment.
2. *The instruction sheets.* These are the pages describing the purpose of the experiment, the theory, the apparatus, and the procedure. Perforations allow these pages to be easily torn out of this book for inclusion in the report.
3. *All original data and observations.* As already noted, these are entered in the blank data tables provided for each experiment. The data table sheets are also perforated so that they can be easily removed from the book.
4. *All the required calculations.* Make these in the spaces provided. The calculation sheets are then detached along their perforations for inclusion in the report. If extra calculation pages are used, include them in the report in the proper order.
5. *Graphs and diagrams, whenever they are required.* Graph paper pages are provided as needed in this manual and are also perforated so that they can be easily detached and inserted in the report.
6. *A summary and discussion of the results.* The summary is included in tabular form under the data. It usually involves a comparison of the computed results with the accepted values together with the percent errors involved. You are encouraged to add a brief discussion of the sources of these errors and any other comments you would like to make about the working of the experiment.
7. *Answers to the questions at the end of the experiment.* The answers are written in the spaces provided after each question. The question sheets are then torn out along the perforations and added to the report. Take care to use complete sentences and to make the answers as clear and readable as possible. Use extra sheets if needed and then include them in the proper order as in the case of the calculations.

UNITS

If the properties of our physical world are to be investigated quantitatively, units must be introduced in terms of which the quantities we wish to measure can be stated. We are all familiar with units of length such as feet, inches, and centimeters, and units of time such as hours and minutes, but because physics is a precise science, we have to look at such things more carefully and be very precise in our definitions.

Although the choice of units is quite arbitrary, two paramount considerations must be observed when such choices are made: (a) The chosen unit must be of a size that is convenient to use for the proposed measurements, and (b) everyone must agree on its definition. This latter requirement is accomplished by international agreement. The most recent conference for this purpose (the fourteenth) was held in 1971, and the so-called fundamental units of length, mass, time, and electric current were there agreed upon. In the case of length, the meter was established as the basic unit and defined in terms of the wavelength of the light emitted by the krypton 86 atom when undergoing a transition between a particular pair of its allowed levels of energy. *Note that such an agreement involves a definition that is accessible to everyone—krypton atoms are all identical, and anyone anywhere in the world can get some krypton and follow the prescribed procedure for determining the specified wavelength. An identical situation arose in the case of the second, the basic unit of time. The standard second has been defined in terms of the period of a certain frequency ob-

---

*Because of some variations subsequently discovered in the measurement of the krypton 86 wavelength, the meter was redefined in 1983 in terms of the velocity of light. The most accurate measurement then known for this quantity, 2.9979249×10^8 m/s, was *defined* as its *exact* value, and this, together with the definition of the second, now serves as the definition of the meter.

served in the cesium atom. The basic unit of mass, on the other hand, is the kilogram, which everybody has agreed is the mass of a platinum-iridium cylinder kept in the vaults of the International Bureau of Weights and Measures near Paris. These units are called *fundamental*, not because physics gives them some fundamentally special status but because we can express all other units in terms of them. Thus, velocity is measured in meters per second (m/s), acceleration in meters per second per second (m/s^2), and force in kilogram-meters per second per second ($kg-m/s^2$). In the case of a unit like force, which is an unwieldy combination of fundamental units, a special name is given to represent that combination. It has become customary to honor famous scientists by using their names for this purpose; thus our unit of force is called the newton (after Isaac Newton) and is defined as a kilogram-meter per second per second.

The meter, the kilogram, and the second suffice to give us all the units we need for measurements in mechanics. But electricity introduces a new physical quantity, electric charge, for which we need an additional fundamental unit. Actually, because a given amount of charge is hard to determine precisely, the International Conference agreed instead on the size of the ampere, the unit of electric current. Clearly, since by definition of current an ampere is a unit of charge going by in a second, and since a second has already been defined, a precise definition of the unit of electric charge is immediately obtainable. It is called the coulomb after the French physicist Charles Augustin Coulomb, just as the name of the unit of current honors André Marie Ampere.

Although the meter, the kilogram, and the second are convenient in size for measurement of the lengths, masses, and times encountered in everyday life, both smaller and larger units are needed in many areas of physics. Accordingly, a system of prefixes indicating multiples by powers of ten is used to express such quantities. These prefixes are listed in Table III at the end of the book. Some common examples are the centimeter

(10^{-2} m), the millimeter (10^{-3} m), the kilometer (10^3 m); and for time, the millisecond (10^{-3} s), the microsecond (10^{-6} s) and, in this day of electronic circuitry, the nanosecond (10^{-9} s) and the picosecond (10^{-12} s). Note that the standard unit of mass is the kilogram, not the gram, although this presents no particular problem, a standard gram being a mass of 10^{-3} kg. Conversion to other units in common use but not in the agreed-upon system, such as the foot, the mile, the minute, and the hour are given in Table II at the end of the book, which points out that in the so-called English system the unit of length is now defined by making the inch *exactly* 2.54 cm.

The system agreed upon at the 1971 International Conference and described earlier is called the SI system, an abbreviation for the full French title "Système Internationale d'Unités." Because the meter, the kilogram, the second, and the ampere (and hence the coulomb) are its fundamental units, the SI system is essentially identical with what was previously called the MKS system (for meter-kilogram-second). The only difference, in fact, lies in the precise definitions of the meter and second described above. These definitions superseded earlier, less precise ones, but the number of wavelengths of the krypton light making up a meter and the number of cycles of the cesium frequency making up a second were chosen to agree with the earlier definitions of these units within the precision already established. Thus, existing secondary standards did not have to be altered following the 1971 conference. For our purposes, the terms MKS system and SI system are synonymous, but we shall use the presently preferred designation SI. We will also follow current practice in using the SI system in this book. There will be some exceptions, however. For example, meter sticks, vernier calipers, and micrometer calipers are calibrated in centimeters, and recording the reading of such instruments in centimeters should certainly be permissible. But conversion to meters will usually be advisable, if not essential, when the data are used in calculations.

MASS AND WEIGHT

One other difficulty that needs to be overcome right from the start is the confusion between mass and weight. *It is imperative to remember that weight is a force, namely, for our purposes in a laboratory on Earth's surface, the force with which objects are attracted to Earth by gravity.* Mass, on the other hand, is an intrinsic property of all objects. Since the force of gravity goes inversely with the square of the distance separating the attracting objects, an object can be made weightless by taking it far away from Earth and all other heavenly bodies, but it never becomes massless. Confusion arises because an object's mass is apparently responsible for its gravitational attraction to other objects, and in fact the gravitational force is proportional to the masses in-

volved. Thus, a mass m will be attracted to Earth with a force W (its weight) that is proportional to m. As the proportionality constant is a known number (the acceleration of gravity g), a measurement of the weight of an object also determines its mass. As a result, scales and balances, which actually measure weight, are often calibrated in grams or kilograms, which are units of mass. Thus, it is all too easy to say that a certain object "weighs so many grams." Great care must be taken not to do this but to say rather that the object's mass is so many grams, even though the instrument making the measurement is actually measuring the force of gravity on the object. The distinction between mass and weight will be carefully observed throughout this book, although sometimes a seeming contradiction may appear. Thus, it is common parlance to speak of "a set of weights," meaning objects of calibrated mass to be used in balances to produce known forces of gravity. We shall adopt such usual terms rather than the somewhat forced "set of calibrated masses," but the distinction between mass and weight must be kept in mind at all times, even when confusing situations such as those just mentioned arise.

PLOTTING OF CURVES

Graphs are of particular importance in physics because they display the relationship between pairs of interdependent quantities in a readily visualized form. Thus, if two quantities x and y have the linear relation

$$y = ax + b \qquad (1)$$

a graph of y against x will be a straight line whose slope is a and whose y intercept (the value of y at which the line crosses the y axis, that is, at $x = 0$) is b. Conversely, if a given theory predicts that a certain physical quantity y depends linearly on another physical quantity x, this conclusion can be tested experimentally by measuring corresponding values of x and y and plotting these results. The plotted points will readily show whether a straight line can be drawn through them, even if various errors cause them to have a "scatter" instead of all lying right on a line. If there is a scatter, the straight line that represents the best average should be drawn as shown in Fig. 1. There are numerous rules for obtaining the line that is a true "best fit" to a given set of experimental points, but a simple determination by visual inspection using a transparent (plastic) straightedge is usually good enough and is all that will be required in these experiments. If the scatter of points is so large that a good decision as to where to draw a straight line through them cannot be made, the conclusion that y depends linearly on x should be seriously questioned. If, on the other hand, the points readily define a straight line, not only does the resulting graph supply evidence that x and y are indeed linearly related, but the value of a in Equation 1 may be obtained by finding the line's slope. Note that

this is a convenient method of getting the average value of a. Moreover, a value for b representing an average result of all the plotted data can be read directly off the graph. This simple procedure is equivalent to the much more tedious one of determining the best values of a and b by a "best fit" calculation.

A linear relation between x and y is easily recognized when the points are plotted, but other relationships are not so obvious. Thus, suppose the relationship to be investigated were

$$y = ax^2 + b \qquad (2)$$

Plotting y against x would give a curve, but it would be very difficult to distinguish this curve from the curve resulting from, say, $y = ax^3 + b$. In fact, the straight line is the only graph that is really obvious. However, if the validity of Equation 2 is to be tested graphically, a new variable $u \equiv x^2$ can be introduced so that Equation 2 becomes

$$y = au + b \qquad (3)$$

Then, if Equation 2 is valid, a plot of y against u (that is, against x^2) will yield a readily recognizable straight line. This procedure may be used in many cases where a new variable u may be substituted for a function of x to produce a linear relation such as Equation 3.

A special case of this procedure arises when y depends exponentially on x, so that

$$y = Ae^{ax} \qquad (4)$$

where e is the base of natural logarithms and is approximately equal to 2.718. This situation occurs often enough to merit special treatment. The first step is to take the natural logarithm of both sides of Equation 4:

$$\ln y = \ln Ae^{ax} = \ln A + \ln e^{ax} = \ln A + ax \qquad (5)$$

Equation 5 is just like Equation 1 except that $\ln y$ rather than y is to be plotted against x. This plot will thus be linear if Equation 4 is valid, and the y intercept (b in Equation 1) will be $\ln A$. Hence A can be found by taking the antilog of the intercept.

Because this situation arises often, special graph paper is printed on which the graduations along the ordinate (y axis) are logarithmically rather than linearly spaced. This means that if a value of y is plotted on the given ordinate scale, the actual position of the point along the y axis will be proportional to the logarithm of y. In other words, plotting a value of y on this special graph paper automatically takes the logarithm, making a

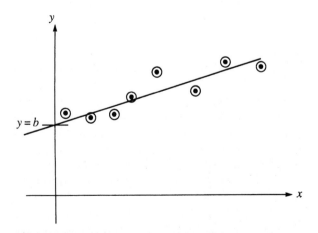

Figure 1   *A "best fit" straight line drawn through a set of experimental points*

separate calculation of $\ln y$ for each value of y unnecessary. However, note that commercial logarithmic graph paper is set up for common (base 10) logarithms rather than natural (base e) logarithms. Taking the common log of both sides of Equation 4 yields

$$\log y = \log Ae^{ax} = \log A + \log e^{ax}$$

$$= \log A + ax \log e$$

$$= \log A + a(0.4343)x \qquad (6)$$

The slope of the resulting straight line is now $0.4343a$ rather than just a, but note that it cannot be read directly from the graph because you have plotted the actual values of y, not their logarithms. The best procedure for determining a is therefore to take a value x_1 of x near one end of the plot, note the corresponding value y_1 of y and substitute in Equation 5 to obtain

$$\ln y_1 = \ln A + ax_1 \qquad (7)$$

Similarly, pick a value x_2 of x near the other end of the plot, and write

$$\ln y_2 = \ln A + ax_2 \qquad (8)$$

where y_2 is the value of y corresponding to $x = x_2$. Subtracting Equation 7 from Equation 8 then yields

$$\ln y_2 - \ln y_1 = a(x_2 - x_1) \qquad (9)$$

Hence,

$$a = \frac{\ln y_2 - \ln y_1}{x_2 - x_1} = \frac{\ln \frac{y_2}{y_1}}{x_2 - x_1} \qquad (10)$$

The special graph paper has not helped us here, and it is indeed necessary to look up the natural logarithm of the ratio $\frac{y_2}{y_1}$, but at least that's better than having to do this for every point measured. If it's easier to obtain the common logarithms of y_1 and y_2, Equation 6 may be used in the foregoing procedure instead of Equation 5 to give

$$a = \frac{\log y_2 - \log y_1}{(0.4343)(x_2 - x_1)} = \frac{\log \frac{y_2}{y_1}}{(0.4343)(x_2 - x_1)} \qquad (11)$$

Note, however, that the value of A may be read off the graph paper directly as the $x = 0$ intercept, for here the fact that the y axis is graduated logarithmically helps you by providing the actual value of y when $x = 0$. Equation 6 shows this to be true, for when $x = 0$, $\log y = \log A$ and therefore $y = A$.

Another popular special case arises when a functional relation of the form

$$y = ax^n \qquad (12)$$

is to be investigated. Although this can be handled by introducing $u \equiv x^n$ as already discussed, such a procedure requires calculating x^n for each value of x. Since n may be any number, positive or negative, such calculations

can get tedious unless a reasonably sophisticated calculator is available. Another very convenient method is again to take the logarithm of both sides of Equation 12. Then,

$$\log y = \log a + n \log x \qquad (13)$$

and a plot of $\log y$ against $\log x$ will produce a straight line with slope n.* You may object that since you must look up the logarithms of all values of both x and y, things haven't been simplified much, but again special graph paper is available that makes this calculation unnecessary. Since $\log y$ is now to be plotted against $\log x$ rather than x, this graph paper has both the ordinate and the abscissa graduated logarithmically. It is therefore called full log or log-log paper, whereas paper with the ordinate graduated logarithmically and the abscissa linearly is called semilog. The logarithmic scales are called 1-cycle, 2-cycle, etc., depending on the number of powers of ten covered on the axis in question. Thus, an axis graduated logarithmically from 1 to 10 is called 1-cycle; from 1 to 100, 2-cycle; etc. Scales of up to 5 cycles are available commercially, and in the case of full log paper there are various standard combinations of numbers of cycles along the ordinate and the abscissa. Appropriate graph paper pages are included in this book as needed; refer to Experiment 31 for an example of semilog paper and to Experiment 41 for the full log type.

In drawing graphs, scales for the coordinate axes should be chosen so that the curve extends over most of the graph sheet and so that decimal parts of units are easily determined. This can be done if each small division is made equal to one, two, five, or ten units. The same scale need not be used for both axes. The independent variable should be plotted along the x axis and the dependent variable along the y axis. Each axis should be labeled with the name of the quantity being plotted and the scale divisions used. The numbers should increase from left to right and from bottom to top. Each graph should have a title indicating what the curve is intended to show.

Each point should be plotted as a dot surrounded by a small circle, which shows where the point is located even if the dot is obscured by the curve drawn through it. A straight line (or smooth curve if a straight-line plot is not being sought) should then be drawn through the dots. The curve need not pass through all the dots but should be drawn so as to fit them as closely as possible, as already mentioned. In general, as many points will lie on one side of the curve as on the other. The extent to which the plotted points coincide with the curve is a measure of the accuracy of the results.

---

*Notice that in this case there is no intercept in the usual sense. $\log y = \log a$ and $y = a$ when $x = 1$. In addition, a procedure similar to that described for the semilog case must be used to find n.

SIGNIFICANT FIGURES

The numbers dealt with in mathematics are exact numbers. When mathematicians write 2, they mean 2.00000 . . . , and all subsequent calculations assume that the 2 means exactly two, not the tiniest fraction more or less. In physics the situation is very different. Many of the numbers dealt with come from measurements of physical quantities, and these can never be exact. For example, suppose that a distance is measured with an ordinary centimeter rule and found to be 5.23 cm. In this measurement, the 3 is an estimate, for the smallest divisions on a centimeter rule are millimeters (tenths of a centimeter). The 3 represents a guess as to where between the 5.2 and 5.3 cm divisions the end of the measured distance lies. The statement that the distance was found to be 5.23 cm does not mean that it is exactly 5.23 cm but merely that it is probably not less than 5.22 cm or more than 5.24 cm. If a high-quality micrometer had been used, the distance might have been found to be 5.2347 cm, where the 7 represents a guess as to where the micrometer's index line fell between the .234 and the .235 divisions. Thus the micrometer yields a much more precise value of the measured length than does the centimeter rule, but it is not exact either. More precise measurement methods might give further decimal places that cannot be determined with the micrometer any more than the 4 and the 7 could be found with the centimeter rule. Thus, when the result of the centimeter rule measurement has been written as 5.23 cm, it doesn't mean that the distance is exactly 5.23 cm or that zeros can be written after the 3. Nothing can be written after the 3 because the instrument being used gives no information as to what to write there.

The 5, the 2, and the 3 in the centimeter rule measurement are called *significant figures* because they each give trustworthy information about the size of the physical quantity being measured. The centimeter rule is quite good enough as a length-measuring instrument to determine that the length in question lies between 5.2 and 5.3 cm, and the 3 in the next place represents a significant guess as to where between 5.2 and 5.3 cm the actual length lies. The centimeter rule measurement is thus *good to three significant figures*, whereas the micrometer measurement gave five significant figures, the micrometer being a much more precise length-measuring instrument than the centimeter rule. The 3 in the centimeter rule measurement and the 7 in the micrometer measurement are less significant than the other figures but are still considered significant because they give some real information about the desired length even though there is some doubt about their actual values. Clearly, however, if there is some doubt about them, any figures that might get written to the right of them in the respective cases would be meaningless. In particular, one should be careful not to write zeros there. If the length measurement made with the centimeter

rule were recorded as 5.230 cm, the zero would be a significant figure and would mean that somehow someone was able to interpolate between the 5.2 and 5.3 divisions to 1/100 rather than just 1/10 of the space between them. Indeed, the micrometer measurement shows that the figure to the right of the 3 should be 4, so that putting a zero there says something that isn't true. Always take care to distinguish between zeros that are significant and those that are not. In general, zeros that merely serve to place the decimal point are not significant. Thus, if the length measurement were to be stated in meters, the two zeros in 0.0523 m would not be significant. They merely place the decimal point appropriately in the three-significant-figure measurement. However, if in measuring the distance with the centimeter rule the end of this distance appeared to fall right opposite the .2 cm division following the 5 cm mark, it would be recorded as 5.20 cm and the zero would be significant. In general, zeros appearing to the right of figures that are already to the right of the decimal point must be regarded as significant, for if they weren't they wouldn't be there. Zeros between other figures and the decimal point should usually be regarded as serving only to place the decimal point. The example of the length of 0.0523 m is typical, there being no doubt that the zeros are not significant. There are some ambiguous cases, however. Suppose that a certain race course is found to be 1.2 km long. As written, this is a two-significant-figure measurement. The same result may be given as 1200 m. Here again the zeros are not significant but must be present in order to locate the decimal point properly. Without the knowledge that the original measurement of 1.2 km contained only two significant figures, however, there is no way to tell whether these zeros are significant or not. In such cases, the experimenter must refer to the measuring instrument to determine how many significant figures are justified.

There is usually no problem in deciding how many significant figures a given measurement should contain, but difficulties arise when these numbers are used in calculations. This is because mathematics assumes that all numbers are exact and thus automatically fills all places to the right of the last significant figure with zeros even though this is physically wrong. The calculations then often produce a great many figures that look as if they were significant but really are not, for clearly no mathematical manipulation can give a result whose precision is greater than that of the quantities put into it. Some examples may serve to show how this problem should be handled.

1. *Addition and subtraction:* When carrying out addition or subtraction by hand, do not carry the result beyond the first column that contains a doubtful figure. This means that all figures lying to the right of the last column in which all figures are significant should be dropped. Thus, in obtaining the sum of these numbers,

806.5		806.5
32.03	they should be written as	32.0
0.0652		0.1
125.0		125.0
		963.6

Note that in dropping nonsignificant figures, the last figure retained should be unchanged if the first figure dropped is less than 5 and should be increased by 1 if the first figure dropped is 5 or greater. This is a normal convention to which this book will adhere.

If an electronic calculator is used, the numbers to be added may be entered without the bother of determining which figures to drop, in which case all figures will appear in the sum. This is like adding the numbers as given on the left in the example above. The result will be 963.5952. You must then look at the data and observe that in two of the numbers being added there is no indication of what the figure in the second decimal place should be. The result must therefore be rounded off to one decimal place by dropping the 952. Since the 9 is equal to or greater than 5, the figure in the first decimal place is raised by one to give 963.6 as before.

2. *Multiplication and division:* The operations of multiplication and division usually produce many more figures than can be justified as significant, so that results must be properly rounded off. The rule is to retain in the result only as many figures as the number of significant figures in the least precise quantity in the data. Suppose the area of a plate is to be measured. A centimeter rule is used to find that the plate has a length of 7.62 cm and a width of 3.81 cm. As in the earlier example with the centimeter rule, these measurements each contain three significant figures, of which the third is doubtful. If the area is now found by multiplying 3.81×7.62 either by hand or with a calculator, 29.0322 cm^2 will be obtained. This number appears to have six significant figures, but the two original quantities have only three each. Therefore, only three significant figures should be retained in the result, which should be written as 29.0 cm^2. With some exceptions, the measurements to be made in the experimental work covered in this book will contain three or four significant figures, so that when an electronic calculator is used, the final result must be rounded off to the number of figures that can be justified by the data as being significant.

THEORY OF ERRORS

All measurements are affected by errors; this means that measurements are always subject to some uncertainty. There are many different types of errors, such as personal, systematic, instrumental, and accidental errors. Personal errors include blunders, such as mistakes in arithmetic, in recording an observation, or in reading scale divisions. Another important kind of personal error is known as personal bias, such as trying to fit the measurements to some preconceived idea, or being prejudiced in favor of the first observation. Systematic errors are characterized by their tendency to be in one direction only, either positive or negative. For example, if a meter stick is slightly worn at one end, and measurements are taken from this end, then a constant error will occur in all these measurements. Instrumental errors are those introduced by slight imperfections in the manufacture or calibration of the instrument. The worn meter stick just mentioned or an electrical meter that has not been properly set to zero with no input are examples of instrumental errors. Note that such errors are usually also systematic. Accidental errors are deviations beyond the control of the observer. These errors are due to jarring, noise, fluctuations in temperature, variations in atmospheric pressure, and the like. Included in this category are variations in observed data due to inherently random processes such as the intersurface actions that produce the force we call friction and the radioactive decay of atomic nuclei. Since the causes just listed for accidental errors are essentially random in nature, all these causes

of data variation are subject to treatment by statistical methods, as will be discussed here.

It will be assumed in these experiments that instrumental errors due to improper calibration, zeroing, etc., have been prevented by proper inspection and adjustment of the equipment and that care has been taken to eliminate systematic errors, personal errors, and personal bias. There remain accidental errors, which make themselves known by causing a spread in the values obtained when a given measurement is repeated several times. Two examples may serve to illustrate how this comes about.

Consider first the distance measurement with the centimeter rule discussed in connection with significant figures. In the measurement of 5.23 cm, the 3 was doubtful, being an interpolation between the 5.2 and 5.3 cm divisions, which are the smallest divisions on the centimeter rule. If a two-significant-figure result were adequate, the distance could have been quoted as 5.2 cm. In this case, if the measurement were repeated many times, even by different experimenters, the likelihood is that 5.2 cm would be obtained each time. No accidental error is revealed because the measuring instrument is not being pushed to the limit of its precision and random processes in the experiment (such as small variations in the length of the rule and/or the distance being measured due to temperature fluctuations) are negligible compared to the smallest scale unit in the measuring instrument (the millimeter divisions on the centimeter rule in this case). The precision here is said to be limited by the scale of the instrument.

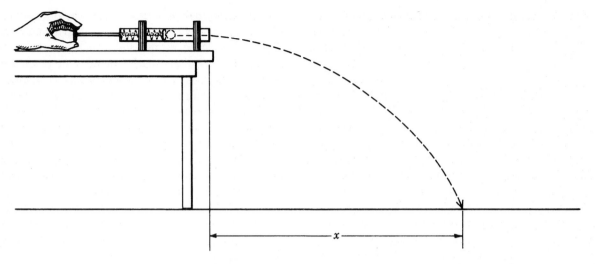

Figure 2 *The range of a spring gun*

However, if the distance measurement is repeated with an estimated interpolation made each time between the 5.2 and 5.3 cm divisions, the same estimate may not always be made. This would be especially true if each measurement were made by a different experimenter who had no knowledge of the others' results. Thus, one might guess 5.22 cm, another 5.21, another 5.24, etc. To handle this situation, a mean or average of the various measured values is calculated. As will be discussed shortly, this average is more accurate than any one of the measurements alone and can in fact be shown to improve in accuracy as the square root of the number of individual measurements made. Clearly, a way of improving the accuracy of experimental data is to measure each quantity many times, and an important matter of judgment in experimental work is to decide on how many times a given quantity is to be measured. In this regard, it must be remembered that to measure something N times takes N times as long as measuring it once, but the accuracy obtained by doing so is only $\sqrt{N}$ times as great. Thus, if a certain measurement takes one minute, making it ten times will take only ten minutes but will yield over three times the accuracy. However, making it a hundred times will take an hour and forty minutes, but this investment in time will only yield another threefold increase in accuracy. Clearly, a compromise based on the accuracy required, the time needed for a particular measurement, and the time available must be reached in each case.

A second example, one dominated by random processes inherent in the experiment itself, is that of the range of a spring-operated gun. The experimental setup is shown in Fig. 2 and will actually be tried in Experiment 5. The spring gun consists of a tube containing a spring-loaded plunger. A small steel ball is placed in the tube against the plunger. The plunger is pulled back a given distance, compressing the spring by a known

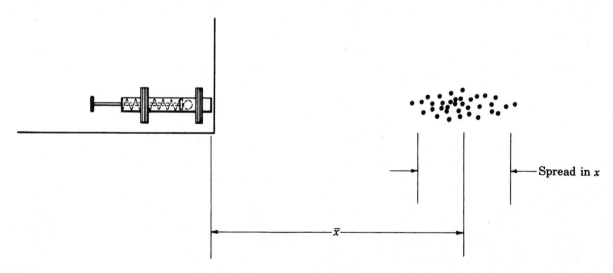

Figure 3 *The range of a spring gun. Plan view*

amount, and is then released sharply, propelling the ball out of the tube. The ball strikes the floor at a horizontal distance x from the end of the tube (the gun's muzzle), this distance being the range in question.

If this experiment is repeated under conditions as identical as possible to those in effect on the first try (the ball is carefully put back in the tube, the plunger is pulled back a distance as close as possible to the distance used the first time, and care is taken to release the plunger in the same way), will the ball strike the floor at exactly the same point? Simple theory predicts that it will, but small variations in the distance the plunger was pulled back, in the state of the spring, and in the condition of the surfaces of the ball and the inside of the tube—all random, uncontrollable effects—will cause the measured range to vary somewhat on subsequent shots. Indeed, no one would really expect successive shots from a gun all to land in precisely the same spot even though the gun was clamped in a fixed position and given the same charge each time. Instead, a spread of impact points would be expected, as shown in the plan view of the spring-gun experiment in Fig. 3. The extent of the spread may be reduced by using great care in the experimental technique (wiping off the ball after each shot, handling it with plastic gloves to prevent getting fingerprints on it, taking care in the measurement of how far the plunger is retracted, and releasing the plunger smartly each time), but the spread can never be reduced to zero. The size of the spread is a measure of the precision of the experiment. An estimate of this precision is very desirable in all experimental work, and the following discussion will show how the extent of the spread can be used to express such an estimate quantitatively.

According to statistical theory, the arithmetic mean or average of a number of observations will give the most probable result. This is clear from the results of the range experiment illustrated in Fig. 3. If a single number is to be quoted as the range of the spring gun, it should be the distance from the gun muzzle to the center of the distribution of impact points. In the absence of any peculiar experimental effects, we expect the distribution of points to be densest near the center, to thin out as we go away from the center, and to be symmetrical (to show as many impact points beyond the center as short of it). Hence, in this normal case, the *average* range $\bar{x}$ is also the *median* (the midpoint of the distribution with as many points with bigger x as with smaller x) and the *most probable* value (the point near which there is the greatest density of points). Thus the first step in data analysis is to find the average of the distances from the gun muzzle to all the individual impact points. This is shown on the left in Table 1. Note that these measurements are made with a meter stick and could therefore be given to one more significant figure by interpolation between the millimeter divisions on the stick. If this order of precision were desired, each value of x would be measured several times by different investigators, each of whom would make an

interpolation, and an average value would be obtained for each x. Then every entry in the left-hand column of Table 1 could be quoted to two decimal places (five significant figures). However, this would take a great deal of time and effort, all of which would be wasted because the spread in the data is several centimeters, making the fifth significant figure in each measurement nonsignificant in the final result. In other words, the random effects in the experiment dominate the picture and limit the usable precision of the measuring instrument.

One obvious way of expressing the extent of the spread in a set of experimental data is to note the deviation of each measurement from the average or arithmetic mean just found. In the example of the spring-gun range experiment, these deviations (differences between each measurement and the average) are tabulated in the middle column of Table 1 and their average is then computed. Note that, in computing this average, no account is taken of the algebraic signs of the deviations. A deviation represents an error—a difference between a particular measurement and the average of all the measurements, this average being the closest available approximation to the true value of the quantity being measured. Which way the deviation lies makes no difference; it is still an error. The average error is a measure of the scatter of the observed values about their average. The average deviation thus found is therefore often called the average error, and for the purposes of the elementary laboratory, it may be taken as the possible error in the mean value. Consequently, the result of the range measurement should be written as 133.9 ± 3.1 cm to show that the true value of the range has a high probability of lying between 130.8 cm ($133.9 - 3.1$ cm) and 137.0 cm ($133.9 + 3.1$ cm). Actually, a statistical analysis shows that if a very large number of range measurements were made, 57.5% of them would lie inside this interval. That is, 57.5% of the impact points would be between 130.8 and 137.0 cm from the gun muzzle.

Statistical theory also presents some other useful ways of stating the accuracy of an experimental result.

Table 1 *The Range of a Spring Gun*

Range, cm	Deviations, cm	Deviations squared, cm^2
134.2	+0.3	0.09
139.5	+5.6	31.36
133.0	−0.9	0.81
136.6	+2.7	7.29
129.4	−4.5	20.25
127.8	−6.1	37.21
130.6	−3.3	10.89
136.5	+2.6	6.76
135.3	+1.4	1.96
131.9	−2.0	4.00
138.1	+4.2	17.64
11⌐1472.9	11⌐33.6	11⌐138.26
$\bar{x} = 133.9$ cm	a.d. = 3.1 cm	12.57
		$\sigma = \sqrt{12.57} = 3.5$ cm

For example, the fact that the average of a set of measurements gets more and more accurate in proportion to the square root of the number of measurements made can be reflected in the stated error by dividing the average error by $\sqrt{N}$, where N is the number of measurements. The result is called the *average deviation of the mean* (A.D.). Thus,

$$\text{A.D.} = \frac{\text{a.d.}}{\sqrt{N}} \tag{14}$$

where a.d. stands for the average deviation *from* the mean, that is, the average error already discussed. The A.D. is a measure of the deviation of the arithmetical mean from the true value and is in this context generally known as the probable error. The significance of the A.D., from probability theory, is that the chances are 50% that the true value of the quantity being observed will lie within $\pm$A.D. of the mean. Thus, in the example of the spring gun, the mean of the measured ranges is 133.9 cm and the average deviation from the mean (the a.d.) is 3.1 cm, which says that on the average the readings differ from the mean (133.9 cm) by ± 3.1 cm. The average deviation *of* the mean (the A.D.) is $3.1\sqrt{11} = 0.9$ cm, which says that the chances that the true value of the range will lie in the interval 133.9 ± 0.9 cm are 50%, while the chances that it will lie outside this interval are also 50%. Because quoting a result as the average of the measurements $\pm$ the A.D. makes it look better than it really is (for a 50-50 chance of the true value lying within this range is really not very good), the A.D. will not be used in this book.

Another (and, from the standpoint of statistical theory, most important) measure of the *dispersion* (scatter of experimental points) is the *standard deviation*. This is defined as the square root of the average of the squares of the individual deviations, or, mathematically, by

$$\sigma = \sqrt{\frac{(x_1 - \bar{x})^2 + (x_2 - \bar{x})^2 + \cdots + (x_N - \bar{x})^2}{N}} \tag{15}$$

where σ is the standard deviation, $x_1, x_2, \ldots, x_N$ are the N individual measurements, and $\bar{x}$ is their average. Note that the signs of the various deviations make no difference in calculating σ since each is squared. An example of a standard deviation is given in the right-hand column in Table 1. Like other measures of dispersion, the standard deviation gives information about how closely the distribution is grouped about the mean. Statistical analysis shows that for a large number of normally distributed measurements, 68.3% of them will fall within the interval $x \pm \sigma$. In the results of our hypothetical range experiment, this is 133.9 ± 3.5 cm, and after a large number of firings we would expect to find that about 68% of the impact points lay between 130.4 and 137.4 cm from the gun muzzle.

When a very large number of measurements of a given quantity are made and variations between the different values obtained are due to truly random effects, a *normal distribution* of these values will be found. The word *distribution* as used here means an expression of the relative frequency with which the different observed values occur. Such an expression often takes the form of a graph in which the number of observed values in a small interval centered on a particular value of x is plotted against x. Thus, suppose in the experiment with the spring gun a very large number of range observations were made. We could, for example, count the number of such observations falling in the interval 127.0 ± 0.5 cm and plot this number as the ordinate of a point whose abscissa was 127 cm. Another point would be the number of observations falling between 127.5 and 128.5 cm plotted with an abscissa of 128 cm, and this process could be continued until we found somewhere beyond 140 cm that there were no more observed points to plot. A smooth curve could then be drawn through the plotted points. According to our earlier discussion, this curve should show a maximum at the mean value $\bar{x} = 133.9$ cm and should fall off symmetrically on either side. Such a curve, called the *normal curve*, was first discovered by a famous French mathematician, Abraham De Moivre (1667–1754), while working on certain problems in games of chance. It was also derived independently by Laplace and Gauss, who made statistical use of it and found that it accurately represents the errors of observation in scientific measurements. The curve is also known as the *normal probability* curve because of its use in the theory of probability, as the *normal curve of error*, and as the *Gaussian curve*. Here error is used to mean a deviation from the true value. Whenever any measurements are made in which there are random fluctuations, the results predicted by the normal curve are found to be valid. Thus it has been found that this curve describes very well many distributions that arise in the fields of the physical sciences, biology, education, and the social sciences.

The mathematical representation of the normal curve is given by the equation

$$y = \frac{N}{\sigma\sqrt{2\pi}}\, e^{-(x - \bar{x})^2/2\sigma^2} \tag{16}$$

where y represents the distribution function. In accordance with the foregoing discussion of how the normal curve is obtained in a physical case, the number $y\Delta x$ is the number of measurements of x (out of a very large total number N) that fall within the very small interval Δx centered on the value of x for which y was computed. Thus suppose we want to know how many measurements of x will fall in the vicinity of a certain value x_0. Putting x_0 into Equation 16 for x, we calculate the corresponding value y_0 for y. Then $y_0\Delta x$ is the number of measurements of x expected to be found in the interval $x_0 - \frac{1}{2}\Delta x$ to $x_0 + \frac{1}{2}\Delta x$. Thus y_0 is to be interpreted as the number of measurements *per unit interval* in x falling in the neighborhood of x_0. Note that $y_0\Delta x$ is the area of a tall, thin rectangle Δx wide located at x_0 on the x axis of the coordinate system in which the normal curve is plotted and extending up to the curve (a height y_0). Because this area represents the number of measurements of x falling within Δx, the area under the complete curve should be equal to N, the total number of measurements made. The factor $1/\sigma\sqrt{2\pi}$ in Equation 16, called a normalizing factor, is chosen to bring this about. It follows that the area under the curve between ordinates erected at some pair of values $x = a$ and $x = b$ is the number of measurements falling between a and b.

The curve is bell-shaped and symmetrical about the line $x = \bar{x}$. It has a maximum for this value of x (bearing out the idea that the mean value of x is also the most probable value) and falls quite rapidly toward the x axis on both sides. For different normal distributions, the curve has the same general shape, but its steepness, height, and location along the x axis will depend on the values of N, $\bar{x}$, and σ. The characteristic properties of the normal curve can be studied very readily by representing it in a new set of variables by means of a mathematical transformation. The first step is to divide Equation 16 through by N. The quantity $(y/N)\Delta x$ is then the fraction of the total number of measurements or the probability of obtaining a measurement in the narrow interval Δx centered on the value of x for which y is computed. Put another way, y/N is the probability *per unit interval in x* of getting a value of x lying in that interval whenever a measurement of x is made. Clearly, the area under the curve of y/N is unity, corresponding to the fact that the probability of getting *some* value of x is 1, or 100%. Next, we take the origin of our transformed coordinate system to be at the arithmetic mean and use the standard deviation as the unit of measurement along our new horizontal axis. This is done by choosing a new variable t related to x by the equation $t = (x - \bar{x})/\sigma$. When this is substituted into Equation 16 with N divided out and the normalizing factor appropriately modified, we obtain in place of $y(x)$ a new function $\phi(t)$ given by

$$\phi(t) = \frac{1}{\sqrt{2\pi}}\, e^{-t^2/2} \qquad (17)$$

This is the *standard form* of the normal curve, where $\phi(t)$ is the distribution and t is the variable. For this distribution, $t = 0$ corresponds to no deviation from the mean, and, because the t axis is calibrated in units of σ, the standard deviation along this axis is equal to one. The advantage of representing the normal curve in the standard form is that the area under the curve between any two values of t may be calculated once and for all. These values are tabulated and can be obtained from tables of probability integrals. Any normal distribution may then be expressed in the standard form, and the required calculations of the characteristics of the distribution can easily be made.

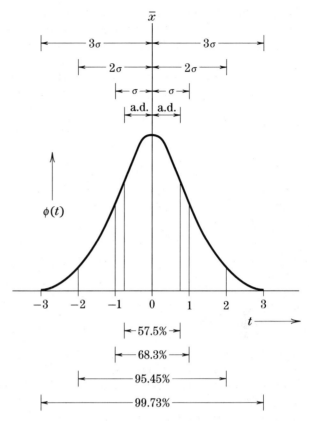

Figure 4   *The percentage distribution of area under the normal curve*

Some interesting properties of this curve, which is shown in Fig. 4, deserve mention. Note that theoretically it extends from $-\infty$ to $+\infty$, but practically it is so close to the axis beyond $t = \pm 3$ that the area under the curve beyond these points is negligible. The variable t is measured in units of σ along the horizontal axis, and the mean, the median, and the most probable value all coincide at the origin ($t = 0$). The total area under the curve is equal to 1. Hence, the area under any portion of the curve represents the relative frequency (expressed as a fraction of unity or as a percent) with which measurements in that interval occur. Numerical values of such areas may be obtained from tables and can be changed into the actual frequencies of occurrence by multiplying by N.

The *percentage distribution* of area under the normal curve is given in Fig. 4, where σ is the unit of measurement. The significance of the values given in the figure is that if the values of x are normally distributed, the probability that a value chosen at random will fall within the range $\bar{x} \pm \sigma$ is 0.683. The probability that it will lie within the range $\bar{x} \pm 2\sigma$ is 0.9545. The probability that it will lie within the range $x \pm 3\sigma$ is 0.9973. Thus the probability that it will lie outside of this range is only 0.0027, or 0.27%. Hence a deviation of 3σ on both sides of the arithmetic mean includes practically the whole of a normal distribution. The probability that any particular value of x will lie in the range $\bar{x} \pm$ a.d. is also easily found from Equation 17 and Fig. 4. The average deviation from the mean (a.d.) is simply the average $\bar{t}$ of t expressed in units of σ, as is readily seen from the definition of t given above. Equation 17 can then be used to calculate t. The result is $\sqrt{2/\pi}$, or 0.798, so that

$$\text{a.d.} = \sigma\bar{t} = \sigma\sqrt{\frac{2}{\pi}} = 0.798\sigma \qquad (18)$$

The area under the curve of Fig. 4 between −0.798 and +0.798 is then found to be 0.575 or 57.5% of the total area. This means that, as noted earlier, 57.5% of a large number of measurements will fall within the interval $\bar{x} \pm \text{a.d.}$ Similarly, the area under the curve between −1 and +1 is found to be 0.683 of the total area so that, also as noted earlier, 68.3% of a large number of measurements will fall within the interval $\bar{x} \pm \sigma$.

PERCENT ERROR

The error in a measured quantity is often conveniently expressed as a percent of the quantity itself. This comes up most often when an accepted value of the quantity in question is known from other, possibly more precise or detailed work. In this case the percent error can be calculated by taking the difference between the result obtained from the experiment (the mean value M of the measurements) and the value M_t accepted as the true value and expressing this difference as a percent of M_t. Thus,

$$\% \text{ error} = \frac{M - M_t}{M_t} \times 100\%$$

For example, suppose the velocity of sound in dry air at 0°C is measured and found to be 333.1 m/s, while the accepted value is 331.4 m/s. The error is 1.7 m/s. The relative error is 1.7/331.4, or 0.005. The percent error is

$$\frac{1.7}{331.4} \times 100\% = 0.5\%$$

There is no definite value for the allowable percent error to be expected in the following experiments. In many cases it is reasonable to expect results within 1%, while in others the error may be 5% or more, depending on the apparatus used. However, all measurements should be made with the greatest care, so as to reduce the error as much as possible.

CALCULATING WITH ERRORS

Whenever an experimental result is used in a calculation, account must be taken of the fact that an error is associated with it. Suppose we have two results x_1 and x_2 with respective standard deviations σ_1 and σ_2. We quote these results as $x_1 \pm \sigma_1$ and $x_2 \pm \sigma_2$ on the basis that the 68% chance that the true values lie within these ranges is good enough.* If the theory of the experiment requires these two quantities to be added, the sum is $x_1 + x_2 \pm \sigma_1 \pm \sigma_2$. It would, of course, be very nice if the error in one measurement were in the opposite direction from and therefore canceled the error in the other, but this happy event can hardly be counted on. It is much safer to assume the worst, that is, that the errors are in the same direction so that the sum becomes $x_1 + x_2 \pm (\sigma_1 + \sigma_2)$. We conclude that to be on the safe side we should add the errors in the individual quantities to obtain the error in the sum. Statistics show, however, that this approach is unduly pessimistic and that, in fact, when standard deviations are being used, the standard deviation σ_s in the sum is the square root of the sum of the squares of the individual standard deviations. Thus,

$$\sigma_s = \sqrt{\sigma_1^2 + \sigma_2^2}$$

The procedure is identical in the case of a subtraction, but care should be taken to note that the percent error can increase tremendously when two quantities of about the same value are to be subtracted one from the other. In such a case, we get $x_1 - x_2 \pm \sqrt{\sigma_1^2 + \sigma_2^2}$, and if x_1 and x_2 are nearly equal, their difference may be smaller than the error. This means that the errors associated with x_1 and x_2 may be large enough so that we cannot tell whether the quantities are equal or slightly unequal and hence whether or not a difference actually exists. Consider, for example, two automobiles driving down the highway with one slowly passing the other. The problem is to measure the passing speed, that is, the difference between the speeds of the individual automobiles. The best way to do this (since we are not interested in the individual speeds) is to measure the relative speed directly, but suppose experimental difficulties made this impossible so that the only data obtainable were the speedometer readings in the two cars. We find that one reads 61 miles per hour and the other 62 miles per hour. Can we conclude that one car is passing the other at the rate of 1 mph? Not really, for automobile speedometers are good to only two significant figures with the last one in doubt. There is therefore an error of about 1 mph in each speedometer, so that their readings should be reported as 61 ± 1 and 62 ± 1 mph, respectively. This being the case, the difference must be given as $1 \pm \sqrt{1^2 + 1^2} = 1 \pm 1.4$ mph. From this result alone we cannot tell whether the cars are proceeding side by side or, if they are not, which is passing the other. Certainly there seems to be a bias in favor of one car passing the other, but a reasonably certain conclusion simply cannot be drawn on

---

* For the experiments covered in this manual, the a.d., which is easier to find than σ, will usually be good enough. As already noted, the A.D. will not be used at all.

the basis of the available data. Thus, if we really want to know which car is passing the other and how fast, a method must be developed for measuring their relative velocity directly.* Although this does not seem to be a difficult task in our example of the cars, the progress of research in physics has often been marked by break-throughs due to someone's inventing a method for direct measurement of a quantity that had been formerly obtainable only as the difference between two other quantities.

If the two quantities $x_1 \pm \sigma_1$ and $x_2 \pm \sigma_2$ are to be multiplied, we have

$$(x_1 \pm \sigma_1)(x_2 \pm \sigma_2) = x_1x_2 \pm \sigma_1x_2 \pm \sigma_2x_1 + \sigma_1\sigma_2$$

On the assumption that the errors are reasonably small, $\sigma_1\sigma_2$ can be dropped based on the fact that the product of two small quantities is negligibly small. Assuming the worst case—that the errors are in the same direction and therefore add—we get

$$(x_1 \pm \sigma_1)(x_2 \pm \sigma_2) \approx x_1x_2 \pm (\sigma_1x_2 + \sigma_2x_1)$$

This result is inconvenient to handle in computation, but note that the percent error calculated on the basis of x_1x_2 being the true value is

$$\% \text{ error} = \frac{\sigma_1x_2 + \sigma_2x_1}{x_1x_2} \times 100\%$$

$$= \left(\frac{\sigma_1}{x_1} + \frac{\sigma_2}{x_2}\right) \times 100\%$$

$$= \frac{\sigma_1}{x_1} \times 100\% + \frac{\sigma_2}{x_2} \times 100\%$$

In other words, the percent error in the product is the sum of the percent errors in each factor. It is easy to show that the same holds true for division.

Raising an experimental quantity to an integer power is treated just like multiplication, for the integer power merely tells how many times the quantity is to be multiplied by itself. Thus, in calculating the area x^2 of a square of measured side x, the exponent 2 is simply an instruction to multiply x by x and has no error associated with it. If the error in x is σ, we have

$$(x \pm \sigma)^2 = x^2 \pm 2\sigma x + \sigma^2 \approx x^2 \pm 2\sigma x$$

and the percent error is

$$\pm \frac{2\sigma x}{x^2} \times 100\% = 2\frac{\sigma}{x} \times 100\%$$

or twice the percent error in the measurement of x. The situation may also be regarded as the product of $x \pm \sigma$ with itself, with error $\pm\sigma x \pm\sigma x$, but note that in this case the two errors σx *must* be considered to add since

they are not separate but identical and hence necessarily in the same direction. In general, it can be shown that if a measured quantity is raised to any power n, integer or fractional, the percent error in the result will be n times the percent error in the quantity. Notice that this is not n times the error but n times the *percent* error. Consequently, great care must be taken in the measurement of a quantity that is to be raised to a power greater than one, for the percent error in the result will be greater than that in the measurement by a factor of that power. A quantity raised to a high power may be thought of as "dominating" the result in that the error in this quantity will be much more important than the error in other quantities that may be multiplied or added in without being raised to such a power.

Conversely, the percent error in quantities raised to a power less than one is less important because only the fraction of that error represented by the less-than-unity value of n appears in the result. For example, in the experiment on the simple pendulum, in which the theory says that the period is proportional to the square root of the pendulum's length, only half the percent error in the length measurement will show up in the calculated period. But in an investigation of the Stefan-Boltzmann law, which states that the power radiated by a hot surface is proportional to the fourth power of the surface temperature, the temperature measurements will be extremely critical.

Finally, the case of the exponential (a mathematical number such as the base e of natural logarithms, with which no error is associated, being raised to a power given by a measured quantity) must be considered. Suppose that the theory under investigation calls for e to be raised to the power x where x is a measurement containing standard error σ. We have

$$e^{x \pm \sigma} = e^x e^{\pm \sigma} = e^x\left(1 \pm \sigma + \frac{\sigma^2}{2} \pm \cdots\right)$$

where we have used the property $e^{a+b} = e^a e^b$ of exponentials and have expanded $e^{\pm\sigma}$ in its power series. Since σ is a reasonably small quantity, terms in σ^2, σ^3, etc., can be neglected to give

$$e^{x \pm \sigma} \approx e^x(1 \pm \sigma) = e^x \pm \sigma e^x$$

The percent error is then $\pm \sigma \times 100\%$, which means that here *the error in the measured quantity is numerically equal to the relative error in the result and hence to the percent error after multiplication by 100%*. Thus, in any experiment in which e is to be raised to a measured quantity, particular care must be exercised in making the measurement in order to keep the error as small as possible. There is, in fact, a famous case (that of the thermionic emission of electrons from a heated metal surface) in which the theory, which predicts that the current will be proportional to some power of the temperature multiplied by an exponential having the temperature in the exponent, has never been really satisfactorily checked.

---

* We could also substitute speedometers of much higher precision, but this is an inelegant approach, as most of the precision will be wasted in getting the desired result.

There is no question about the exponential dependence, but it so dominates the result that even very careful experimental data has not been able to verify by what, if any, power of the temperature the exponential should be multiplied. Such are the challenges, the frustrations, and the fascination of experimental physics. It is hoped that the student will in some measure taste all three while performing the work described in this manual.

Measurement of Length, Mass, Volume, and Density

1

Physics is a quantitive experimental science and as such is largely a science of measurement. Many measuring instruments of great accuracy have been developed to meet the requirements of the physics laboratory. The measurement of length is of fundamental importance in scientific work; hence, it is fitting to begin experimental work with this type of measurement.

Mass is another fundamental quantity that must often be measured, and mass determinations will also be done in this experiment. Finally, if the dimensions of a regular geometrical object have been measured, its volume may be calculated; and if its mass is also known, its density may be found. Density is one of the important characteristics of materials. The engineer uses tables of densities to determine the weights of bridges and other structures that cannot be weighed. From the dimensions the volume can be computed and then from tables of densities the total weight can be calculated.

In this experiment the dimensions of various objects will be measured by means of a meter ruler, vernier caliper, and micrometer caliper, and the probable error in these measurements will be determined. There will also be an opportunity to compare the English and metric units of length. The mass of each object will be found with a triple-beam balance, and the density of the material of which each object is made will be obtained.

THEORY

It is important to be able to use the proper instruments when measuring various physical quantities. The simplest way to measure length is to use an ordinary meter stick or an English ruler. For ordinary measurements, a length is determined by a rough comparison with the scale—the result is given at a glance and in round numbers. For a more precise determination, the scale must be very accurately made and must be read to a fraction of its smallest scale division.

As discussed in the Introduction, when an instrument is used to the limit of its precision, certain errors occur that cannot be eliminated; these are called accidental errors. Thus, when a series of measurements of a physical quantity is made, the individual results usually differ among themselves because of the accidental errors involved. The best value of the quantity measured, that is, the most probable value, is the arithmetic mean (the average of the values obtained).

The precision of measurement can usually be increased by using more complex and more accurate equipment and by taking precautions to eliminate errors as much as possible. However, you should always try to make measurements with the greatest accuracy attainable with the given apparatus.

The Vernier Caliper   When a measurement is made with an ordinary meter stick, it is necessary to estimate the tenths of a millimeter, the millimeter being the smallest division on the scale. A vernier helps in reading accurately the fractional part of a scale division. It is a small auxiliary scale that slides along the main scale. The grad-uations of the vernier are different than those of the main scale, but they are simply related.

The vernier caliper, which is illustrated in Fig. 1.1, consists of a fixed part and a movable jaw. The fixed part includes a stem, on which is engraved the main scale, and a fixed jaw attached to the stem. The movable jaw is free to slide on the fixed stem and has a vernier scale engraved on it. The fixed scale is divided into centimeters and millimeters. The vernier is divided so that ten divisions on it cover the same interval as nine divisions on the main scale. Hence, the length of each vernier division is $\frac{9}{10}$ the length of a main-scale division. When the jaws are closed, the first line at the left end of the vernier, called the zero line or the index, coincides with the zero line on the main scale. However, the first vernier division is 0.1 mm away from the first main-scale division, the second vernier division is 0.2 mm away from the second main-scale division, and so on. If the jaws are slightly opened, it is easy to tell what fraction of a main-scale division the vernier index has moved by noting which vernier division coincides with a main-scale division.

A measurement is made with the vernier caliper by closing the jaws on the object to be measured and then reading the position where the zero line of the vernier falls on the main scale. This setting of the zero line of the vernier on the main scale is an incomplete measurement, however, because no attempt was made to estimate to a fraction of a main-scale division. The fractional part of a main-scale division is obtained by noting which line on the vernier coincides with a line on the main scale. Several possible vernier readings are illustrated in Fig. 1.2.

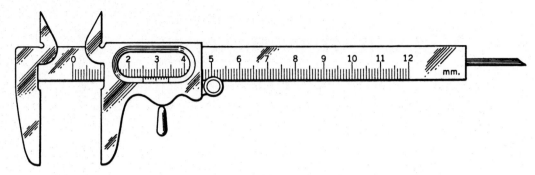

Figure 1.1 *The vernier caliper*

Frequently a vernier does not read zero when the jaws are completely closed. In such cases a zero correction must be applied to every reading; this correction may be either positive or negative. If the errors in the direction of increasing main-scale readings are called positive, then the zero correction is always made by subtracting the zero reading from the final reading.

Vernier scales are attached to many different kinds of instruments, and not all of them divide a main-scale division into ten equal parts. In any given case, the number of vernier divisions equals the number of equal parts into which a main-scale division is to be divided. The vernier works because the total number of vernier divisions spans a lesser number (usually one less) of main-scale divisions.

The Micrometer Caliper The micrometer caliper (see Fig. 1.3) is an instrument used for the accurate measurement of short lengths. Essentially, it consists of a carefully machined screw mounted in a strong frame. The object to be measured is placed between the end of the screw and the projecting end of the frame, called the anvil. The screw is then advanced until the object is gripped gently between the two jaws of the instrument. Most micrometers are provided with a ratchet arranged to slip on the screw as soon as a light and constant force is exerted on the object. By using the ratchet, it is possible to tighten up the screw by the same amount each time and also to avoid using too great a force. If this arrangement is absent, great care should be taken not to force the screw, for the instrument may be easily damaged.

The micrometer caliper used in this experiment consists of a screw with a pitch of 0.5 mm, a longitudinal scale engraved along a barrel containing the screw, and a circular scale engraved around a thimble which rotates with the screw and moves along the scale on the barrel. The longitudinal scale is divided into millimeters. The circular scale has 50 divisions. Since the pitch of the screw is 0.5 mm, which is the distance advanced by the screw in turning through one revolution, it is clear that rotating the thimble through one scale division will cause the screw to move a distance of $\frac{1}{50}$ of 0.5 mm, or 0.01 mm. Hence, readings may be taken directly to one hundredth of a millimeter, and by estimating tenths of a thimble-scale division, they may be taken to one thousandth of a millimeter.

The micrometer is read by noting the position of the edge of the thimble on the longitudinal scale and the position of the axial line of the barrel on the circular scale, and adding the two readings. The reading of the main scale gives the measurement to the nearest whole main-scale division; the fractional part of a main-scale division is read on the circular scale. Since two revolutions of the screw are required to make it advance a distance of 1

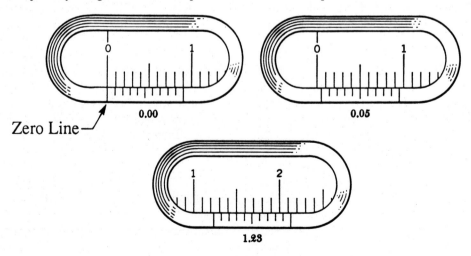

Zero Line

0.00

0.05

1.23

Figure 1.2 *Vernier readings*

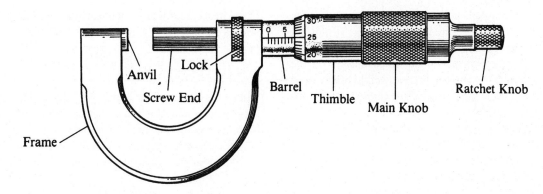

Figure 1.3 *The micrometer caliper*

mm, it is necessary always to be careful to note whether the reading on the circular scale refers to the first half or the second half of a millimeter. (In Fig. 1.3 the reading is 7.75 mm.)

The micrometer should be checked for a zero error, for it may not read zero when the jaws are completely closed. In such cases a zero correction has to be applied to every reading and may be either positive or negative. The value of the zero reading is obtained by rotating the screw until it comes in contact with the anvil and then noting the reading on the circular scale. Great care should be taken when doing this to ensure that both the screw end and anvil surfaces are completely clean.

The Triple-Beam Balance The masses of various objects will be determined in this experiment with a triple-beam balance, so called because it has three beams, each provided with a sliding weight, as shown in Fig. 1.4. These weights have different sizes, the largest sliding along a scale with a notch every 100 g, the next on a scale with a notch every 10 g, and the smallest on a scale with 1-g main divisions and 0.1-g subdivisions. The unknown mass is placed on the tray, the 100-g and 10-g weights set in the appropriate notches (determined by

trial), and the smallest slider adjusted for balance. The triple-beam balance can thus be read to 0.1 g, or, if one is willing to interpolate between the subdivisions, to about .05 g. However, great care is required to make the second decimal place really significant. In Fig. 1.4, the balance is set to 371.4 g.

In this and all further work in physics, be very careful to distinguish between mass and weight, which are not the same despite the very common confusion between the two. Mass is the property of material bodies that makes them hard to accelerate. Even if a body is taken out into deep space far from the earth, the sun, and the other planets, it will not become massless—it will still have inertia; that is, a force will still be required to change its velocity. Weight, on the other hand, is simply the force of gravity with which the earth attracts a body. Since this force depends on the distance between the earth and the body, the body will indeed become weightless if taken far away (meaning many thousands of miles). The confusion arises because apparently the property of material bodies called mass is responsible for their mutual gravitational attraction, the two being in fact proportional. Thus, the weight of a body (that is, the force with which it is attracted to the earth) is proportional to

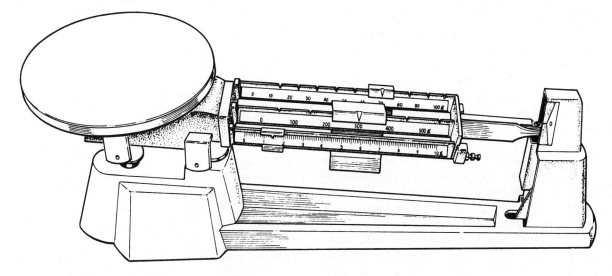

Figure 1.4 *The triple-beam balance*

its mass; if its weight is measured, its mass is also known. Scales and balances measure weight.

The triple-beam balance compares the weight of the unknown object with that of the fixed masses on the beams. However, because mass and weight are proportional, the scales on the beams can be calibrated in grams (a unit of mass) rather than in newtons (a unit of force), even though the weight—that is, the force of gravity—is what is actually being measured. There is nothing wrong with this—a measuring instrument's scale can always be graduated in units proportional to those it is actually measuring—but a physicist is nevertheless careful not to forget the distinction between mass and weight. For a further discussion of this matter, see the Mass and Weight section in the Introduction.

In the present experiment the masses of several objects will be determined and their volumes calculated from the measurements of their dimensions. The density

of the material of which each is made, defined as the material's mass per unit volume, is then calculated from

$$D = \frac{M}{V}$$

where M is the mass in grams, V is the volume in cubic centimeters, and D is the density, whose units are thus seen to be grams per cubic centimeter.

For the metal cylinders and the wire (which may be considered a long, thin cylinder), the volume is given by

$$V = \pi r^2 L = \frac{\pi d^2 L}{4}$$

where r is the cylinder's radius, d is the diameter ($2r$), and L is the cylinder's length. For the irregular body, the volume will be determined by measuring the volume of liquid that it displaces.

APPARATUS

1. Triple-beam balance
2. English ruler
3. Metric ruler (meter stick)
4. Vernier caliper
5. Micrometer caliper
6. Three cylinders of different metals

   a. aluminum
   b. brass
   c. iron

7. Length of copper wire
8. An irregular solid
9. A 250-mL graduated cylinder

PROCEDURE

1. Measure the length of the piece of copper wire with the metric ruler. Read the position of both ends of the wire, estimating to the nearest tenth of a millimeter. Record both readings in centimeters and read to 0.01 cm. Make four independent measurements—that is, use a different part of the ruler for each measurement—and record all the readings.

2. Measure the length of the wire with the English ruler. Read the position of both ends of the wire, estimating to the nearest tenth of the smallest scale division. Record both readings, expressed in inches, to the nearest 0.01 in. Make four independent measurements—that is, use a different part of the ruler for each measurement—and record all the readings.

3. Determine the zero reading of the vernier caliper. This is the reading when the jaws of the instrument are in contact with each other. Be sure that the jaws are clean so that no grit or other foreign matter gets between them and prevents true contact. Record the zero reading to 0.01 cm. Make four independent determinations of this reading—that is, open and close the jaws before each setting—and record your results.

4. Measure the length and diameter of each cylinder with the vernier caliper. These measurements are made by closing the jaws of the caliper on the length or diameter of the cylinder being measured and reading the posi-

tion where the zero line of the vernier falls on the main scale. The fractional part of a main-scale division is obtained by noting which line on the vernier coincides with a line of the main scale. Record the reading in centimeters and read to 0.01 cm. Make four independent measurements—that is, open and close the jaws before each setting. This is most conveniently done by measuring the diameter of each cylinder in turn once, then measuring the length of each in turn once, and repeating the complete sequence four times. Record all readings.

5. Determine the zero reading of the micrometer caliper, that is, the reading when the surfaces of the anvil and the screw end are in contact. Be sure these surfaces are very clean, since even a small speck of dust can give a false reading. The screw end may be brought almost into contact with the anvil by turning the thimble directly, but actual contact must always be made by turning the ratchet slowly until it clicks several times. If the ratchet is absent, great care must be taken not to force the screw; merely let the screw end approach the anvil very slowly and stop turning the screw as soon as the two surfaces touch. Record the value of this reading in centimeters and read to 0.0001 cm, estimating to one tenth of the smallest scale division. Make four independent determinations of the zero reading—that is, open and close the instrument before each setting—and record the readings.

6. Measure the diameter of the copper wire with the micrometer caliper. The measurement is made by placing the wire between the screw end and the anvil and advancing the screw until the wire is gripped between the anvil and screw-end surfaces. Again, the ratchet should be used. If the instrument has no ratchet, care should be taken not to force the screw. Record the reading in centimeters and read to 0.0001 cm, estimating to one tenth of the smallest scale division. Make six independent measurements of the diameter of the wire—that is, open and close the caliper before each setting, and make the measurements at different points along the wire's length. Also, spin the wire between measurements so that different diameters are measured each time. Record the six readings so obtained.

7. Determine the mass of each cylinder, the irregular solid, and the copper wire using the triple-beam balance.

8. Determine the volume of the irregular solid by the displacement method. Partly fill the graduated cylinder with water and read the water level. Filling the cylinder to an exact graduation division will prove convenient in the subsequent calculations. Record the water level. Suspend the solid by means of a thread and lower it into the cylinder until it is completely submerged. Read and record the water level now observed. When the first reading is subtracted from the second, the volume of the irregular solid is obtained in cubic centimeters.

DATA _____

Measurement of Wire Length Using a Metric Ruler

| Ruler readings | | Lengths | Deviations | Deviations squared |
Left end	Right end			
Average values				

Value of σ _____

Measurement of Wire Length Using an English Ruler

| Ruler readings | | Lengths | Deviations | Deviations squared |
Left end	Right end			
Average values				

Value of σ _____

Number of centimeters in 1 inch

    Calculated value _____

    Accepted value _____

Difference between calculated and
    accepted values _____

Percent error _____

Length and Diameter of Metal Cylinders Using a Vernier Caliper

	Vernier caliper readings				Average
	1	2	3	4	
Zero reading					
Length, aluminum cylinder					
Length, brass cylinder					
Length, iron cylinder					
Diameter, aluminum cylinder					
Diameter, brass cylinder					
Diameter, iron cylinder					

Diameter of Copper Wire Using a Micrometer Caliper

	Micrometer readings						Average
	1	2	3	4	5	6	
Zero reading					✕	✕	
Reading with wire							
Diameter of wire							

Determination of Density

Object used	Mass, g	Length, cm	Diameter, cm	Volume, cc (cm³)	Density computed	Density (from Table IV)	Percent error
Aluminum cylinder							
Brass cylinder							
Iron cylinder							
Copper wire							
Irregular solid		✕	✕				

CALCULATIONS

1. From the data of Procedure 1, compute the length of the copper wire for each set of measurements. Calculate the average value of the length by finding the arithmetic mean of the lengths obtained, to the proper number of significant figures.

2. From the data of Procedure 2, compute the length of the copper wire for each set of measurements. Calculate the average value of the length by finding the arithmetic mean of the lengths obtained, to the proper number of significant figures.

3. Compute the number of centimeters in one inch from your measurements in Procedures 1 and 2. Use the calculated average values of the wire's length in making the computations. Compare your result with the accepted value by finding the difference between them and the percent error.

4. From the data of Procedure 3, compute the average value of the zero reading of the vernier caliper. From the data of Procedure 4, compute the average value of the setting of the vernier caliper for the length and diameter of each cylinder. Making sure to include its algebraic sign properly, use the average zero reading to obtain the measured length and diameter of each cylinder and insert these values in the table of data for the density determinations.

5. From the data of Procedure 5, compute the average value of the zero reading of the micrometer caliper. From the data of Procedure 6, compute the average value of the setting of the micrometer caliper for the diameter of the copper wire. Calculate the diameter of the copper wire by inserting the zero correction with the proper algebraic sign and enter your result in the table of data for the density determination.

6. From the data of Procedures 1 and 2, calculate the deviations of the lengths obtained from the average value of the length. Compute the average deviation from the mean (the a.d.) and the standard deviation σ for both sets of measurements. (See the Theory of Errors in the Introduction.) The a.d. should be entered in your tables as the average of your deviations in Procedures 1 and 2, respectively. The standard deviation is obtained in each case by taking the square root of the averages of the squares of the individual deviations.

7. Calculate the error to be associated with your result for the number of centimeters in 1 in. found in Calculation 3 and note whether the accepted value falls within the range of this error. Use both the a.d.'s and the standard deviations found in Calculation 6 and see whether the accepted value falls within both the error ranges so obtained.

8. Compute the volume of each object measured.

9. Calculate the density of each substance used.

10. Compare your results with those given in Table IV at the end of the book and find the percent error. *Note:* If the instructor has designated the substance of which the irregular solid is made as an unknown, try to identify it by comparing your computed value with the list in Table IV.

QUESTIONS _____

1. Why are several observations taken for each measurement?

2. (a) What is the smallest part of a centimeter that can be read or estimated with a meter stick? (b) What is the smallest part of a centimeter that can be read or estimated with your vernier caliper? (c) Which readings are more reliable? Why?

3. What is the smallest part of a centimeter that can be read or estimated with your micrometer caliper? This represents the sensitivity of the micrometer, for the sensitivity of a measuring instrument is the value of the smallest quantity that can be read or estimated with it.

4. State the number of significant figures in the data of Procedures 1, 4, and 6.

5. (a) What is the significance of the average deviation from the mean (the a.d.)? (b) What is the significance of the standard deviation σ?

6. In measuring the length and diameter of a cylinder, which dimension should be measured more carefully? Why? *Hint:* Remember that the length and diameter are being measured in order to determine the volume.

7. Why was the micrometer used instead of the vernier caliper in determining the diameter of the copper wire?

8. (a) What is the volume in cubic millimeters of the largest cylinder you measured? (b) What is its volume in liters? (c) What is its mass in kilograms?

9. A thin circular sheet of copper has a diameter of 30.0 cm and a thickness of 1 mm. Find the weight of the sheet in newtons.

10. A certain graduated cylinder has an inside diameter of 4.00 cm. (a) The graduations on its side are labeled "cc" (cubic centimeters). How far apart are these graduations? (b) The cylinder is partially filled with water, and a solid sphere of radius 1.20 cm is then totally submerged therein. Through what distance does the water level rise when the sphere is so immersed?

Addition of Vectors. Equilibrium of a Particle 2

When a system of forces, all of which pass through the same point, acts on a body, it may be replaced by a single force called the resultant. The purpose of this experiment is to show that the magnitude and direction of the resultant of several forces acting on a particle may be determined by drawing the proper vector diagram and that the particle is in equilibrium when the resultant force is zero.

THEORY

A scalar is a physical quantity that possesses magnitude only; examples of scalar quantities are temperature, mass, and density. A vector is a quantity that possesses both magnitude and direction; examples of vector quantities are velocity, acceleration, and force. A vector may be represented by drawing a straight line in the direction of the vector, the length of the line being made proportional to the vector's magnitude. The sense of the vector (for example, whether it is pointing toward the right or toward the left) is indicated by an arrowhead placed at the end of the line.

Vectors may be added either graphically or analytically. Note carefully, however, that words like "add" and "sum," when used in connection with vectors, do not mean the same as they do in common parlance about numbers. Vectors are not added by simply adding their magnitudes. Instead, the sum, or more properly, the *resultant* of two or more vectors, is the single vector that produces the same effect. For example, if two or more forces act at a certain point, their resultant is the force that, if applied at that point, has the same effect as the two separate forces acting together. Thus, in Fig 2. l, the resultant of two forces **A** and **B** is shown. The equilibrant, defined as the force equal and opposite to the resultant, is also shown. Note that if the equilibrant is regarded as a third force (in addition to **A** and **B**), the sum of **A**, **B**, and the equilibrant is zero.

The operation of adding vectors graphically consists of constructing a figure in which a straight line is drawn from some point of origin to represent the first vector, the length of the line being proportional to the magnitude of the vector and the direction of the line being the same as the direction of the vector. From the arrowhead (nose) end of this line and at the proper angle with respect to the first vector, another line is drawn to represent the second vector, and so on with the remaining vectors. The resultant is the vector drawn from the origin or tail of the first vector to the nose of the last vector (see Fig. 2.2). If a closed polygon is formed—that is, if the nose of the last vector falls upon the tail of the first—then the resultant is zero. If the vectors represent forces, they are in equilibrium.

If no more than two vectors are to be added, the parallelogram method may be used. This consists of constructing both vectors with their tails at the origin and completing the parallelogram, as shown in Fig. 2.3. The diagonal of the parallelogram is then the resultant. Note, however, that this method is really identical to the previously described method of graphical addition by constructing the vectors nose-to-tail. Thus, the side of the parallelogram opposite **A** is parallel to and has the same

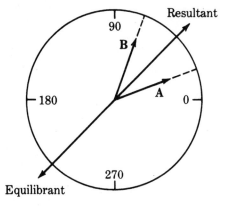

Figure 2.1 *The resultant and equilibrant for two forces*

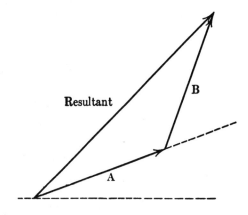

Figure 2.2 *Graphical addition of vectors*

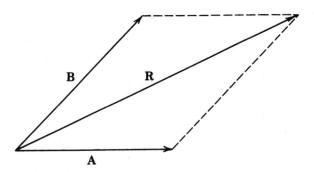

Figure 2.3   *Parallelogram method of adding vectors*

length as **A**. Hence, it is really a reconstruction of **A** with its tail on the nose of **B**. The resultant **R** is then drawn from the tail of **B** to the nose of **A** as before. Similarly, the side of the parallelogram opposite **B** is equal to **B** and is therefore a reconstruction of **B** with its tail on the nose of **A**. Here the resultant may be considered as being drawn from the tail of **A** to the nose of **B**. The parallelogram method thus demonstrates the commutative property of vector addition: in vector addition, as in algebraic addition, the order in which the quantities are added doesn't matter. This property may be stated mathematically as

$$\mathbf{A} + \mathbf{B} = \mathbf{B} + \mathbf{A}$$

and may seem obvious, but bear in mind that mathematical processes do exist for which it does not hold.

Vectors may also be added analytically, and this is in fact the preferred method since it does not require the making of precise drawings and does not involve the inaccuracies inherent in the measurements made on them. The method is illustrated in Fig. 2.4, in which the vectors of Fig. 2.2 have been placed on the usual cartesian (x–y) coordinate system. In Fig. 2.4, the vectors **A** and **B** have been resolved into their x and y components. For example, the x component A_x of the vector **A** is the projection of **A** on the x axis and is equal to $A \cos \theta_A$, where A is the

magnitude of **A**. Similarly, the y component of **A** is the projection of **A** on the y axis and is given by $A \sin \theta_A$.

It is easy to show how a vector is represented by its components. Fig. 2.4 introduces the two *unit* vectors **i** and **j**, which are defined as vectors of unit magnitude and with directions along the x and y axes, respectively. Then $\mathbf{i}A_x$ is a vector along the x axis having magnitude A_x, and $\mathbf{j}A_y$ is a vector along the y axis having magnitude A_y. Fig. 2.4 shows that, by the parallelogram method of vector addition,

$$\mathbf{A} = \mathbf{i}A_x + \mathbf{j}A_y \tag{2.1}$$

By the same arguments, vector **B** is given by

$$\mathbf{B} = \mathbf{i}B_x + \mathbf{j}B_y \tag{2.2}$$

where B_x and B_y, the x and y components of **B**, have the values $B \cos \theta_B$ and $B \sin \theta_B$, respectively.

Finally, Fig. 2.4 shows that the x component R_x of the resultant **R** is just the sum of the two x components A_x and B_x, while R_y is the sum of A_y and B_y. Thus, the analytical method consists of finding the x and y components of all the vectors to be added, adding the x components algebraically (paying due attention to sign) to get the resultant's x component, and adding the y components to get the resultant's y component. In mathematical terms,

$$R_x = A_x + B_x \tag{2.3}$$

and

$$R_y = A_y + B_y \tag{2.4}$$

With the x and y components of **R** known, the magnitude and direction of this vector are easily found if desired. Inspection of Fig. 2.4 shows that

$$R^2 = R_x{}^2 + R_y{}^2 \tag{2.5}$$

and

$$\tan \theta = \frac{R_y}{R_x} \tag{2.6}$$

However, since a vector is completely specified by its components, it is often left in this form, the magnitude and direction being calculated only if they are needed.

The conclusion that the components of a resultant are each the algebraic sum of the respective components of the vectors to be added may be arrived at analytically as well as by inspection of Fig. 2.4. Thus, in the addition of vectors **A** and **B**, these vectors may be given by Equations 2.1 and 2.2 so that

$$\mathbf{R} = \mathbf{A} + \mathbf{B} = \mathbf{i}A_x + \mathbf{j}A_y + \mathbf{i}B_x + \mathbf{j}B_y \tag{2.7}$$

Writing **R** as $\mathbf{i}R_x + \mathbf{j}R_y$ and grouping terms appropriately in Equation 2.7 leads to

$$\mathbf{i}R_x + \mathbf{j}R_y = \mathbf{i}(A_x + B_x) + \mathbf{j}(A_y + B_y) \tag{2.8}$$

This can be true only if $R_x = A_x + B_x$ and, separately, $R_y = A_y + B_y$ in agreement with Equations 2.3 and 2.4.

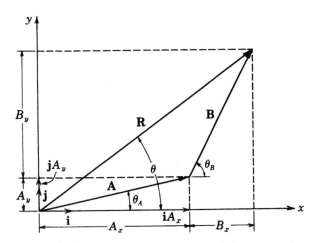

Figure 2.4   *Addition of vectors by components*

The apparatus used in this experiment (see Fig. 2.5) consists of a horizontal force table graduated in degrees and provided with pulleys that may be set at any desired angles. A string passing over each pulley supports a weight holder upon which weights may be placed. A pin holds a small ring that acts as the particle and to which the strings are attached. When a test for equilibrium is to be made, the pin is removed; if the forces are in equilibrium, the particle will not be displaced.

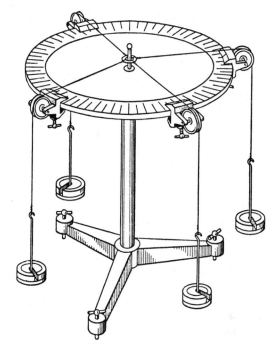

Figure 2.5 *Force table*

APPARATUS

1. Force table
2. Four pulleys
3. Four weight hangers
4. Set of slotted weights, including:
   four 100-g weights
   four 50-g weights

    two 20-g weights
    two 10-g weights
5. Set of small weights (5-, 2-, 2-, 1-g)
6. Protractor
7. Metric ruler
8. Thin string for suspending weight hangers

PROCEDURE

1. Mount a pulley on the 20° mark on the force table and suspend a mass of 100 g over it. Mount a second pulley on the 120° mark and suspend a mass of 200 g over it. Draw a vector diagram to scale, using a scale of 0.2 N/cm, and determine graphically the direction and magnitude of the resultant by using the parallelogram method. Note that your weights are labeled according to their masses in grams but that you are dealing with forces measured in newtons. You will have to make the appropriate conversion in your calculations. For more information on this matter, see the Mass and Weight section in the Introduction.

2. Check the result of Procedure 1 by setting up the equilibrant on the force table. This will be a force equal in magnitude to the resultant, but pulling in the opposite direction. Set up a third pulley 180° from the calculated direction of the resultant and suspend masses over it giving a weight equal to the magnitude of the resultant. Cautiously remove the center pin to see if the ring remains in equilibrium. Before removing the pin, make sure that all

the strings are pointing exactly at its center; otherwise, the angles will not be correct.

3. Mount the first two pulleys as in Procedure 1, with the same weights as before. Mount a third pulley on the 220° mark and suspend a mass of 150 g over it. Draw a vector diagram to scale and determine graphically the direction and magnitude of the resultant. This may be done by adding the third vector to the sum of the first two, which was obtained in Procedure 1. Now set up the equilibrant on the force table and test it as in Procedure 2.

4. Clamp a pulley on the 30° mark on the force table and suspend a mass of 200 g over it. By means of a vector diagram drawn to scale, find the magnitude of the components along the 0° and the 90° directions. Set up these forces on the force table as they have been determined. These two forces are equivalent to the original force. Now replace the initial force by an equal force pulling in a direction 180° away from the original direction, that is, by suspending the 200-g mass over a pulley clamped at the 210° mark. Test the system for equilibrium.

CALCULATIONS _____

1. Calculate the resultant in Procedure 1 by solving for the third side of the force triangle algebraically. Find the magnitude of the resultant vector by using the law of cosines and the direction from the law of sines. Compare your result with that obtained in Procedure 1. How do these predicted results compare with your experimental finding in Procedure 2?

2. Calculate the resultant in Procedure 1 by using the analytical method of adding vectors. Compare this result with those already obtained in Procedures 1 and 2 and in Calculation 1.

3. Calculate the resultant in Procedure 3 by the analytical method. Note that you have already found the components of two of the force vectors in Calculation 2 and need only find and add the components of the third. Compare your result with that obtained in Procedure 3.

QUESTIONS _____

1. State how this experiment has demonstrated the vector addition of forces.

2. In Procedure 3, could all four pulleys be placed in the same quadrant or in two adjacent quadrants and still be in equilibrium? Explain.

3. State the condition for the equilibrium of a particle.

4. The forces used in this experiment are the weights of known masses, that is, the forces exerted on these masses by gravity. Bearing this in mind, explain the function of the pulleys.

5. The analytical method of adding vectors expressed in terms of their components may be applied to vectors in three dimensions, for which graphical work is inconvenient. Find the magnitude of the resultant of the vectors $\mathbf{A} = \mathbf{i}12 - \mathbf{j}37 + \mathbf{k}58$ and $\mathbf{B} = \mathbf{i}5 + \mathbf{j}30 - \mathbf{k}42$, where $\mathbf{i}$, $\mathbf{j}$, and $\mathbf{k}$ are unit vectors along the x, y, and z axes, respectively.

6. The x and y components of a certain force are measured and found to be 68 ± 3 and 42 ± 2 N, respectively. Calculate the direction and magnitude of this force, expressing your result to the proper number of significant figures and showing the error in both magnitude and angle. Note that the error in the latter is obtained from the error in its tangent. Devise a procedure for doing this and show your work.

Equilibrium of a Rigid Body 3

When a rigid body is acted upon by a system of forces that do not all pass through the same point, a change may be produced in the angular (rotational) velocity of the body as well as in its linear (translational) velocity. Under certain conditions the body will be in equilibrium— that is, there will be no tendency for either its translational or rotational motion to change. If it is at rest, it will remain at rest. This experiment presents a study of the conditions for the equilibrium of a rigid body under the action of several forces.

THEORY

In the preceding experiment, the equilibrium of a particle was investigated, and it was found that the particle was in equilibrium (had no tendency to change its translational velocity) when the resultant force acting on it was zero. In that case, the particle was a body so small that it could be considered a point and therefore the question of where on the body the various forces were applied did not arise. Moreover, rotation was not considered, since rotation of a very small body is of little consequence.

More generally, bodies of appreciable extent must be considered. Then the points on a body at which forces are applied become important, and rotation of the body as a unit is a new motion that must be considered in addition to its translational motion, that is, its displacement as a whole from one location to another. The bodies in question are assumed to be rigid, which means that their parts do not change their distances from one another. Internal forces (between atoms and molecules) hold a rigid body together. These forces maintain all parts in fixed relative positions and are assumed strong enough to do this no matter what external forces may be applied. The body then remains rigid and unbroken, and the internal forces need be considered no further.

A force's tendency to produce rotation of an extended rigid body is called the torque or the moment of the force. It is equal to the product of the force's magnitude and the perpendicular distance from the axis of rotation to the force's line of action. This definition is illustrated in Fig. 3.1. Here a force $\mathbf{F}$ is shown applied to a disk wheel at point P a distance r from the wheel's axis. The dashed line is the line of action of force $\mathbf{F}$ and is simply a line constructed through P along the direction of $\mathbf{F}$. It therefore includes the arrow that represents $\mathbf{F}$ and is really no more than an extension of that arrow in both directions. The perpendicular distance from the axis of rotation to the line of action of $\mathbf{F}$ is shown as l in Fig. 3.1, and the magnitude of the torque is Fl. The distance l is called the lever arm or moment arm of the force $\mathbf{F}$ about the wheel's axis.

Since torque has a direction associated with it, it is a vector, and in general this vector lies along the axis of rotation, that is, perpendicular to the plane determined by the line of action of the force and the lever arm. In Fig. 3.1 this plane is the plane of the page, and the torque vector may be imagined as being perpendicular to the page and lying along the wheel's axis, which is shown in Fig. 3.1 as a point representing the end view of the axis. This will be the situation in all cases encountered in elementary mechanics, so that all torque vectors will be parallel to each other and only their sense need be considered when they are added. The usual convention that torques tending to produce counterclockwise rotation are taken to be positive whereas those tending to produce clockwise rotation are taken to be negative will be adhered to throughout this book.

An important problem arises when the lever arm of the force of gravity on (the weight of) a rigid body is to be found about some given axis. The force of gravity acts on every little piece of the body and is therefore distributed over the extent of the body rather than being

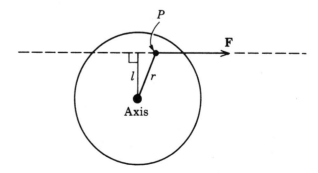

Figure 3.1 *The torque or moment of a force*

applied at one particular point. However, it is found that the body can be balanced, meaning that it can be supported at a single point about which (if the point is properly chosen) it will show no tendency to rotate. But if the body has no tendency to rotate, the total torque on it must be zero. Moreover, if the body is just being supported, the upward supporting force must be exactly equal to its weight. Fig. 3.2 now shows that if the entire weight W of the body is considered to act at the balance point, then the lever arms of the weight and the supporting force will be the same about any arbitrarily chosen axis, and their respective torques will be equal and opposite. The net torque will be zero and there will be no tendency for the body to rotate. Since, in fact, the body is observed not to rotate when supported at its balance point, the actual situation is equivalent to the hypothetical one in which the entire weight of the body is considered to be applied at the balance point. In fact, correct results will be obtained if the weight is taken to act at the balance point, and this point is therefore called the *center of gravity*. Thus, the moment or lever arm of the weight of a rigid body about any given point may be calculated by assuming the body's entire weight to be concentrated at its center of gravity.

The foregoing shows that for a rigid body to be in equilibrium, it is not enough for the resultant force to be zero. That merely guarantees that the body's translational velocity will not change. If the body's rotational velocity is not to change either, the net torque due to all the applied forces must also be zero. There are thus two conditions for the equilibrium of a rigid body: (1) the vector sum (the resultant) of all the forces acting on the body must be zero and (2) the algebraic sum of all the torques about any axis must be zero. The first condition means that the sum of the forces in any direction must be equal to the sum of the forces in the opposite di-

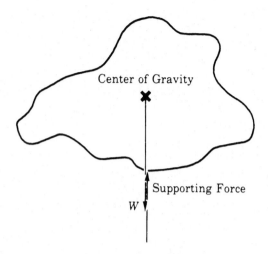

Figure 3.2   *Center of gravity*

rection. The second condition means that the sum of the torques producing clockwise rotation around any point must be equal to the sum of the torques producing counterclockwise rotation around the same point. The present experiment illustrates these principles using a meter stick as the rigid body. The arrangement is shown in Fig. 3.3. The meter stick is supported at an axis of rotation near its center, and various forces are applied to it by weights and the supporting stand. The torque due to each force is calculated and the unknown force or distance computed by application of the two conditions of equilibrium. The weight and location of the center of gravity of the meter stick are also found, and the torque due to this weight is calculated as if the weight were concentrated at this point.

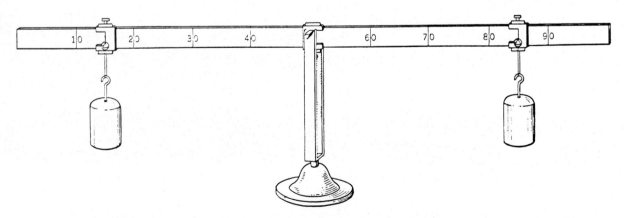

Figure 3.3   *A meter stick as a rigid body in equilibrium*

APPARATUS

1. Meter stick
2. Three meter stick knife-edge clamps
3. Stand for supporting the meter stick with a clamp
4. Set of hooked weights
5. Triple-beam balance
6. Body of unknown mass (about 100 g)

PROCEDURE

1. Weigh the meter stick and record the weight. Note that weight is a force and should be expressed in newtons even though your scales may read in grams. See the Mass and Weight section in the Introduction.

2. Weigh the three meter stick clamps together and compute their average weight.

3. Find the center of gravity of the meter stick to the nearest 0.5 mm by balancing it in one of the clamps.

4. Put another clamp near one end of the stick and hang a 100-g weight from it. Slide the stick through the supporting clamp until the position of balance is found. Record the position of the axis of rotation—that is, the point of support—and the position of the weight.

5. Put another clamp near the other end of the stick and suspend a 200-g weight from it. Leaving the other

weight in place, find the new point of balance by again sliding the meter stick through the supporting clamp. Record the position of the 200-g weight and of the new point of balance.

6. Remove the weights and clamps. Clamp the meter stick in the support clamp at its center of gravity. Put another clamp near one end of the stick and suspend the body of unknown weight from it. Slide another clamp, with a 200-g weight attached, along the other end of the stick until the position of equilibrium is found. Record the positions of both weights.

7. Weigh the body of unknown mass used.

DATA

Weight of the meter stick _____

Weight of the three meter stick clamps _____

Average weight of the meter stick clamps _____

Center of gravity of the meter stick _____

Position of the 100-g weight _____

Position of the axis of rotation _____

Position of the 200-g weight _____

New point of balance _____

Position of the unknown mass _____

New position of the 200-g weight _____

Actual weight of the unknown mass _____

Weight of the meter stick computed by the method of moments _____

Weight of the unknown mass computed by the method of moments _____

CALCULATIONS

1. Compute the weight of the meter stick from the data of Procedure 4 by the method of moments. Compare your result with that obtained by direct weighing of the stick. In particular, note whether the two measurements agree within the errors associated with each.

2. Using the point of support as the axis in Procedure 5, compute the moment of force of each of the weights and also of the meter stick, assuming its weight to be concentrated at its center of gravity. Add all these moments together, paying attention to their algebraic signs. Compare this net torque with zero, noting in particular whether zero lies within the error associated with your result.

3. Compute the weight of the body used in Procedure 6 by the method of moments. Compare the measured weight of this body with the computed weight, noting in particular whether the two weights agree within the experimental errors involved.

QUESTIONS _____

1. State the two conditions for the equilibrium of a rigid body and tell how this experiment demonstrates their validity.

2. In Calculation 2, why was the supporting force exerted on the meter stick by the stand not considered?

3. Why was the meter stick clamped at its center of gravity in Procedure 6? Was there any advantage in doing this?

4. Using the data of Procedure 5, assume the axis of rotation to be at one end of the stick and compute the moments of all the forces about this axis. The total upward force at the point of support is equal to the sum of all the downward forces. Add all the positive moments together and all the negative moments together and then find their algebraic sum. What conclusion do you draw from the result?

5. A meter stick is pivoted at its 50-cm mark but does not balance because of nonuniformities in its material that cause its center of gravity to be displaced from its geometrical center. However, when weights of 150 and 200 g are placed at the 10-cm and 75-cm marks, respectively, balance is obtained. The weights are then interchanged and balance is again obtained by shifting the pivot point to the 43-cm mark. Find the mass of the meter stick and the location of its center of gravity.

Uniformly Accelerated Motion. The Atwood Machine 4

In accordance with Newton's first law of motion, when the resultant of all the forces acting on a body is zero, if the body is at rest it will remain at rest, and if it is in motion, it will continue to move with constant speed in a straight line. Newton's second law of motion describes what happens if the resultant is different from zero. This law states that if an unbalanced force is acting on a body, it will produce an acceleration in the direction of the force, the acceleration being directly proportional to the force and inversely proportional to the mass of the body. If the acceleration is constant, the body is said to be moving with uniformly accelerated motion. The purpose of this experiment is to measure the acceleration of a given mass produced by a given force and to compare it with that calculated from Newton's second law of motion.

THEORY

The Atwood machine consists of two weights connected by a light, flexible string that passes over a light pulley, as shown in Fig. 4.1. (Fig. 4.2 and Fig 4.3 show two models of Atwood machines.) The pulley should be as nearly frictionless as possible. The machine is used in measuring the acceleration produced by an arbitrarily chosen force acting upon a given mass. Once the mass and the force have been chosen, the acceleration produced is determined by Newton's second law of motion

$$F = ma \qquad (4.1)$$

where F is the net force in newtons acting on a body, m is the mass of the body in kilograms, and a is the acceleration in meters per second per second.

In the Atwood machine, the total mass that is being accelerated is the sum of the two masses. The driving force, which is expressed in newtons, is the difference in the weights on the two ends of the string. The important feature of this device is that a large mass (the sum) is being accelerated by a small force (the weight difference) so as to produce accelerations that are considerably smaller than the acceleration of gravity and therefore easier to measure. Moreover, the total mass being accelerated may be kept constant while the force is varied by transferring masses from one side of the moving system to the other. Before measurements can be made, however, the force due to friction in the pulley bearing must be determined. This is done by transferring masses from the ascending side to the descending side until the mass on the descending side moves downward with constant velocity when given a very slight push. The difference between the weights at the two string ends is then just compensating for the opposing frictional force and may be taken as equal to that force.

For convenience, the distance through which the descending mass falls should be the same in all of the observations. The starting point is taken as the position of the moving system in which one of the masses rests on the floor. As this mass ascends, the other will descend an equal distance, and the stopping point is taken as the instant at which this mass strikes the floor. The distance traversed should be about 1.5 m. The time required for the mass to move through this distance is measured with a stopwatch.

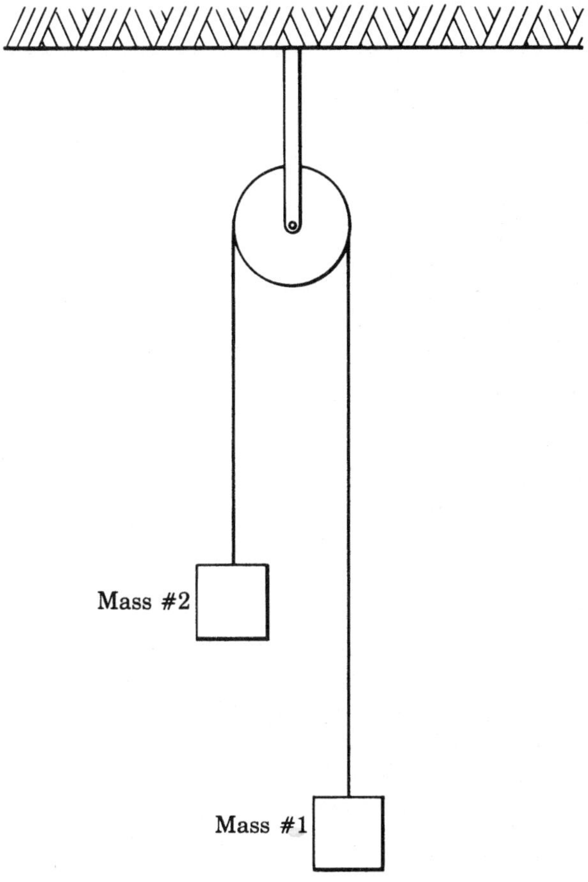

Figure 4.1 *The Atwood machine*

25

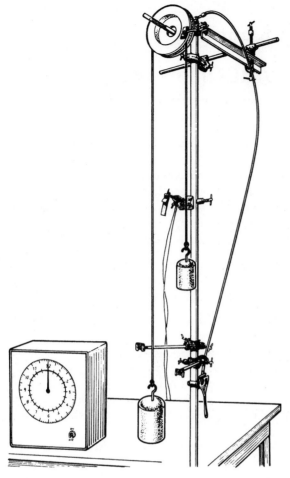

Figure 4.2 *Demonstration Atwood machine with electric timing device*

In uniformly accelerated motion, the velocity is increased by the same amount in each succeeding second. The distance traveled is equal to the average velocity multiplied by the time. Since the system starts from rest and experiences a uniform acceleration, the final velocity will be twice the average velocity. From the time taken to acquire this final velocity, the corresponding acceleration can be computed. The equations of motion involved in uniformly accelerated motion follow. The distance is given by

$$s = \bar{v}t \tag{4.2}$$

where s is the distance in meters, $\bar{v}$ is the average velocity in meters per second, and t is the time in seconds. The average velocity is given by

$$\bar{v} = \frac{v_1 + v_2}{2} \tag{4.3}$$

where $\bar{v}$ is the average velocity; v_1 is the initial velocity, which is zero in this case; and v_2 is the final velocity. The final velocity is given by

$$v_2 = v_1 + at \tag{4.4}$$

where v_2 is the final velocity; v_1 is the initial velocity (zero in this case); a is the acceleration in meters per second per second; and t is the time in seconds.

APPARATUS

1. Pulley clamped to a vertical rod
2. Two weight holders
3. Set of slotted weights
4. Set of small weights, including
     one 10-g weight
     one 5-g weight

    five 2-g weights
    one 1-g weight
5. String (strong fishing line)
6. Stopwatch or stop clock
7. 2-m stick

PROCEDURE

1. Using a total mass of about 2 kg, determine the force of friction in the machine by transferring masses from the ascending side to the descending side until the mass on the descending side moves downward with constant velocity when given a very slight push. Record the mass on the descending side and the mass on the ascending side. Be sure to include the mass of each weight hanger. Arrange the weights so as to have five 2-g weights on the ascending side when the machine is in the balanced condition. CAUTION: Always stand clear of the suspended weights because the string may break.

2. Transfer two more grams from the ascending side to the descending side, and thus determine the accelera-

tion produced by a net force of 0.004 kg × 9.8 m/s² = 0.0392 N. Begin the observation when the ascending mass is on the floor, starting the stopwatch at the instant when you let the weights go, and stopping it at the instant when the descending mass strikes the floor. Make and record four independent observations.

3. Measure the distance traversed.

4. Repeat Procedure 2, using accelerating forces of 0.008 × 9.8, 0.012 × 9.8, 0.016 × 9.8, and 0.02 × 9.8 N, by transferring two additional grams each time. Make four independent observations for each accelerating force.

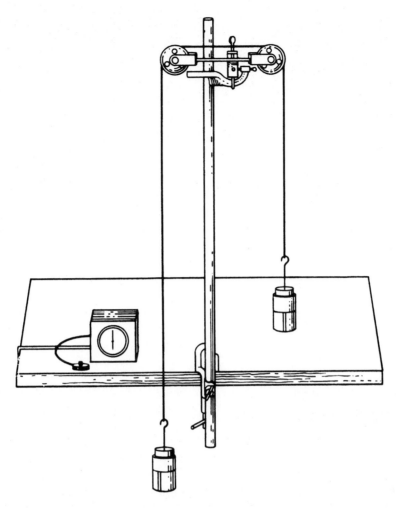

Figure 4.3   *Two-pulley Atwood machine allowing the use of small, light pulleys without the risk of collision between the rising and falling weights*

DATA _____

Distance traversed	_____	Ascending mass, for constant velocity	_____
Descending mass, for constant velocity	_____	Force of friction	_____

Descending mass	Ascending mass	Net force	Time 1	2	3	4	Average

Net force	Average velocity	Final velocity	Acceleration experimental value	Acceleration from $F = ma$	Percent error

CALCULATIONS

1. Calculate the force due to friction by subtracting the value of the ascending mass from that of the descending mass as found in Procedure 1 and multiplying this difference by the acceleration of gravity. Record your result, being careful to use proper units. Note that all forces should be expressed in newtons.

2. Compute the average time taken with each accelerating force. From the known distance and the time taken in each case, compute the average velocity corresponding to each accelerating force.

3. Calculate the final velocity for each set of observations.

4. Calculate the acceleration produced by each accelerating force from the values of the final velocity and the time.

5. Compute the theoretical value for the acceleration from Newton's second law of motion. Assuming this to be the correct value, compute the percent error of the observed value for each case.

QUESTIONS _____

1. (a) What time would be required for a mass to fall freely through the distance traversed by the masses in your experiment? (b) Could you have measured this time accurately with your stopwatch? (c) Given your answers to parts (a) and (b) and the fact that investigators at the time the Atwood machine was invented (by the Reverend George Atwood in 1784) had no fast electronic timing equipment at their disposal, state in your own words what this machine accomplishes.

2. What is the advantage of transferring masses from one side to the other, instead of adding masses to one side?

3. What is the difference between uniform motion (constant velocity) and uniformly accelerated motion?

4. Why should the pulley be as light as possible as well as nearly frictionless?

5. If you gave the system an initial velocity different from zero, how would this affect your results?

6. Compute the a.d. for each of your sets of time measurements, which will give you an error to be associated with each experimental value of the acceleration. Does the accepted value determined from Newton's second law fall within your limits of error in each case? If not, try to explain why not.

7. Using Equations 4.2, 4.3, and 4.4, derive an expression for s as a function of t for this case of uniformly accelerated motion.

Projectile Motion 5

The motion of a projectile is a more general case of a freely falling body. It's more general because the projectile, instead of simply being dropped, is launched with an initial velocity in some arbitrary direction. Because the force of gravity is directed vertically down, the motion takes place in a plane determined by this vertical direction and the direction in which the projectile is launched. The motion is thus limited to two dimensions; that is, we will have x and y components but no z components. Because we throw things all the time, and because here on Earth things are inevitably subject to the force of gravity, projectile motion is an important application of Newton's second law. Although air resistance must be taken into consideration for fast-moving projectiles, we can neglect it in the present experiment while still gaining insight into what happens when we throw something or fire a gun. In addition, we will have the opportunity to study a case in which random error (caused by small, uncontrollable variations in the firing of the launcher) will be apparent.

THEORY

Projectile motion may be studied very effectively by noting that in the two-dimensional plane in which the motion takes place, the horizontal and vertical directions are at right angles to each other so that vectors along these respective directions are independent. We thus really have two independent motions—one of constant speed in the horizontal direction (there being no component of force and hence no acceleration in this direction) and the other of constant acceleration in the vertical direction due to gravity. If the launcher is fixed at some angle θ with the horizontal as shown in Fig. 5.1, the projectile has a vertical component $v_0 \sin \theta$ of its initial velocity v_0

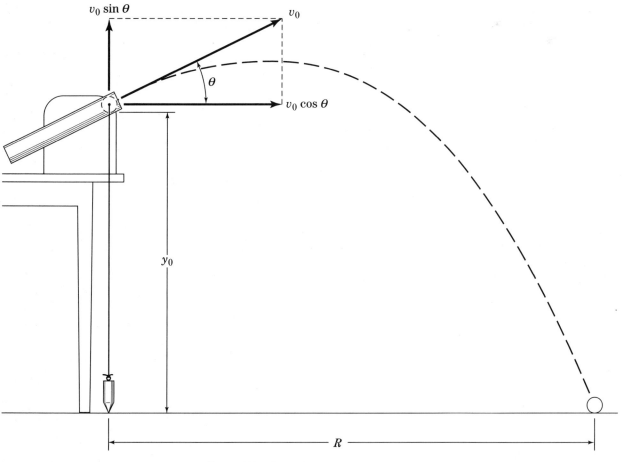

Figure 5.1 *Setup for range measurements*

in the vertical direction, and the motion in this direction is simply that of a freely falling body with acceleration $-g$ and initial velocity $v_0 \sin \theta$. There is also a horizontal velocity component $v_0 \cos \theta$, and as there is no force component in this direction, the horizontal motion is merely one of constant speed at this value. To state this situation in quantitative terms, we choose an $x - y$ coordinate system with the x axis horizontal, the y axis vertical, and the origin at the launcher's exit. We can then write for the horizontal direction

$$x = (v_0 \cos \theta)t \qquad (5.1)$$

where t is the time in seconds measured from the instant ($t = 0$) at which the projectile leaves the launcher. Similarly, for the vertical direction

$$y = (v_0 \sin \theta)t - \frac{1}{2}gt^2 \qquad (5.2)$$

Now we are interested in the range R, the horizontal distance from the point where the projectile leaves the launcher to the point where it strikes level ground a distance y_0 below the launching point, as Fig. 5.1 shows. We can therefore find the flight time t_f by substituting $-y_0$ for y in Equation 5.2. The result is

$$t_f = \frac{1}{g}\left[v_0 \sin \theta \pm \sqrt{v_0^2 \sin^2 \theta + 2gy_0}\right] \qquad (5.3)$$

where the solution of the quadratic equation has given rise to the + or −. However, as the minus sign would produce a negative time, we shall reject that solution. Equation 5.1 will then give the range R for x if we put the value t_f from Equation 5.3 in for t. We thus have

$$R = \frac{v_0 \cos \theta}{g}\left[v_0 \sin \theta + \sqrt{v_0^2 \sin^2 \theta + 2gy_0}\right] \qquad (5.4)$$

In the present experiment we shall investigate two values of the angle θ: $0°$, which is the case of the launcher being fired horizontally, and $45°$. It is readily seen that if $\theta = 0°$, $\sin \theta = 0$, $\cos \theta = 1$, and Equation 5.4 becomes

$$R = v_0 \sqrt{\frac{2y_0}{g}} \qquad (\theta = 0°) \qquad (5.5)$$

For $\theta = 45°$, $\sin \theta = \cos \theta = 1/\sqrt{2}$ and we get

$$R = \frac{v_0^2}{2g}\left[1 + \sqrt{1 + \frac{4gy_0}{v_0^2}}\right] \qquad (\theta = 45°) \qquad (5.6)$$

We shall fire ten shots from the launcher at each of these angles, obtaining the initial or muzzle velocity v_0 in each case by electronically measuring the time the projectile takes to pass through two precisely spaced light gates mounted at the launcher's exit. We will use carbon paper to mark the spot where the projectile hits the floor and can thus measure the range R for each shot. The purpose of the experiment will then be to see if the predicted values of R obtained from Equations 5.5 and 5.6 agree with the range measurements.

APPARATUS

1. Launcher with projectile ball
2. Mounting stand for launcher
3. Light-gate assembly
4. 2-meter stick
5. Surveyor's measuring tape

6. Metric steel scale
7. Plumb bob with string
8. Carbon paper
9. White paper and masking tape
10. Level

PROCEDURE

1. When you come into the laboratory you will find the launcher mounted on its stand, which is, in turn, clamped to the bench and directed to fire over empty floor space. You will also find the light-gate assembly attached to the launcher and connected to the electronic timer. Note that the two light gates, each consisting of a light source and a photocell, are mounted a fixed distance apart on a bracket attached to the launcher and holding the gates so that the ball must pass through first one and then the other, interrupting the light beam and triggering the photocell as it does so. Interruption of the beam in the first light gate starts the timer, and interruption in the second one stops it. The ball's initial velocity, v_0, with which it leaves the launcher, can then be found by dividing the distance between the light gates by the elapsed time shown on the timer. Examine the apparatus before you start work, and have the instructor explain the operation of the launcher, the light gates, and the timer to you.

2. For the first part of the experiment we will fire the launcher horizontally (angle of elevation $\theta = 0$). The launcher is clamped to its stand in such a way that by loosening the clamping screws, it may be swung to any angle. If it is equipped with its own small plumb line and angle scale, you may use this to set θ to zero, but greater precision can be obtained by laying your level along the flat top of the launcher and setting it accordingly. Tighten the clamping screws after you have set θ.

3. Measure the distance between the light beams in the two light gates with the steel scale. This will be approximately 10 cm and is critical for your determination of v_0, so be careful to make this measurement as precise as possible. In some cases this distance has been set very accurately by the manufacturer and will be given to you by your instructor. Record your result in the space provided.

4. Use the 2-meter stick to measure the height above the floor of the ball at the point where it leaves the launcher. Measure to the *bottom* (not the center) of the ball, as it is the bottom of the ball that strikes the floor. Your result will then be y_0, the actual vertical distance through which the ball falls. Record this value in your data sheet.

5. Hang the plumb bob on a string passing over the launcher through the point where the ball makes its exit. Lower the plumb bob until it almost touches the floor. Then take the surveyor's tape and, using masking tape, fasten the surveyor's tape to the floor with its end directly under the point of the plumb bob. Extend the surveyor's tape in the direction in which the ball will go. *Note:* More precision may be obtained if you position, say, the 10-cm mark rather than the tape end under the plumb bob's point. If you do this, make sure the tape extends in the direction in which the ball will go and don't forget to subtract the 10 cm when using the tape to make your range measurements.

6. Get the launcher ready for firing by inserting the ball in the launcher's muzzle and pushing it back against the spring with the ramrod provided. If the launcher has an adjustment for the spring tension, use the long-range setting. Push the ball into the launcher until the trigger is engaged. Turn on the light-gate timer and set it to zero. Fire the launcher and note about where the ball hits the

floor. Also note the timer reading. Repeat this firing several times for practice to familiarize yourself with the operation of the timer and to see about where the ball lands.

7. Using the masking tape, fasten a sheet of white paper to the floor so that the ball will land near its center. Place a sheet of carbon paper face down over the white paper so that the point where the ball strikes will be marked.

8. You are now ready to take data. Fire ten shots, recording the timer reading and measuring the range with the surveyor's tape each time. When you are finished, remove the white paper with the landing marks on it from the floor for inclusion in your report.

9. Loosen the clamping screws and reset the launcher to an angle of elevation of 45°. Make this setting either with the launcher's plumb line and angle scale or, if your level has a bubble glass at 45°, by again laying the level along the flat top surface of the launcher. Then repeat Procedures 4–8. *Note:* If the launcher pivots about the point where the ball leaves it, you will get the same result for y_0 here as you did in Procedure 4, but you should nevertheless check this. When you have finished, remove all masking tape from the floor, pick up the surveyor's tape, place the plumb bob on the bench, and switch off the timer.

DATA _____

Distance between the light gates _____   Height y_0 of launching point above
Angle of elevation $\theta = 0°$ the floor _____

Shots	Timer reading	Initial velocity v_0	Deviations	Range R, measured	Deviations
1					
2					
3					
4					
5					
6					
7					
8					
9					
10					
	Averages				

Calculated value of R _____   Percent error _____

Angle of elevation $\theta = 45°$ Height y_0 of launching point above
                                                                  the floor   _____

Shots	Timer reading	Initial velocity v_0	Deviations	Range R, measured	Deviations
1					
2					
3					
4					
5					
6					
7					
8					
9					
10					
	Averages				

Calculated value of R _____   Percent error   _____

CALCULATIONS

1. From the data of Procedure 8, calculate the initial velocity v_0 for each shot and find the average value of v_0.

2. Calculate the deviation of each of your values of v_0 from the average and then calculate the average deviation (a.d.). Express your result for v_0 as the average ± the a.d.

3. From the range data taken in Procedure 8, calculate the average range of the projectile. Take this to be your measured value of R.

4. Calculate the deviation of each of your range measurements from your average found in Calculation 3 and then calculate the average deviation (a.d.) in the measured range. Express your result for R as the average value ± the a.d.

5. Substitute the average value of v_0 found in Calculation 1 and the value of y_0 measured in Procedure 4 in Equation 5.5 to obtain the calculated value of R for the case of $\theta = 0$.

6. Treat your calculated value of R as the true value and find the percent error in your measured value. Note whether the true value of R falls within ± the a.d. of the measured value.

7. Find the possible error in the calculated value of R due to the average deviation in v_0 . Does your result here affect your work in Calculation 6?

8. Repeat Calculations 1–7 for the case of $\theta = 45°$ using Equation 5.6 and the data of Procedure 9. Note that in repeating Calculation 7 with this equation it will only be necessary to substitute your value of v_0 + the a.d. and your value of v_0 – the a.d. to get the possible spread in the calculated value of R.

QUESTIONS

1. Briefly discuss the errors you encountered in this experiment, noting in particular whether your calculated value of the range fell within the error associated with the measured value. Note also how much uncertainty the error in your determination of v_0 introduced into the calculated value of R and justify the fact that in using Equations 5.5 and 5.6 you attached no error to the values of the acceleration of gravity g or the height y_0 of the launch point above the floor.

2. Using the deviations in the range data found in Calculation 4 for the case of $\theta = 0$ and in Calculation 8 for the case of $\theta = 45°$, find the standard deviation σ for your two measured ranges. See Equation 15 in the Introduction. Do your calculated values of R fall within $\pm\sigma$ of the respective average values of the measured range?

3. Derive Equation 5.3. Explain the significance of the minus sign in this equation and tell why we were justified in rejecting it.

4. Show how Equations 5.5 and 5.6 are obtained from Equation 5.4 in the respective special cases.

The Ballistic Pendulum 6

The principle of conservation of momentum follows directly from Newton's laws of motion. According to this principle, if there are no external forces acting on a system containing several bodies, then the momentum of the system remains constant. In this experiment, the principle is applied to the case of a collision, using a ballistic pendulum. A ball is fired into the pendulum's bob, and the initial velocity of the ball is determined in terms of the masses of the ball and the bob and the height to which the bob rises after impact. This is a particularly interesting experiment because it had an important application—namely, the determination of the muzzle velocities of guns in the days before sophisticated electronics made direct determination of such high velocities possible. Remember that momentum is the product of mass and velocity, and you've been taught that it is conserved. In particular, it remains unchanged in a collision because, unlike energy, there is no other form such as heat into which it can be converted.

Thus, if we fire a small bullet with high velocity into a massive block, the momentum stays the same but the block with the bullet in it has much greater mass than the bullet alone and therefore moves off with a correspondingly smaller velocity, which can be measured by elementary means. In the case of the ballistic pendulum, the massive block is the pendulum bob, and we measure its relatively small velocity by noting how high the bob rises in its swing, determining from this data the kinetic energy the bob and the embedded projectile had right after impact, and obtaining therefrom the required velocity. Conservation of momentum along with our weighing of the projectile and pendulum bob to determine their masses will then enable us to find the muzzle velocity. In our case, in our modern electronic age, we will be able to measure this velocity directly, and the purpose of the present experiment will be to show that our method of finding it using conservation of momentum really works.

THEORY

The momentum of a body is defined as the product of the mass of the body and its velocity. Newton's second law of motion states that the net force acting on a body is proportional to the time rate of change of momentum. Hence, if the sum of the external forces acting on a body is zero, the linear momentum of the body is constant. This is essentially a statement of the principle of conservation of momentum. Applied to a system of bodies, the principle states that if no external forces act on a system containing two or more bodies, then the momentum of the system *as a whole* does not change.

In a collision between two bodies, each one exerts a force on the other. These forces are equal and opposite, and if no other forces are brought into play, the total momentum of the two bodies is not changed by the impact. Hence, the total momentum of the system after collision is equal to the total momentum of the system before collision. During the collision the bodies become deformed and a certain amount of energy is used to change their shape. If the bodies are perfectly elastic, they will recover completely from the distortion and will return all of the energy that was expended in distorting them. In this case, the total kinetic energy of the system also remains constant. If the bodies are not perfectly elastic, they will remain permanently distorted, and the energy used up in producing the distortion is not recovered.

Inelastic impact can be illustrated by a device known as the ballistic pendulum, once used to determine bullet speeds. If a bullet is fired into a pendulum bob and remains imbedded in it, the momentum of the bob and bullet just after the collision is equal to the momentum of the bullet just before the collision. The velocity of the pendulum before collision is zero, so that the total momentum of the system consisting of the bullet and bob is then just that of the bullet alone. After the collision, the pendulum and the bullet move with the same velocity. Thus, conservation of the system's momentum requires

$$mv_0 = (M + m)V \qquad (6.1)$$

where m *is* the mass of the bullet in kilograms, v_0 is the velocity in meters per second of the bullet just before the collision, M is the mass of the pendulum bob, and V is the common velocity of the bob and bullet just after the collision.

As a result of the collision, the pendulum with the imbedded bullet swings about its point of support, and the center of gravity of the system rises through a vertical distance h. From a measurement of this distance, it is possible to calculate the velocity V. The kinetic energy of the system just after the collision must be equal to the increase in potential energy of the system as the pendulum stops at its highest point. This follows from the law of conservation of energy; here we assume that the loss of energy due to friction at the point of support is negligible. The energy equation gives

$$\frac{1}{2}(M + m)V^2 = (M + m)gh \qquad (6.2)$$

where M is the mass of the pendulum bob, m is the mass of the bullet, V is their common velocity just after the collision, g is the known value of the acceleration of gravity, and h is the vertical distance through which the center of gravity of the system rises. The left-hand side of Equation 6.2 represents the kinetic energy of the system just after the impact. When the pendulum stops at the top of its swing, all this kinetic energy has been changed to the potential energy given by the right-hand side of this equation. Solving it for V yields

$$V = \sqrt{2gh} \qquad (6.3)$$

after which Equation 6.1 can be used to determine v_0. We shall also use the electronic arrangement described in Procedure 1 of Experiment 5 to measure our launcher's muzzle velocity directly and will then be able to see how well this application of the law of conservation of momentum can serve to determine this velocity in the absence of fancy electronics.

The apparatus used in this experiment includes a spring gun or launcher that fires a steel ball into a massive bob hollowed out to receive the ball and provided with a catch to retain the ball in the bob once it has entered. The bob is suspended by a light rod pivoted at its upper end on a bearing that has been made as frictionless as possible. The height to which the center of gravity of the loaded pendulum rises may be determined in either of two ways, depending on the type of apparatus you have. In the version shown in Fig. 6.1(a), the pendulum carries a pawl that engages a series of teeth on a curved track and prevents the pendulum from swinging back, thus catching it at its highest point. The distance risen is then easily measured. In another version, the pendulum pushes a light pointer along an angular scale so that this pointer indicates the maximum angle ϕ through which the pendulum has swung. Fig. 6.1(b) shows this form of the apparatus. Note that the distance h through which the center of mass has risen must be obtained from the formula $h = r(1 - \cos \phi)$, which follows from a consideration of Fig. 6.2. Here r is the distance from the pivot to the center of mass of the bob *with the ball in it*. With either apparatus, if the center of mass has not been marked on your pendulum, you will have to find it by balancing the pendulum and ball on a narrow, round rod or suspending it on a string. Good results have been obtained by these simple means, but try to be as precise as possible. Remember that you can get good results without highly sophisticated equipment, but you have to be careful.

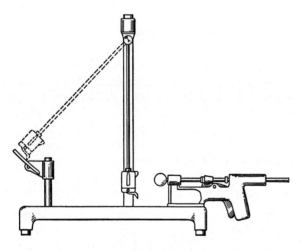

Figure 6.1(a) *The Blackwood ballistic pendulum using pawl and ratchet track*

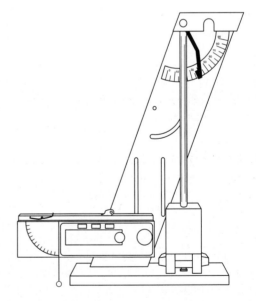

Figure 6.1(b) *The Blackwood ballistic pendulum using pointer and angle scale*

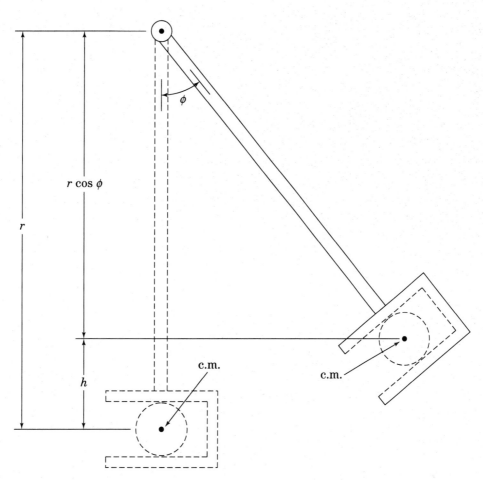

Figure 6.2   *Determination of* h *in the ballistic pendulum*

APPARATUS

1. Blackwood ballistic pendulum
2. Spring gun or launcher with projectile ball
3. Light-gate assembly
4. Triple-beam balance
5. Metric steel scale or ruler
6. Level

PROCEDURE

1. We will first find the initial velocity of our projectile by timing it over a short distance between two light gates as shown in Fig. 6.3 and described in Procedure 1 of Experiment 5. If you have not done this experiment, have your instructor explain the operation of the launcher, the light gates, and the timer to you. The pendulum is not used in this step and must be removed from the apparatus by unscrewing the pivot rod. In the form of the apparatus using a pawl and ratchet track, the pendulum is removed by loosening the thumbscrew holding the supporting needle bearing. In addition, the ratchet track may also be in the way and must be removed. Your instructor will supply the proper tool for doing this. He or she will also show you how to install the light-gate assembly on the launcher. Finally, check that the apparatus is level; if it is not, place sheets of paper under the feet as required.

2. Measure the distance between the light beams in the two light gates with the steel scale. This will be approximately 10 cm and is critical for your determination of v_0, so be careful to make this measurement as precise as possible. In some cases this distance has been set very accurately by the manufacturer and will be given to you by your instructor. Record this parameter in the space provided. Then mount the light-gate assembly on the launcher.

3. Weigh the pendulum and the ball separately and record their respective masses in the spaces provided in the data sheet.

4. Get the launcher ready for firing by inserting the ball in the launcher's muzzle and pushing it back against the spring with the ramrod provided. If the launcher has an adjustment for the spring tension, use the long-range setting. Push the ball into the launcher until the trigger is

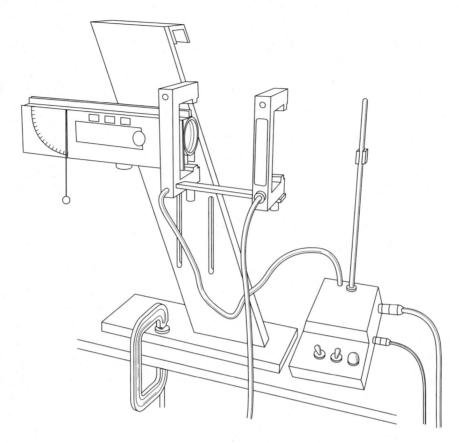

Figure 6.3 *Launcher with light gates for muzzle velocity measurements*

engaged. Turn on the light-gate timer and set it to zero. Place an appropriate backstop in front of the launcher and pull the trigger. Note and record the timer reading. Repeat this procedure four more times to obtain five readings of the timer, and record your results.

5. Remove the light-gate assembly from the apparatus and reinstall the pendulum, making sure that it swings freely. If you have an apparatus using a pointer and angular scale, check that the pointer is in a position to be pushed up the scale when the pendulum swings. If you have an apparatus using a pawl and ratchet track, reinstall the latter and make sure the pawl catches it when the pendulum swings up. Then load the launcher, let the pendulum hang straight down, and when it is at rest, pull the trigger, thus firing the ball into the bob and causing the pendulum to swing up with the ball inside it. Record the notch on the ratchet track reached by the pawl if you are using this form of the apparatus, or the angle to

which the pointer has been pushed if you are using the other form.

6. Repeat Procedure 5 four more times, recording either the notch reached by the pawl or the indicated angle each time.

7. If you are using the apparatus with the pawl and ratchet track, calculate an average value of the position of the pendulum on the track from the data of Procedure 5. Then set the pendulum with the pawl engaged in the notch that corresponds most closely to the average just found. Using the metric steel scale, which you should read to 0.1 mm, measure the vertical distance h_1 from the base of the apparatus to the center of gravity marked on the pendulum.

8. With the pendulum hanging straight down, measure the vertical distance h_2 from the base of the apparatus to the center of gravity. Again use the steel scale and measure to 0.1 mm.

DATA

Distance between the light gates _____

Timer readings	1. _____	2. _____	3. _____	4. _____	5. _____
Initial velocity v_0	1. _____	2. _____	3. _____	4. _____	5. _____
Deviations	1. _____	2. _____	3. _____	4. _____	5. _____

Average value of v_0 _____   Average deviation _____

Mass of the pendulum bob _____   Mass of the ball _____

Distance r from the pendulum pivot to
the center of mass of the pendulum
with the ball in the bob _____

Ballistic pendulum with pawl and ratchet track:

Highest notch reached   Trials

Average _____   1. _____   2. _____   3. _____   4. _____   5. _____

Average height h_1 of pendulum's center   Vertical distance h through which the
of gravity in its highest position _____   pendulum's center of gravity rose _____

Height h_2 of the pendulum's center of
gravity in its lowest position _____

Ballistic pendulum with pointer and angle scale:

Value of angle ϕ 1. _____   2. _____   3. _____   4. _____   5. _____

Value of height h 1. _____   2. _____   3. _____   4. _____   5. _____

Deviations   1. _____   2. _____   3. _____   4. _____   5. _____

Average value of h _____   Average deviation (a.d.) _____

Velocity of pendulum and ball just   Difference between this value of v_0
after the collision _____   and that measured directly _____

Velocity v_0 of the ball just before the   Percent difference _____
collision _____

CALCULATIONS _____

1. From the data of Procedure 4, calculate your launcher's initial velocity v_0 for each of the five shots and calculate the average.

2. Calculate the deviation of each of your values of v_0 from the average and then calculate the average deviation (a.d.). Enter these results in the spaces provided for them in the data table and note that you should express this directly measured value of v_0 as the average ± the a.d.

3. (a) Pendulum with pawl and ratchet track: From the data of Procedures 5–8, calculate the vertical distance *h* through which the center of gravity of the loaded pendulum rose as a result of the collision.

(b) Pendulum with pointer and angle scale: From the data of Procedures 5 and 6, calculate the height *h* for each of your five shots and the average value of *h*. Then calculate the deviations from this average and the average deviation (a.d.).

4. Compute the value of *V*, the common velocity of the pendulum bob and ball just after the collision, by using Equation 6.3.

5. Calculate the velocity of the ball before the collision by substituting the value of *V* found in Calculation 4 and the measured values of the masses of the pendulum bob and of the ball in Equation 6.1.

6. Assuming the directly measured value of v_0 to be the true value, compare it with the value obtained in Calculation 5 by finding the difference between them and expressing it as a percent. Note also whether your value of v_0 from Calculation 5 falls within the range of error found in Calculation 2 for the directly measured value.

7. (a) Pendulum with pawl and ratchet track: Estimate your error in the measurement of h and find the resulting error to be associated with the value of v_0 obtained in Calculation 5. Does your directly measured value of v_0 fall within this error range?

   (b) Pendulum with pointer and angle scale: Use the error found in your measurement of h in Calculation 3(b) to find the error to be associated with the value of v_0 from Calculation 5. Does your directly measured value of v_0 fall within this error range?

QUESTIONS _____

1. Using the data of your experiment, calculate the kinetic energy of the ball just before impact from the value of the velocity of the ball obtained with the ballistic pendulum and the mass of the ball.

2. Calculate the kinetic energy of the pendulum bob and ball just after impact from the values of their common velocity and their masses.

3. (a) Using the results of Questions 1 and 2, calculate the fractional loss of energy during this inelastic impact. Express it in percent. (b) What became of the energy lost?

4. (a) Compute the ratio of the mass of the pendulum bob to the total mass of the bob and the ball. Express it in percent. (b) How does this ratio compare with the fraction of energy lost during the impact?

5. Compare the momentum and the kinetic energy of an automobile weighing 2000 lb and moving with a velocity of 60 mi/h with the momentum and the kinetic energy of a projectile weighing 70 lb and moving with a velocity of 2500 ft/s.

6. A bullet of mass 10 g is fired horizontally into a block of wood of mass 2 kg and suspended like a ballistic pendulum. The bullet sticks in the block and the impact causes the block to swing so that its center of gravity rises 0.1 m. Find the velocity of the bullet just before the impact.

7. Derive an equation giving the gun's muzzle velocity v_0 in terms of h and the masses of the ball and pendulum by combining Equations 6.1 and 6.3. Comment on the dependence of v_0 on the ratio of the two masses.

Uniform Circular Motion 7

Whenever a body moves in a circular path, a force directed toward the center of the circle must act on the body to keep it moving in this path. That force is called centripetal force. The reaction, which is equal and opposite, is the pull of the body on the restraining medium and is called centrifugal force. The purpose of this experiment is to study uniform circular motion and to compare the observed value of the centripetal force with the calculated value.

THEORY

If a body moves with constant speed in a circle, it is said to be moving with uniform circular motion. Even though the speed is constant, the velocity is continuously changing because the direction of the motion is continuously changing. Thus, such a body has an acceleration. It can be shown that the direction of the acceleration is always toward the center of the circle (because it is only the direction and not the magnitude of the velocity that is changing) and that its magnitude is given by

$$a = \frac{v^2}{r}$$

where v is the speed of the body in meters per second and r is the radius of the path in meters.

A force is necessary to produce this acceleration. Because it must be in the same direction as the acceleration—namely, toward the center of the circle as noted above—it is called *centripetal force*. Newton's second law now requires that the magnitude of this force be equal to the mass times the acceleration produced, so that in our present case

$$F = m\frac{v^2}{r}$$

where F is the force in newtons, m is the mass of the rotating body in kilograms, and v and r are the same as before. But by Newton's third law of motion, an equal and opposite force is exerted by the body on the restraining medium. This reaction is called *centrifugal force*.

The centripetal force can also be expressed in terms of the angular speed, since

$$v = r\omega \quad \text{and} \quad \omega = 2\pi f$$

where v is the linear speed and r the radius of the path as before, ω is the angular speed in radians per second, and f is the number of revolutions per second. Thus,

$$F = mr\omega^2 \quad \text{or} \quad F = 4\pi^2 f^2 rm$$

where F is the centripetal force in newtons as already described.

In one form of apparatus used for this experiment (see Fig. 7.1) a stretched spring provides the centripetal force necessary to keep a mass rotating in a circle of a particular measurable radius. The rotator is adjusted to a rotational speed such that the back of the mass inside it just touches the sensing probe of a sensitive indicator. The indicator's pointer then rises to stand opposite an index ridge at the end of the shaft on which the apparatus turns. As this index and the pointer are near the center of

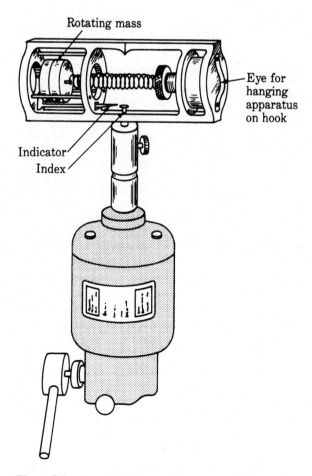

Figure 7.1 *Centripetal force apparatus, motor-driven form*

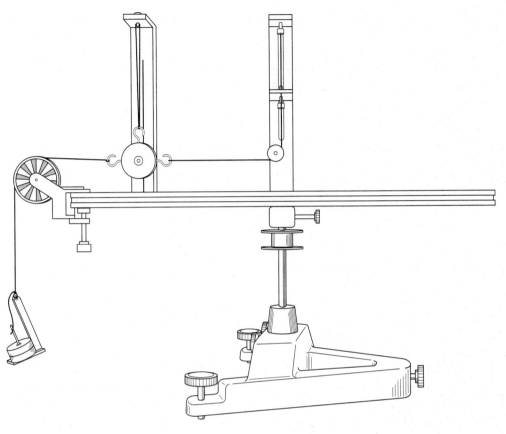

Figure 7.2   *Centripetal force apparatus, manual form*

rotation, they are easily visible regardless of the rotational speed. This speed is kept constant at the value just described and is measured by timing some convenient number of turns. The force required to stretch the spring by the same amount is subsequently measured by hanging weights from it.

Another type of apparatus is shown in Fig. 7.2. Again, a stretched string provides the centrifugal force to keep the rotating mass moving in a circle, but the complete assembly is mounted on a horizontal bar (the rotating platform) mounted on a vertical shaft carried in very good bearings. A knurled portion of the shaft provides a convenient grip for the experimenter's fingers, and you'll find with a little practice that you can keep the assembly rotating at a constant speed quite accurately. Note that at the bottom of the spring there is a small pink disk, and because the spring and disk are at the center of rotation, the position of the latter is easily observed even with the

apparatus in motion. The mass to be kept rotating at a particular radius is suspended by threads from a post mounted on the rotating platform at that radius. Weights hung on a thread passing over a pulley at the end of the platform are used to bring the rotating mass to the selected radius with the apparatus at rest, and the marker ring on the central post is moved to the position now occupied by the pink disk. The weights are then removed and the apparatus rotated to a speed that again brings the pink disk into the marker ring. The centripetal force exerted by the spring must then be the same as that exerted by the weight when the apparatus wasn't turning, and the rotating mass must be back at the selected radius. The apparatus parameters have been chosen so that no rotational speed will be high enough to make counting the turns difficult. Be careful, however, to count only one end of the platform as it comes around.

APPARATUS

1. Centripetal force apparatus
2. Motor with coupler and means for counting turns (Procedure A only)
3. Stopwatch or stop clock
4. Supporting rods, hook, right-angle clamp, and bench clamp (Procedure A only)

5. Weight hanger and weights
6. Vernier caliper (Procedure A only)
7. Level (Procedure B only)
8. Triple-beam balance (Procedure B only)

PROCEDURE A Motor-Driven Apparatus _____

1. When you come into the laboratory, examine the rotating assembly carefully in order to understand how it works. Do not change the spring tension adjustment; it has already been set by your instructor. Note and record the value of the rotating mass. This value either is marked on the mass itself or will be given to you.

2. Set up the bench clamp, a vertical rod, right-angle clamp, horizontal rod, and hooked collar so as to suspend the apparatus vertically. Hook the weight hanger to the eye on the rotating mass and add weights carefully until this mass is pulled down just enough to touch the sensitive indicator and bring the pointer to the index. Record the mass required to do this. Don't forget to include the mass of the weight hanger.

3. With the weights still on the hanger so that the rotating mass is in the position established in Procedure 2, use the vernier caliper to measure the mass's radius of rotation. The measurement is made from the point on the shaft above the index (the axis of rotation) to the line engraved around the mass, which marks the location of its center of gravity in that direction.

4. Unhook the apparatus from the weight hanger and supporting rods and mount it on the motor shaft by means of the coupling provided. Mount the motor unit securely on the bench using the bench clamp, and check that all mounting clamps and screws are tight.

5. Make sure the speed control is at its lowest position and turn on the motor. As you gradually increase the speed, note that until the rotating mass has moved out to the radius at which it engages the indicator, the apparatus is highly unbalanced and will shake quite badly. Increase the speed until you see the pointer rise, but be careful not to increase it too much. CAUTION: It is dangerous to let the apparatus rotate at excessive speeds. Best results will be obtained if you get the pointer to rise just above the index and then slow the rotation very carefully, making very small speed changes and watching the results. Your aim is to keep the motor speed at the value for which the pointer stands exactly opposite the index, but this adjustment is very critical, and you may want to practice awhile before attempting to take data. Holding a sheet of white paper behind the apparatus as you watch the pointer and index will help you see them more clearly.

6. When you have the speed set at the correct value so that the pointer is opposite the index, measure this speed by timing a counted number of turns. If you are using a revolution counter, note its reading and then engage it and start the stop clock simultaneously. Disengage it and simultaneously stop the clock after about a minute and record the number of turns and the elapsed time. The setup shown in Fig. 7.1 uses a variable-speed motor with an 18:1 reduction gear on one end and a pointer mounted on the slowly turning shaft thus provided. Revolutions of this shaft are easily counted visually. In performing this experiment, it is a good plan to have one observer pay strict attention to the proper adjustment of the speed while the other counts turns and operates the stop clock.

7. Repeat Procedures 5 and 6 three more times so as to have a total of four measurements of the rotation speed. Record the number of revolutions counted and the elapsed time in each case.

PROCEDURE B Manual Apparatus _____

1. When you come into the laboratory, examine the rotating assembly carefully in order to understand how it works. Make sure that it is level by placing your level on the rotating platform and adjusting the leveling screws in the base until the platform's top surface remains exactly horizontal no matter which way it is turned. Observe that the rotating mass consists of three parts and is suspended by two threads from a supporting post that may be set at any point along the rotating platform. A centimeter scale on this platform and a vertical line on the post allow the distance of the post from the center of rotation to be measured quite accurately. Unhook the complete rotating mass from the threads, weigh it, and record your result. Then remove the two outer sections, weigh the central section alone, and record this result also. You will thus have two possible values for the rotating mass.

2. Replace the central piece of the complete rotating mass on the supporting threads and note that in addition to the hook by which you hang this mass, it has two other hooks, one facing toward the center of the apparatus and the other facing outward. A piece of thread about 30 cm long should run from the inner hook, under the pulley on the central post, and up to the bottom of the pink indicator disk. Check that this thread is in place (install it yourself if it is not) and that the run from the hook to the pulley is horizontal. If it is not, the height of the bracket holding the supporting threads may be adjusted until it is.

3. Run another thread from the outer hook on the rotating mass over the pulley at the end of the platform and down to a weight hanger. Place 150 g on the hanger and record the total mass (weights plus the hanger mass) so suspended. This will pull the rotating mass out and stretch the spring.

4. Carefully loosen the clamp fixing the supporting post to the platform and slide it along its track until the threads supporting the rotating mass are exactly vertical. You can do this by lining up the two threads with each other and with the vertical line scribed on the post. Be sure the post itself is vertical by sliding it so that its bottom edge bears on the track on the platform's lower edge. Clamp the post in position and read the radius at which you have set it from the scale on the platform. Record this value. Finally, move the indicator ring on the central post so that the pink indicator disk is centered in it and fix the ring in this position.

5. Now remove the weight hanger, weights, and thread from the outer hook on the rotating mass. The spring will pull this mass in toward the center of the apparatus, and your plan is to spin the assembly up to the speed at which the rotating mass returns to the position in which it was held by the weights with the apparatus at rest. Do this by carefully turning the knurled portion of the supporting shaft with your fingers, gradually increasing the rotational velocity until the pink indicator disk is pulled back down into the indicator ring. Practice for a while until you can keep this velocity constant at the

value for which the pink disk is centered in the ring. Then start the stop clock as one end of the platform comes around and stop it after ten turns. Measure the time for ten turns in this manner four times and record your results. Note that it is a good plan in making these measurements to have one observer pay strict attention to keeping the speed constant at the proper value while the other counts turns and operates the stop clock.

6. Repeat Procedures 3–5 with 100 g on the weight hanger and again using 50 g.

7. Reassemble the two outer portions of the rotating mass to the central piece and repeat Procedures 3–6 using this larger rotating mass.

DATA

Value of the rotating mass   _____

Mass hung on the spring   _____

Total mass necessary to stretch the spring (Procedure A only)   _____

Measured value of the centripetal force   _____

Difference between calculated and measured value   _____

Radius of rotation   _____

Centripetal force calculated from the theory   _____

Percent discrepancy   _____

Trial	Number of turns	Time interval	Rotation speed	Deviation
1				
2				
3				
4				
Averages				

Value of the rotating mass   _____

Mass hung on the spring   _____

Total mass necessary to stretch the spring (Procedure A only)   _____

Measured value of the centripetal force   _____

Difference between calculated and measured value   _____

Radius of rotation   _____

Centripetal force calculated from the theory   _____

Percent discrepancy   _____

Trial	Number of turns	Time interval	Rotation speed	Deviation
1				
2				
3				
4				
Averages				

Value of the rotating mass　　_____　　Radius of rotation　　_____

Mass hung on the spring　　_____　　Centripetal force calculated from the

Total mass necessary to stretch the
　spring (Procedure A only)　　_____　　theory　　_____

Measured value of the centripetal force　　_____　　Percent discrepancy　　_____

Difference between calculated and
　measured value　　_____

Trial	Number of turns	Time interval	Rotation speed	Deviation
1				
2				
3				
4				
Averages				

Value of the rotating mass　　_____　　Radius of rotation　　_____

Mass hung on the spring　　_____　　Centripetal force calculated from the

Total mass necessary to stretch the
　spring (Procedure A only)　　_____　　theory　　_____

Measured value of the centripetal force　　_____　　Percent discrepancy　　_____

Difference between calculated and
　measured value　　_____

Trial	Number of turns	Time interval	Rotation speed	Deviation
1				
2				
3				
4				
Averages				

Value of the rotating mass　　_____　　Radius of rotation　　_____

Mass hung on the spring　　_____　　Centripetal force calculated from the

Total mass necessary to stretch the
　spring (Procedure A only)　　_____　　theory　　_____

Measured value of the centripetal force　　_____　　Percent discrepancy　　_____

Difference between calculated and
　measured value　　_____

Trial	Number of turns	Time interval	Rotation speed	Deviation
1				
2				
3				
4				
Averages				

Value of the rotating mass   _____   Radius of rotation   _____

Mass hung on the spring   _____   Centripetal force calculated from the

Total mass necessary to stretch the   theory   _____

     spring (Procedure A only)   _____   Percent discrepancy   _____

Measured value of the centripetal force   _____

Difference between calculated and

     measured value   _____

Trial	Number of turns	Time interval	Rotation speed	Deviation
1				
2				
3				
4				
Averages				

CALCULATIONS

1. From the data of Procedures 5–7, calculate the speed of rotation for each reading you have taken. Note that dividing the number of turns by the elapsed time gives you revolutions per second, that if you used a reduction gear you must multiply by the gear ratio to get the rotational speed of the apparatus, and that you must then multiply by 2π to get your result in radians per second.

2. Compute the average of your four speed measurements, the deviation of each from the average, and the average deviation (a.d.) for each set of readings. Note that you will have only one set if you used the motor-driven apparatus (Procedure A) but six sets if you used the manual apparatus (Procedure B).

3. Calculate the centripetal force from the theory using the average speed of rotation obtained in Calculation 2. If you are using Procedure B, you will have to do this and the next three Calculations six times.

4. *Procedure A:* Find the directly measured value of the centripetal force by adding the value of the rotating mass given you in Procedure 1 to the mass found in Procedure 2 and multiplying by the acceleration of gravity to get the force stretching the spring in newtons. Record this result in the space provided in the data sheet. *Procedure B:* The mass whose weight stretches the spring is just the mass hung on the weight hanger in Procedures 3, 6, and 7 plus the mass of the weight hanger. Multiply by the acceleration of gravity to get the force stretching the spring in newtons in each case and record your three results in the appropriate spaces.

5. Compare the calculated value of the centripetal force with the value measured directly by subtracting one from the other and computing the percent discrepancy.

6. Using the a.d. found in Calculation 2 as the error in your rotational speed measurement, find the resulting error in your calculated value of the centripetal force and see whether the directly measured value falls within the limits of error so established.

QUESTIONS _____

1. State what the experiment checks.

2. (a) How does the centripetal force vary with the speed of rotation for a constant radius of the path? (b) How does it vary with the radius of the path for a constant speed of rotation?

3. Distinguish between centripetal force and centrifugal force. Explain in what direction each force is acting and on what it is acting.

4. Calculate at what speed Earth would have to rotate in order that objects at the equator would have no weight. Assume the radius of Earth to be 6400 km. What would be the linear speed of a point on the equator? What would be the length of a day (time from sunrise to sunset) under these conditions?

5. Engines for propeller-driven aircraft are limited in their maximum rotational speed by the fact that the tip speed of the propeller must not approach the speed of sound in air (Mach 1). Taking 6 ft as a typical diameter for a propeller of a light airplane and 1100 ft/s as the speed of sound, find the upper limit on the rpm (revolutions per minute) of the propeller shaft.

Friction 8

Whenever a body slides along another body, a resisting force known as the force of friction is called into play. This is a very important force and serves many useful purposes. For example, a person could not walk without it, nor could a car propel itself along a highway without the friction between the tires and the road surface. On the other hand, friction is very wasteful; it reduces the efficiency of machines because work must be done to overcome it, and this energy is wasted as heat. The purpose of this experiment is to study the laws of friction and to determine the coefficient of friction between two surfaces.

THEORY

Friction is the resisting force encountered when one surface slides over another; this force acts along the tangent to the surfaces in contact. The force necessary to overcome friction depends on the nature of the materials in contact, their roughness or smoothness, and on the normal force, but not on the area of contact, within wide limits, or the speed of the motion. It is found experimentally that the force of friction is directly proportional to the normal force. The constant of proportionality is called the coefficient of friction.

When the contacting surfaces are actually sliding one over the other, the force of friction is given by

$$F_r = \mu_k N \qquad (8.1)$$

where F_r is the force of friction and is directed parallel to the surfaces and opposite to the direction of motion, N is the normal force, and μ_k is the coefficient of friction. The subscript k stands for *kinetic,* meaning that μ_k is the coefficient that applies when the surfaces are moving one with respect to the other. It is therefore more precisely called the coefficient of *kinetic* or *sliding* friction. Note carefully that F_r is always directed *opposite* to the direction of motion. This means that if you reverse the direction of sliding, the frictional force reverses too. In short, it is always against you. Friction is therefore called a *nonconservative force* because energy must be used to overcome it no matter which way you go. This is in contrast to a so-called *conservative force* such as gravity, which is against you on the way up but with you on the way down. Thus, energy expended in lifting an object may be regained when the object redescends, but the energy used to overcome friction is dissipated—that is, lost or made unavailable—as heat. As you will see later in your study of physics, the distinction between conservative and nonconservative forces is a most important one, fundamental to our concepts of heat and energy.

A method of checking the proportionality of F_r and N and of determining the proportionality constant μ_k is to have one of the surfaces in the form of a plane placed horizontally with a pulley fastened at one end. The other surface is the bottom face of a block that rests on the plane and to which is attached a cord that passes over the pulley and carries weights. These are varied until the block moves at constant speed after having been started with a slight push. Since there is thus no acceleration, the net force on the block is zero, which means that the frictional force is equal to the tension in the cord. This tension, in turn, is equal to the total weight attached to the cord's end. The normal force between the two surfaces is equal to the weight of the block and can be increased by placing weights on top of the block. Thus, corresponding values of F_r and N can be found, and plotting them will show whether F_r and N are indeed proportional. The slope of this graph gives μ_k.

When a body lies at rest on a surface and an attempt is made to push it, the pushing force is opposed by a frictional force. As long as the pushing force is not strong enough to start the body moving, the body remains in equilibrium, which means that the frictional force automatically adjusts itself to be equal to the pushing force and thus to just balance it. There is, however, a threshold value of the pushing force beyond which larger values will cause the body to break away and slide. We conclude that in the static case (body at rest) the frictional force automatically adjusts itself to keep the body at rest up to a certain maximum. But if static equilibrium demands a frictional force larger than this maximum, static equilibrium conditions will cease to exist because this force is not available, and the body will start to move. This situation may be expressed in equation form as

$$F_r \leq \mu_s N \qquad \text{or} \qquad F_{r_{max}} = \mu_s N \qquad (8.2)$$

where F_r is the frictional force in the static case, $F_{r_{max}}$ is the maximum value this force can assume, and μ_s is the coefficient of *static* friction. It is found that μ_s is slightly larger than μ_k, which means that a somewhat larger force is needed to break a body away and start it sliding than is needed to keep it sliding at constant speed once it is in

55

motion. This is why a slight push is necessary to get the block started for the measurement of μ_k.

One way of investigating the case of static friction is to observe the so-called *limiting angle of repose,* defined as the maximum angle to which an inclined plane may be tipped before a block placed on the plane just starts to slide. The arrangement is illustrated in Fig. 8.1. The block has weight W whose component $W \cos \theta$ (where θ is the plane angle) is perpendicular to the plane and is thus equal to the normal force N. The component $W \sin \theta$ is parallel to the plane and constitutes the force urging the block to slide down the plane. It is opposed by the frictional force F_r, and as long as the block remains at rest, F_r must be equal to $W \sin \theta$. If the plane is tipped up until at some value θ_{max} the block just starts to slide, then $F_{r_{max}} = W \sin \theta_{max}$. But $F_{r_{max}} = \mu_s N = \mu_s W \cos \theta_{max}$. Hence, $W \sin \theta_{max} = \mu_s W \cos \theta_{max}$ or

$$\mu_s = \frac{\sin \theta_{max}}{\cos \theta_{max}} = \tan \theta_{max} \qquad (8.3)$$

Thus, if the plane is gradually tipped up until the block just breaks away and the plane angle is then measured,

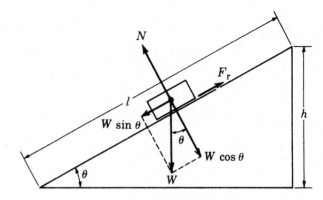

Figure 8.1   *The inclined plane*

the coefficient of static friction is equal to the tangent of this angle, which is called the *limiting angle of repose.* It is interesting to note that W canceled out in the derivation of Equation 8.3, so that the weight of the block doesn't matter.

APPARATUS

1. Board with pulley at one end
2. Rods, bench clamp, and right-angle clamp for supporting the inclined plane
3. Wood block with a cord attached to it
4. Glass block
5. Set of known weights
6. Weight hanger
7. Triple-beam balance
8. Protractor
9. Lintless dust cloth or paper wipers

PROCEDURE

1. Weigh the wood block and record the weight in newtons.

2. Place the board in a horizontal position on the laboratory table with its pulley projecting beyond the table's edge. Be sure that the surfaces of both the board and the wood block are clean, dry, and free of any dust or grit. Wipe them off if necessary with a clean, dry, lintless cloth or paper wiper. *After this has been done, do not touch these surfaces with your hands.* Handle the block with the cloth or a wiper and set it down *only* on the clean board. Begin the experiment by setting the block on the board with its largest surface in contact with the board's surface. Run the cord attached to the block over the pulley and attach it to the weight hanger. Place some weights on the hanger and slowly increase the load until it is just sufficient to keep the block sliding slowly with constant speed after it has been started with a very small push. Record this load. Don't forget to include the weight of the hanger.

3. Repeat Procedure 2 placing masses of 200, 400, 600, 800, and 1000 g successively on top of the wood

block. Record the load needed to produce constant speed in each case.

4. Turn the wood block on its side and repeat Procedure 2 with a mass of 400 g on top of the block. Record the load needed.

5. Again turn the wood block with the largest surface in contact with the plane and place 400 g on top of the block. Gradually increase the load on the hanger until the block just starts to move, without any initial push. Be careful to place the weights on the hanger gently so as not to jerk the cord. Notice whether this time the block moves with uniform speed or whether it is being accelerated. Record the load needed under these conditions. Repeat this procedure twice to obtain three independent measurements of the required load.

6. Set up the board as an inclined plane. Place the wood block on the plane with its largest surface in contact, and gradually tip the plane up until the block just breaks away and starts to slide down. Be very careful to tip the plane slowly and smoothly so as to get a precise value of the angle with the horizontal at which the block just breaks away. This is the limiting angle of repose θ_{max}. Measure it by means of a protractor and record the result obtained in three separate trials. These trials should be independent, meaning that in each case the plane should be returned to the horizontal, the block placed on

it, and the plane carefully tipped up until the limiting angle of repose is reached.

7. Repeat Procedure 6 using the glass block. Record the limiting angle of repose obtained in three independent trials.

DATA _____

Weight of wood block _____

Position of block	Mass placed on the block	Total normal force	Force to keep block moving uniformly
Flat	0 g		
Flat	200 g		
Flat	400 g		
Flat	600 g		
Flat	800 g		
Flat	1000 g		
On side	400 g		

Coefficient of kinetic friction μ_k from
   graph _____

Coefficient of kinetic friction μ_k from
   Procedure 4 _____

Total normal force for Procedure 5 _____

Trial	Force to start block moving	μ_s	Deviation
1			
2			
3			
Averages			

Block used	Trial 1			Trial 2			Trial 3			Averages	
	θ_{max}	μ_s	Deviation	θ_{max}	μ_s	Deviation	θ_{max}	μ_s	Deviation	μ_s	Deviation
Wood											
Glass											

Difference between your two values
   of μ_s _____

Percent error in agreement of the two
   values of μ_s _____

CALCULATIONS _____

1. From the data of Procedures 2 and 3, plot a curve using the values of the total normal force as abscissas and the values of the force of friction as ordinates. See if your curve is a straight line and obtain the coefficient of kinetic friction μ_k for wood on wood by finding its slope.

2. Use the data of Procedure 4 to calculate the coefficient of kinetic friction for the case of the wood block sliding on its side. Record your result and see how it compares with the value of μ_k obtained from your graph.

3. From the data of Procedure 5, compute the coefficient of static friction μ_s for wood on wood from each of your three trials. Calculate an average value of μ_s, the deviation for each trial, and the a.d. Record your result on the data sheet.

4. From the data of Procedure 6, calculate μ_s for wood on wood from each of your three trials. Calculate an average value of μ_s, the deviation for each trial, and the a.d. Record your result on the data sheet. Compare this value of μ_s with that obtained in Procedure 5 by finding the difference between the two values and the percent error in their agreement. See whether your two values agree within their limits of error.

5. From the data of Procedure 7, calculate μ_s for glass on wood for each of your three trials and obtain the average value, the deviation for each trial, and the a.d. Record your result on the data sheet.

QUESTIONS _____

1. Explain in your own words why it is necessary that the block move at constant velocity in Procedures 2–4.

2. (a) How does the coefficient of friction depend upon the normal force between the surfaces in contact? (b) How does it depend upon the area of the surfaces in contact?

3. How does the coefficient of static friction compare with the coefficient of kinetic friction for the same surfaces, areas, and normal forces? In this connection explain what happened in Procedure 5.

4. Using an average value of the limiting angle of repose obtained from your data in Procedure 6 and the value of μ_k from your graph, find the acceleration of the block after it just breaks away in Procedure 6.

5. Calculate the force needed to pull a mass of 20 kg at a uniform slow speed up a plane inclined at an angle of 30° with the horizontal if the coefficient of kinetic friction is 0.20.

6. Suppose the force in Question 5 is a push applied horizontally on the mass instead of a pull directed parallel to the plane. (a) Calculate the value this force must have to keep the mass moving up the plane at constant speed. (b) Is there an angle for the plane at which no horizontal force can keep the mass moving at constant speed? If so, what is it?

Simple Machines and the Principle of Work

A simple machine is a device by which a force applied at one point is changed into another force applied at a different point for some useful advantage. The pulley system, the jack, the lever, and the inclined plane are well-known examples of such machines. The purpose of this experiment is to study an inclined plane, a wheel and axle, and a pulley system and to determine their characteristics.

THEORY

A simple machine is a device by means of which a force applied at one point (called the input) gives rise to another force acting at some other point (the output). The purpose of the machine is to make the new force different in magnitude and/or direction from the applied force in such a way as to make some particular piece of work possible or at least easier. We must recall that in physics the word *work* means a force acting through a distance and is given by the product of that force and the distance through which it acts. The units of work are thus foot-pounds or newton-meters (joules) or dyne-centimeters (ergs). Clearly, a given amount of work may be accomplished by a small force moving through a large distance or a large force moving through a small distance; in either case, the product of force and distance will be the same. A simple machine may be thought of as a device for changing the "mix" of force and distance. Ideally (that is, in the absence of losses due to friction), the work put into a simple machine will be the same as that received from it, the purpose of the machine being to increase the force at the expense of distance or vice versa. The machine may also change the direction of the force in the interests of convenience.

A common example is that of the 2000-lb car that must be raised 6 in. off the road in order to change a tire. The work required is 1000 ft·lb, but the average person cannot do this directly. To lift the car at all requires a 2000-lb effort, which is beyond most people's capacity. The fact that the car need be raised only 6 in. is no help—it might as well be 6 mi!

However, by the use of a jack, the required 2000-lb force is easily developed, the car gets raised the necessary 6 in., and the 1000 ft·lb of work gets done. The operator accomplishes this by applying a force of, say, 50 lb at the jack handle, the jack multiplying this force by 40. But the operator has to pump the handle up and down many times. By the time the car has been raised 6 in., the handle will have been pushed down through a total distance of $6 \times 40 = 240$ in. or 20 ft. The work of $20 \times 50 = 1000$ ft·lb at the input is the same as that of $(\frac{1}{2}) \times 2000 = 1000$ ft·lb at the output, but the proportion of force and distance involved has been drastically altered. The operator's hand has had to move through a much larger distance than the car has been lifted, but this is usually considered a small price to pay for the ability to do the job at all.

The ratio of the force delivered by a simple machine at its output to the force applied at its input is called its *mechanical advantage*. In the above example of an ideal machine in which there are no frictional losses, this ratio is the same as the ratio of the distance moved at the input to the distance moved at the output. In any practical machine, however, frictional losses do exist and must be supplied by additional work at the input. Thus, in the jack, frictional forces in its mechanism might require that a force of 60 lb rather than 50 be exerted on the handle in order to develop the 2000-lb lifting force at the output. The *actual mechanical advantage* (A.M.A.) is then only 2000/60, or 33.3 instead of 40. The handle must nevertheless be pumped through the same total of 20 ft to raise the car 6 in. as before. Thus, under the ideal conditions of no friction, the mechanical advantage the jack would have remains the ratio of the total distance moved by the handle to the distance the car is raised: $20 \div \frac{1}{2}$, or 40. This is called the *ideal mechanical advantage* (I.M.A.) and is given for any simple machine by the ratio of the distance moved at the input to the distance moved at the output. In equation form

$$\text{I.M.A.} = \frac{s}{h} \tag{9.1}$$

where s is the distance through which the force applied at the machine's input moves and h is the distance through which the force delivered by the machine moves. The actual mechanical advantage is given by

$$\text{A.M.A.} = \frac{P}{F} \tag{9.2}$$

where P is the output force exerted by the machine (such as the weight of the car being lifted) and F is the applied force (such as the force applied to the jack handle). Since in most cases the primary purpose of the simple machine is to multiply force, the actual mechanical advantage will be the machine's most important characteristic.

Ideally, the same amount of work is delivered at the output of a simple machine as is put in at the input, but in all actual cases extra work must be put in to make up for frictional losses in the machine. We are therefore interested in the machine's efficiency, which is a measure of how much of the work put in is actually delivered at the output. Specifically, the efficiency of a machine is defined as the ratio of the work delivered at the machine's output to the work done at the input. Thus,

$$\text{Efficiency} = \frac{\text{output work}}{\text{input work}} = \frac{hP}{sF} \quad (9.3)$$

Note that for the ideal machine the efficiency is 1, or 100%, since all of the work put in is delivered at the output. For practical machines, the efficiency will be a fraction less than 1 and may be expressed as a percentage by multiplying by 100. Note also that since the I.M.A. is s/h and the A.M.A. is P/F, the efficiency can be expressed as

$$\text{Efficiency} = \frac{\text{A.M.A.}}{\text{I.M.A.}} \quad (9.4)$$

In the example of the jack, in which frictional losses required the input force to be 60 lb rather than 50 lb, the above considerations show that the efficiency is $2000(\frac{1}{2})/60(20) = 33.3/40 = .83$, or 83%.

In this experiment an inclined plane, a wheel and axle, and a pulley system will be studied. In all of these cases a known load is raised against the gravitational force, and the applied force is measured. The ideal mechanical advantage is easily derived in each case as follows:

Inclined Plane When a block rests on an inclined plane as shown in Fig. 9.1, its weight W, which acts vertically downward, may be resolved into two components, one perpendicular to the plane and one parallel to it. The component of the weight perpendicular to the plane is $W \cos \theta$, and the component parallel to the plane is $W \sin \theta$, where θ is the angle made by the plane with the horizontal. The plane also exerts a force on the block. If the contacting surfaces are frictionless, this force (designated by N in Fig. 9.1) is perpendicular to the plane and is equal to the component of the block's weight perpendicular to the plane. In the absence of friction, a force $W \sin \theta$ directed up the plane will then keep the block in equilibrium, either at rest or moving up or down the plane with uniform velocity. Thus the plane enables us to raise the block to height h with a force $W \sin \theta$ instead of a force W, which would be required to lift the block directly. The plane allows the required force to be reduced by a factor of the sine of the plane angle. However, the distance through which the block must be moved to get it to the top of the plane is l, so that the work done is $(W \sin \theta)l$. Since, as Fig. 9.1 shows, $l \sin \theta = h$, this work is the same as Wh, the work required to lift the block directly. Notice that for the inclined plane

$$\text{I.M.A.} = \frac{l}{h} = \frac{l}{l \sin \theta} = \frac{1}{\sin \theta}$$

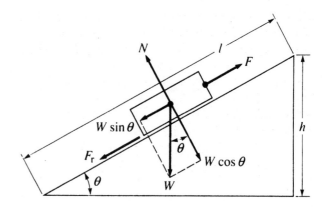

Figure 9.1 *The inclined plane as a simple machine*

Thus the ideal mechanical advantage of an inclined plane is the reciprocal of the sine of the plane angle (the cosecant of the plane angle).

If there is friction between the block and the plane, the force F_r of friction will be directed down the plane when the block is being pulled upward by the applied force F shown in Fig. 9.1. Hence, the value of F needed to keep the block moving up the plane at constant speed will be $W \sin \theta + F_r$, where F_r is the force of kinetic or sliding friction and is given by

$$F_r = \mu_k N = \mu_k W \cos \theta$$

Then

$$F = W \sin \theta + \mu_k W \cos \theta$$

and

$$\text{A.M.A.} = \frac{W}{W \sin \theta + \mu_k W \cos \theta}$$

$$= \frac{1}{\sin \theta + \mu_k \cos \theta} \quad (9.5)$$

Moreover, the input work is $Fl = Wl (\sin \theta + \mu_k \cos \theta)$, so that the efficiency is

$$\text{Efficiency} = \frac{Wh}{Fl} = \frac{Wh}{Wl(\sin \theta + \mu_k \cos \theta)}$$

$$= \frac{\sin \theta}{\sin \theta + \mu_k \cos \theta}$$

$$= \frac{1}{1 + \mu_k \cot \theta} \quad (9.6)$$

Thus, the efficiency is reduced below 100% by the addition of the μ_k term in the denominator. Only in the case of $\mu_k = 0$ (no friction) do we get a 100% efficient inclined plane.

Wheel and Axle Fig. 9.2 is an end view of a wheel and axle. The load W is supported by a cord wrapped around an axle whose radius is r, while the operator applies force F to another cord wrapped around a wheel of radius R fixed on the axle. The load W exerts a clockwise torque Wr about the axis of the system, and the operator exerts a counterclockwise torque FR about this axis by pulling with force F on the cord wrapped around the wheel. If there is no friction in the bearings, equilibrium will exist with W at rest or ascending or descending at constant speed if the two torques are equal. Thus, $W = (R/r)F$ and the ideal mechanical advantage is the ratio R/r of the wheel radius to the axle radius. That this is the I.M.A. as defined in Equation 9.1 may be seen from the fact that when the wheel and axle turn through one revolution, a length of the load-supporting cord equal to one axle circumference gets wrapped around the axle, so that the load is raised by a distance $2\pi r$. Similarly, the operator must draw a length $2\pi R$ of cord off the wheel. Therefore, the input force F moves through this distance. Hence,

$$\text{I.M.A.} = \frac{s}{h} = \frac{2\pi R}{2\pi r} = \frac{R}{r}$$

If there is friction in the bearings, the operator must exert an additional torque beyond FR to overcome this friction and must therefore pull with a force greater than F. The A.M.A. and the efficiency are thus reduced accordingly.

Pulley System A typical pulley system is shown in Fig. 9.3. In the absence of friction in the pulleys, the ten-sion in the cord is the force F with which the operator pulls on the end. But in effect, three strands of the cord pull on the load W, two via the pulley that is attached to W and one that is attached to W directly. Thus, the upward pull on W is $3F$. At equilibrium (W at rest or moving up or down at constant speed), $3F$ must be equal to W, and the ideal mechanical advantage is 3. If there is friction in the pulleys, a force larger than $W/3$ must be exerted on the cord by the operator and the actual mechanical advantage becomes less than 3. However, in either case, lifting the weight W through 1 ft requires that each strand attached to it be shortened by 1 ft, which means that 3 ft of cord must be pulled out of the system. Thus, force F must move through a distance three times that through which the load is raised, confirming that I.M.A. = 3. An interesting point to note here is that in any pulley system consisting of a set of pulleys attached to a fixed support and another set attached to and moving with a load, the ideal mechanical advantage is equal to the number of strands of the cord attached to the load either directly or via the load-end pulleys. A pulley system using just one cord is, of course, assumed. Note that the uppermost pulley in Fig. 9.3 contributes nothing to the mechanical advantage. It merely gives us the convenience of pulling down on the cord at F rather than having to pull up.

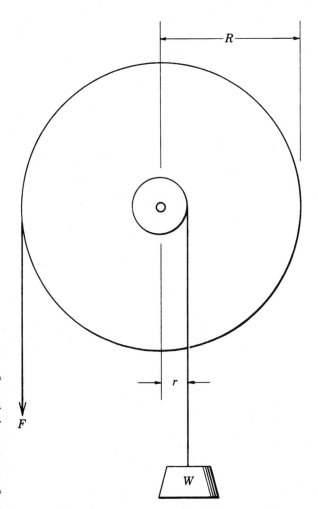

Figure 9.2 *The wheel and axle*

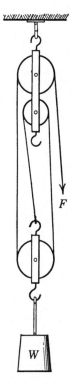

Figure 9.3 *Pulley system*

APPARATUS

1. Board with pulley at one end
2. Rods, bench clamp, and right-angle clamp for supporting the inclined plane
3. Wood block with a cord attached to it
4. Set of known weights
5. Weight hanger
6. Triple-beam balance
7. Meter stick
8. Protractor
9. Vernier caliper
10. Wheel and axle
11. Pulley system
12. Spring balance
13. Lintless dust cloth or paper wipers

PROCEDURE

Inclined Plane

1. Weigh the wood block and record the weight.

2. Measure the length of the board and record this value.

3. Adjust the board as an inclined plane and set it at an angle of 10° to the horizontal with the pulley at the upper end. Be careful about measuring the angle so that precise values of the sine and cosine may be obtained.

4. Clean the surface of the plane and the bottom of the block with the lintless dust cloth or paper wipers. *After this has been done, do not touch these surfaces with your hands.*

5. Lay the block on the plane, run the cord over the pulley, and attach the weight hanger to the cord's end. Place some weights on the hanger and slowly increase this load until it is just sufficient to keep the block sliding slowly up the plane with constant speed after it has been started with a very small push. Record the load so determined.

6. Measure the height *h* of the inclined plane. Be careful in doing this not to include the thickness of the board. Thus, if you measure *h* from the laboratory bench surface, measure it to the edge where the *bottom* surface of the board meets the board's upper end. Record the value of *h*.

7. Set the plane at an angle of 20° and repeat Procedures 3, 5, and 6. Again, use care in measuring the plane angle and the height.

8. Set the plane at an angle of 30° and repeat Procedures 3, 5, and 6, once more using care to obtain precise measurements of your new values of θ and *h*.

Wheel and Axle

9. Wind one string several times around the axle of the wheel-and-axle assembly. Wind the other string several times in the reverse direction around the wheel.

10. Hang a 500-g mass on the string coming from the axle. This is the load to be lifted. Attach the weight hanger to the string coming from the wheel and place weights on it until the load moves upward at a small, constant speed. Record the total weight needed to do this.

11. With the vernier caliper, measure the diameter of the wheel up to the surface on which the string is wound. Similarly, measure the diameter of the axle. Record these values.

Pulley System

12. With a spring balance, measure the force necessary to lift a mass of 1 kg with the pulley system set up in the laboratory. Observe the force while the mass is being lifted at a slow, uniform speed and record the value of this force.

13. Count the number of supporting strands in the pulley system and record it.

DATA

Inclined Plane

Weight of wood block _____   Length of board used as plane _____

	Angle of the inclined plane		
	10°	**20°**	**30°**
Force needed to pull block up plane			
Height *h* of the inclined plane			
Sine of the plane angle θ			
Cosine of the plane angle θ			

Wheel and Axle

Load used in the wheel-and-axle
experiment   _____

Force needed to lift this load   _____

Diameter of the wheel   _____

Diameter of the axle   _____

Pulley System

Load on the pulley system   _____

Force needed to lift this load   _____

Number of supporting strands   _____

Mechanical Advantage and Efficiency

Machine	Actual mechanical advantage	Ideal mechanical advantage	Efficiency	Coefficient of kinetic friction
Inclined plane at 10°				
Inclined plane at 20°				
Inclined plane at 30°				
Wheel and axle				
Pulley system				

CALCULATIONS _____

1. From the data of Procedures 2, 3, and 6, calculate the sine of the plane angle by dividing the height of the plane by its length. Compare this result with the value given in a table of trigonometric functions or by your calculator for the sine of 10°. Your two values should agree to three significant figures. Record your calculated value in the space provided on the data sheet.

2. Using the formula $\sin^2\theta + \cos^2\theta = 1$, calculate the value of cos 10° from your value of sin 10°. Check your result with the tabulated value of cos 10° and enter it on the data sheet.

3. Repeat Calculations 1 and 2 for the 20° and 30° plane angles using the data obtained in Procedures 7 and 8.

4. From the data of Procedures 1, 3, and 5, determine the actual mechanical advantage of the inclined plane set at an angle of 10°. The ideal mechanical advantage is the reciprocal of the sine of the plane angle already found. Record this I.M.A. and calculate and record the 10° inclined plane's efficiency.

5. Use either Equation 9.5 or 9.6 to compute the coefficient of kinetic friction between the block and the plane.

6. Repeat Calculations 4 and 5 for the inclined plane set at 20° using the data from Procedure 7.

7. Repeat Calculations 4 and 5 for the inclined plane set at 30° using the data from Procedure 8.

8. From the data of Procedures 9–11, determine the actual mechanical advantage, the ideal mechanical advantage, and the efficiency of the wheel and axle used.

9. From the data of Procedures 12 and 13, determine the actual mechanical advantage, the ideal mechanical advantage, and the efficiency of the pulley system used.

QUESTIONS

1. Should the coefficients of kinetic friction obtained for the three angular settings of the inclined plane be the same? Are they?

2. The instructions for Calculation 1 suggest that your calculated value of the sine of the plane angle is better than the value obtained by looking up the sine of this angle as measured by the protractor. Why should this be true?

3. (a) State how the mechanical advantage of the inclined plane varies with the inclination of the plane. (b) Does this apply to both the ideal and the actual mechanical advantage?

4. (a) How does the efficiency of the inclined plane vary with the inclination of the plane? (b) Explain the reason for this.

5. How does the uppermost pulley in the system of Fig. 9.3 affect the A.M.A. of this machine? How does it affect the efficiency? Discuss briefly why you think (or do not think) it is worth having.

6. The block and tackle of Fig. 9.3 is used to lift a stone weighing 300 lb. The force required is 125 lb. Calculate the actual mechanical advantage and the efficiency of this simple machine.

7. A body weighing 400 lb is pulled up an inclined plane 15 ft long and 5 ft high. The force required to pull the body up the plane at a slow uniform speed is 200 lb. Compute the ideal mechanical advantage, the actual mechanical advantage, and the efficiency of the plane under these conditions. Also compute the coefficient of kinetic friction between the body and the plane.

8. Calculate the amount of work done in drawing a mass of 20 kg at a uniform slow speed up a plane 2 m long, making an angle of 30° with the horizontal, if the coefficient of kinetic friction is 0.20.

Young's Modulus 10

All bodies are deformed in some way by the application of a force. An elastic body is a body that changes in size or shape upon the application of a distorting force, but returns to its original condition when that force is removed. But there is a limit to the magnitude of the force that may be applied if the body is to return to its original condition. This is called the elastic limit. A greater force will cause a permanent distortion or even a break. In this experiment, a steel wire is stretched by the application of a force, and the resulting changes in length are measured. From these measurements, Young's modulus for a steel wire is calculated.

THEORY

When a wire is stretched by a mass hanging from it, the wire is said to experience a tensile strain, which is defined as the elongation per unit length. The force acting on the wire is a measure of the tensile stress, which is defined as the force per unit area over which it is acting.

Hooke's law states that as long as a body is not strained beyond its elastic limit, the stress is proportional to the strain. Thus, the ratio of stress to strain is a constant and is called a modulus of elasticity. For the case of a wire stretched by a force, this constant is called Young's modulus. Thus,

$$Y = \frac{\text{force per unit area}}{\text{elongation per unit length}} = \frac{F/A}{e/L} = \frac{FL}{eA}$$

where Y is Young's modulus in newtons per square meter, F is the force in newtons, L is the length of the wire in meters, e is the elongation in meters, and A is the area of cross section of the wire in square meters.

The arrangement of the apparatus used in this experiment is shown in Fig. 10.1. The elongations produced in this experiment are very minute; hence, a method called an "optical lever" is used to magnify them. This consists of a vertical mirror mounted on a horizontal T-shaped base having pointed legs at the three extremities. The wire whose elongation is to be measured is clamped at its upper end and has a small cylinder attached to it near the lower end. This cylinder is free to move up and down through a hole in a shelf supporting the mirror and its base. The two front feet of the base rest on the shelf and form an axis of rotation for the mirror. The rear foot rests on the cylinder and moves up and down with it due to the elongation of the wire. A laser is arranged to shine a narrow light ray on the mirror, which reflects it to a vertical strip of paper attached to a wall or screen two or three meters away (see Fig. 10.1). The spot where the ray strikes the paper when the wire is unstretched is marked. As the wire is stretched by applying a load, the angular movement of the mirror rotates the reflected ray so that the spot moves up the paper strip. The elongation can be readily computed from the relation

$$e = \frac{Sd}{2D} \tag{10.1}$$

where e is the elongation in meters, D is the distance in meters from the mirror to the paper strip, d is the perpendicular distance from the moving foot of the optical lever

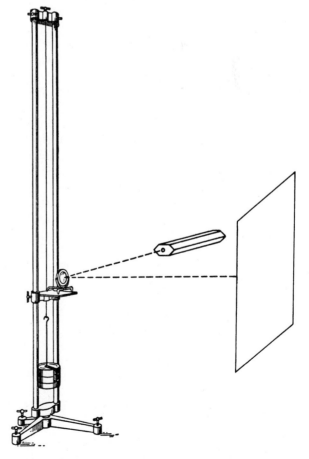

Figure 10.1 *Young's modulus apparatus*

71

to the line joining the other two feet, and S is the distance along the paper strip from the mark made before the load was applied to the position of the spot after elongation. This relation is derived by using the principle of optics that the reflection of a beam of light falling on a plane mirror is turned through an angle twice as large as the angle through which the mirror is rotated, the incident beam remaining fixed.

APPARATUS

1. Young's modulus apparatus
2. Laser with support stand
3. Mirror mounted on T-base (optical lever)
4. Roll of white calculator paper and masking tape
5. Slotted weights (ten l-kg weights)
6. Micrometer caliper
7. Vernier caliper
8. 2-m stick

PROCEDURE

1. Measure the length of the steel wire used, that is, the distance from the lower end of the upper chuck to the top of the cylinder at the lower end of the wire.

2. Measure the diameter of the wire in five places, using the micrometer caliper.

3. Measure the distance of the moving foot of the optical lever from the axis of rotation. This may be done by pressing the pointed legs on a piece of white paper and constructing the required distance. Use the vernier caliper for this measurement.

4. Carefully place the optical lever with its two front feet in the groove provided on the rigid support and its rear foot on the cylinder fixed to the wire.

5. Apply a load of 1 kg to the wire to take up any slack.

6. Mount the laser so that its beam strikes the mirror and is reflected to a wall or screen about 2 m away. The position of the laser, the cylinder on which the optical lever's moving foot rests, and the orientation of the mirror should be adjusted so that the feet of the optical lever are approximately in a horizontal plane and the path of the ray from the mirror to the wall or screen is nearly horizontal. CAUTION: Although the low-power laser used here cannot hurt you even if its beam falls directly on your skin, *do not look into the beam at any time.*

7. Tape a strip of white calculator paper to the wall or screen and position it vertically with the laser light spot falling near its center. Mark the position of the spot on the paper.

8. With the 2-m stick, carefully measure the distance from the mirror to the mark just made. A convenient way of doing this is to make the distance exactly 2 m by positioning the apparatus so that the stick just fits between the wall or screen at the position of the mark and the mirror.

9. Add a second kilogram to the load on the wire and mark the new position of the laser light spot on the paper.

10. Add 1 kg at a time up to a total of 10 kg, marking the spot position after each added load.

11. Remove the load, 1 kg at a time, marking the spot position for each load. *Note:* The values of the load to be used apply only to a steel wire. If a different kind of wire is used, the proper values will be suggested by the instructor. Label each of your marks with the pertinent load value and indicate whether the load was increasing or decreasing.

12. Turn off the laser and with the vernier caliper measure the distance from the zero mark made with the initial 1-kg load to each of the marks made as the load was increased. These are your values of S for an increasing load. Measure the distance from the zero mark made when you had only the l-kg load left at the end of the run to each of the marks made as the load was decreased. These are your values of S for a decreasing load. Record all these measurements in the data table.

DATA

Length of wire used   _____   Diameter of wire   _____

Micrometer readings:

Trials

1. _____   2. _____   3. _____   4. _____   5. _____

Average reading   _____   Zero reading   _____

Cross-sectional area of the wire   _____   Young's modulus (computed value)   _____

Distance from front legs to the rear leg
   of the optical lever   _____   Young's modulus (from Table V at the
     end of the book)   _____

Distance from mirror to scale   _____   Percent error   _____

Total load, kg	Value of S (increasing load)	Value of S (decreasing load)	Average value of S	Elongation e
2				
3				
4				
5				
6				
7				
8				
9				
10				

CALCULATIONS _____

1. Compute the average of the readings for the diameter of the wire. This will be the value of the diameter when properly corrected for the zero reading of the micrometer.

2. Calculate the cross-sectional area of the wire in square meters.

3. Compute the elongation corresponding to each load using the average value of S obtained in each case. Note that as the 1-kg load corresponds to your zero mark, the elongation calculated with the total load of 2 kg is that produced by a load of 1 kg; the elongation calculated with the total load of 3 kg is that produced by 2 kg; etc.

4. Plot a curve on the graph paper using loads in newtons as abscissas and the resulting elongations in meters as ordinates. Note that the load producing a given elongation is that due to the number of kilograms on the weight hanger minus the 1 kg used for the zero reading. Thus, the maximum load appearing as an abscissa in this graph is 9 kg × 9.8 = 88.2 N.

5. Compute Young's modulus for the wire from the slope of your graph. Compare your result with the value given in Table V at the end of the book and find the percent error.

QUESTIONS

1. State what the curve shows and hence what the experiment checked.

2. Why wasn't the whole length of the wire measured?

3. Define elasticity, elastic limit, stress, and strain.

4. (a) State Hooke's law. (b) Does your curve agree with it? (c) How would the curve look if the elastic limit were exceeded?

5. Derive Equation 10.1 from a consideration of the apparatus geometry and the principle of optics stated in the Theory section.

6. (a) Look up what is meant by the *bulk modulus* B of a solid and state the definition of this quantity. (b) If you can, derive a relation between the bulk modulus and Young's modulus.

Rotational Motion 11

The motion of the flywheel of a steam engine, an airplane propeller, and any rotating wheel are examples of a very important type of motion called rotational motion. If a rigid body is acted upon by a system of torques, the body will be in equilibrium as far as rotational motion is concerned if the sum of the torques about any axis is zero. This means that if the body is at rest, it will remain at rest; if it is rotating about a fixed axis, it will continue to rotate about the same axis with uniform angular speed.

However, if an unbalanced torque is acting on the body, it will produce an angular acceleration in the direction of the torque, the acceleration being proportional to the torque and inversely proportional to the moment of inertia of the body about its axis of rotation. The purpose of this experiment is to study rotational motion, to observe the effect of a constant torque upon a body that is free to rotate, and to determine the resulting angular acceleration and the moment of inertia of that body.

THEORY

Rotation refers to motion about some axis in space. In describing this type of motion, the angular displacement, the angular velocity, and the angular acceleration must be given. The unit of angular measurement usually used is the radian. This is the angle subtended at the center of a circle by an arc equal in length to the radius. Clearly, since the circumference of a circle is 2π times its radius and an angle of 360° is once around the circle, 2π radians = 360°, π radians = 180°, and 1 radian = 57.296°, or 57° 17′ 45″. The angular velocity is the time rate of change of angular displacement. It is equal to the angle through which the body rotates divided by the time during which the rotation takes place. The angular acceleration is the time rate of change of angular velocity.

The equations of motion for uniformly accelerated rotational motion are similar to those for uniformly accelerated linear motion except that angular quantities replace the linear ones already considered. Thus, the angular displacement is given by

$$\theta = \bar{\omega}t$$

where θ is the angular displacement in radians, $\bar{\omega}$ is the average angular velocity in radians per second, and t is the time in seconds. The average angular velocity is given by

$$\bar{\omega} = \frac{\omega_1 + \omega_2}{2}$$

where $\bar{\omega}$ is the average angular velocity, ω_1 is the initial angular velocity, and ω_2 is the final angular velocity. The final angular velocity is given by

$$\omega_2 = \omega_1 + \alpha t$$

where α is the angular acceleration in radians per second per second, t is the time in seconds, and ω_1 and ω_2 have the meanings already assigned.

The corresponding equations involved in uniformly accelerated linear motion are

$$s = \bar{v}t$$

$$\bar{v} = \frac{v_1 + v_2}{2}$$

and

$$v_2 = v_1 + at$$

where s is the distance, $\bar{v}$ is the average velocity, v_1 is the initial velocity, v_2 is the final velocity, a is the acceleration, and t is the time.

Using the radian as the unit of angular measurement gives rise to some very simple relations between the linear quantities and the angular quantities. Suppose s represents the distance a particle moves around the circumference of a circle of radius r; let v be the linear speed and a the linear or tangential acceleration of the particle. Then the angular displacement θ, the angular velocity ω, and the angular acceleration α are given by the relations

$$s = r\theta \tag{11.1}$$

$$v = r\omega \tag{11.2}$$

$$a = r\alpha \tag{11.3}$$

When a torque is applied to a body free to rotate about a fixed axis, the body will acquire an angular acceleration given by the relation

$$\tau = I\alpha \tag{11.4}$$

where τ is the sum in meter-newtons of all the torques about the fixed axis of rotation, I is the moment of inertia in kilogram-meters^2 of the body about the same axis, and α is the angular acceleration in radians per second per second.

Notice that moment of inertia here plays the same role as the mass does in linear motion. It is the inertia of a body as regards rotation about an axis. The moment of inertia of a body with respect to an axis is the sum of the products obtained by multiplying the mass of every particle of the body by the square of its distance from the axis. For a simple geometrical solid it can be readily determined by the use of integral calculus. For example, the moment of inertia of a uniform cylinder, or disk, with respect to its longitudinal axis is given by

$$I = \frac{1}{2}MR^2 \qquad (11.5)$$

where M is the mass of the cylinder or disk; R is its radius in meters; and I is the moment of inertia in kilogram-meters^2.

One form of apparatus used in this experiment is shown in Fig. 11.1. It consists of a disk mounted on a shaft that is carried in a very good bearing in the base plate so that the disk is free to rotate with a minimum of friction. On its underside (and thus not visible in the figure), the disk carries a drum with three diameters, around any one of which a string may be wound so as to exert a torque on the rotating assembly. The string passes over a pulley and has a weight attached to the end. When this weight is released, it descends with a constant linear acceleration, while the rotating assembly turns with a constant angular acceleration. This situation is illustrated in a simplified form in Fig. 11.2. Note that in the actual apparatus the disk is mounted in a horizontal plane so that a second disk, a ring, and a plate may be loaded onto it to vary the moment of inertia of the complete assembly. A pulley is thus

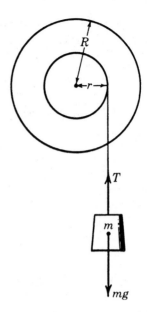

Figure 11.2 *Forces acting on a falling mass*

required to change the string direction from horizontal to vertical. This pulley is chosen to be very light and small so as to contribute a negligible moment of inertia, and its contribution to the friction in the apparatus will be included with the friction in the bearing supporting the disk. As we will neglect the effect of the pulley, Fig. 11.2 is drawn as if the disk of radius R were in a vertical plane so that the string, which is wound around a drum of radius r, can be shown going straight down to the falling mass m. We see that there are two forces acting on this mass: the

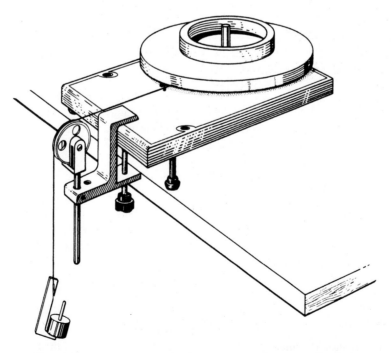

Figure 11.1 *Apparatus for the study of rotational motion, version 1*

downward force of gravity—that is, the mass's weight mg—and the upward pull of the string—namely, its tension T. The resultant force acting on m is thus

$$F = mg - T$$

But by Newton's second law of motion, $F = ma$. Therefore,

$$mg - T = ma \qquad (11.6)$$

where a is the acceleration of the falling mass in meters per second per second.

From Fig. 11.2 we see that the torque acting on the rotating assembly is

$$\tau_a = Tr \qquad (11.7)$$

However, a second torque τ_f due to friction in the bearings is also acting, so that the resultant torque τ is the difference $\tau_a - \tau_f$. Note that the frictional torque is subtracted because friction always operates in the direction opposing the motion.

After an appropriate rearrangement of terms, substitution in Equation 11.4 then yields

$$\tau_a = I\alpha + \tau_f \qquad (11.8)$$

Equation 11.8 shows that the angular acceleration of the rotating assembly depends on the applied torque, the

frictional torque, and the moment of inertia. Assuming the last two remain constant during the experiment, we note that a plot of τ_a against α should be a straight line with slope I and intercept τ_f. With the apparatus of Fig. 11.2, three different drum radii are available, so that τ_a can be varied by changing both the drum radius r and the mass m. The resulting measured values of α will allow an experimentally determined plot of Equation 11.8 to be made. This plot will serve as both a check on the theory (as it should be a straight line) and a means for finding both the moment of inertia I of the rotating assembly and the frictional torque. The experiment also allows for the setting up of various rotating assemblies, all of whose moments of inertia can be calculated from their dimensions, so that a further check on the theory can be made by comparing the calculated and measured values of I in each case.

A somewhat simpler form of this apparatus is shown in Fig. 11.3. It consists of the same rotating platform that was used in Experiment 7 with the equipment used for the centripetal force measurements in that experiment removed. A drum is mounted on the shaft under the platform, but only one diameter is provided, so that the torque exerted by the string can be varied only by varying the mass m. In other respects, however, the two apparatuses are similar.

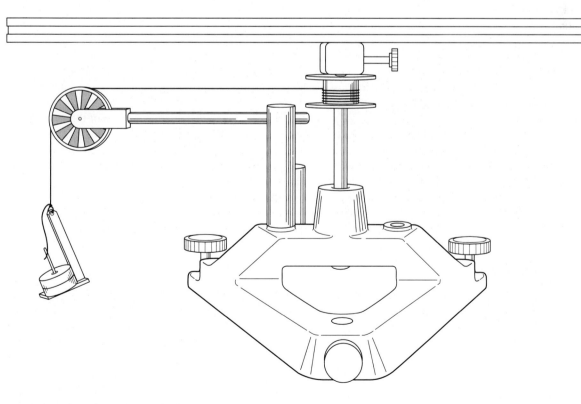

Figure 11.3   *Apparatus for the study of rotational motion, version 2*

APPARATUS

1. Rotational motion apparatus, consisting of a main disk and drum, base plate with bearing and shaft, auxiliary disk, steel ring, and rectangular plate (Fig. 11.1), or a base plate with bearing, shaft, and drum, a platform resembling a large bar, a disk, and a steel ring (Fig. 11.3).
2. Pulley on rod with mounting clamp
3. Weight hanger with 20-, 50-, and 100-g weights
4. Stopwatch or stop clock
5. Meter stick
6. Vernier caliper
7. Triple-beam balance
8. Level
9. String

PROCEDURE (Version 1)

1. Set the base plate of the apparatus on the bench and mount the pulley on one end as shown in Fig. 11.1. Position the base plate so that when the string hangs down over the pulley, it will clear the edge of the bench. Place the shaft in the bearing and level the apparatus using the level provided.

2. Observe the three-diameter drum attached to the main disk and measure each diameter with the vernier caliper. Measure the main disk diameter with the meter stick and weigh this disk on the triple-beam balance. Record all these measurements.

3. Wind the string on the largest-diameter drum in a single layer. Choose a length of string so that when the weight hanger is attached to the free end, the main disk is placed on the shaft, and the string is passed over the pulley, the weight hanger will strike the floor on the way down with two or three turns of the string still left on the drum.

4. For the first set of measurements use a 50-g weight. Assemble the apparatus as shown in Fig. 11.1 with the weight at its highest convenient position. Measure the distance the weight has to fall, that is, the distance from the bottom of the weight hanger to the floor. Choose the weight's highest position to give some convenient value for this measurement. One meter is an obvious choice if the bench is high enough. Always start the weight from this position.

5. Find the time required for the 50-g weight to reach the floor. With the string wound around the largest drum and the weight at the top of its path as described in Procedures 3 and 4, release the disk and start the stopwatch at the same instant; make sure you give no impulse to the disk as you release it so that it truly starts from rest. Stop the watch at the instant the weight strikes the floor.

6. Repeat Procedure 5 twice to obtain three independent observations of the time, and record these results.

7. Repeat Procedures 4–6 using the 100-g weight.

8. Repeat Procedures 3–7 with the string wound on the intermediate-diameter drum.

9. Repeat Procedures 3–7 with the string wound on the smallest-diameter drum.

10. Measure the diameter of the auxiliary disk with the meter stick and weigh it on the triple-beam balance. Record these measurements and then place this disk on the shaft on top of the main disk.

11. Repeat Procedures 3–9 with the auxiliary disk on top of the main disk so that the rotating assembly consists of the two disks together.

12. Weigh the steel ring on the triple-beam balance. Then use the meter stick or, if possible, the vernier caliper, to measure its inside and outside diameters. Record these measurements, remove the auxiliary disk from the apparatus, and replace it with the steel ring. Notice that the ring has two locating pins, which fit into holes in the upper surface of the main disk so that the ring will be properly centered.

13. Repeat Procedures 3–9 with the steel ring replacing the auxiliary disk on the main disk so that the rotating assembly is the combination of this disk and the ring.

14. Measure the length and width of the rectangular plate with the meter stick and then weigh the plate on the triple-beam balance. Record these measurements and then substitute the plate for the ring on top of the main disk.

15. Repeat Procedures 3–9 with the rectangular plate in place on the main disk so that the rotating assembly is the combination of the disk and the plate.

PROCEDURE (Version 2)

1. Set the base unit on the bench, noting that the shaft and drum have already been assembled in the bearing. Make sure that the shaft turns freely. Then mount the rotating platform on the shaft's top end. Place the level on the platform's upper surface and turn the platform to various positions, observing the level each time. Adjust the leveling screws in the base unit so that the level shows the platform to be exactly horizontal in any orientation.

2. Mount the pulley on the base unit by inserting its mounting rod in the holder provided, as shown in Fig.

11.3. Orient the holder so that a string wound around the drum on the apparatus shaft will come straight out to the pulley. If necessary, move the base plate so that when the string hangs down over the pulley, it will clear the edge of the bench. Check that the platform is still level.

3. Remove the platform from the shaft and take out the screws holding the shaft coupling block on the platform's bottom with the special tool provided. Weigh the platform and measure its overall length with the meter stick. Reassemble the platform to its coupling block and replace it on the vertical shaft, making sure that you tighten the set screw against the shaft's flat. Then measure the diameter of the drum with the vernier caliper. Record all these measurements in the data table, noting that you should enter the mass and length of the platform in place of the mass and radius of the other apparatus's main disk. Note also that you can use the spaces for the diameter and radius of the largest drum for the diameter and radius of your one drum and that the spaces for the diameter of the main disk and the diameter and radii of the two other drums will not be used with the present apparatus.

4. Follow Procedures 3–7 for the apparatus of Fig. 11.1, noting that the present apparatus has only one drum diameter, so that you will use six values for the mass *m* to obtain six torques and so that it is the rotating platform rather than the main disk that is on the shaft. When working with this apparatus, treat this platform as the main disk and the one disk you have as the auxiliary disk in the earlier instructions.

5. Follow Procedures 10 and 11 for the apparatus of Fig. 11.1, using the one disk you have and slipping it onto the shaft end *in place of* (not on top of) the rotating platform. Again, note that there is only one drum diameter to be used.

6. Follow Procedures 12 and 13 for the apparatus of Fig. 11.1, noting that the steel ring goes on top of the one disk left on the apparatus from the last procedure and that again there is only one drum around which to wind the string. For this apparatus the steel ring fits in a groove in the disk rather than being provided with the locating pins mentioned in Procedure 12. In this case you must check the exact centering of the ring by seeing that it does not oscillate back and forth as the apparatus is turned.

DATA _____

Mass of the main disk and drum or of the large bar _____

Diameter of the main disk _____

Diameter of the largest (or single) drum _____

Diameter of the intermediate drum _____

Diameter of the smallest drum _____

Distance weight falls _____

Radius of the main disk or length of the large bar _____

Radius of the largest (or single) drum _____

Radius of the intermediate drum _____

Radius of the smallest drum _____

Moment of inertia of the main disk and drum or of the large bar from the slope of the line on the graph _____

Frictional torque from the *y* intercept on the graph _____

Moment of inertia of the main disk and drum or of the large bar calculated from the formula $\frac{1}{2}MR^2$ or $\frac{1}{12}ML^2$ _____

Percent difference between the measured and calculated values of this moment of inertia _____

Mass of the auxiliary disk _____

Diameter of the auxiliary disk _____

Radius of the auxiliary disk _____

Distance weight falls _____

Moment of inertia of the main and auxiliary disks or of your one disk from the slope of the line on the graph _____

Moment of inertia of the auxiliary disk alone (apparatus of Fig. 11.1 only) _____

Frictional torque from the *y* intercept on the graph _____

Moment of inertia of the auxiliary disk calculated from the formula $\frac{1}{2}MR^2$ _____

Percent difference between the measured and calculated values of the moment of inertia of the auxiliary disk _____

Mass of the steel ring _____

Inner diameter of the steel ring _____

Outer diameter of the steel ring _____

Distance weight falls _____

Inner radius of the steel ring _____

Outer radius of the steel ring _____

Moment of inertia of the main disk and steel ring from the slope of the line on the graph _____

Moment of inertia of the steel ring alone _____

Frictional torque from the *y* intercept on the graph _____

Moment of inertia of the steel ring calculated from the formula $\frac{1}{2}M(R_1^2 + R_2^2)$ _____

Percent difference between the measured and calculated values of the moment of inertia of the steel ring _____

Mass of the rectangular plate _____

Length of the rectangular plate _____

Width of the rectangular plate _____

Distance weight falls   _____

Moment of inertia of the main disk
and rectangular plate from the
slope of the line on the graph   _____

Moment of inertia of the rectangular
plate alone   _____

Frictional torque from the y intercept
on the graph   _____

Moment of inertia of the rectangular
plate calculated from the formula
$\frac{1}{12}M(a^2 + b^2)$ _____

Percent difference between the
measured and calculated values
of the moment of inertia of the
rectangular plate   _____

Note: Data for the rectangular plate is obtained with the apparatus of Fig. 11.1 only.

Main Disk Alone or Rotating Platform

Apparatus of Fig. 11.1	Large-diameter drum		Medium-diameter drum		Small-diameter drum	
	50-g mass	100-g mass	50-g mass	100-g mass	50-g mass	100-g mass
Apparatus of Fig. 11.3	20-g mass	40-g mass	60-g mass	80-g mass	100-g mass	120-g mass
Time of fall of mass: 1.						
2.						
3.						
Average						
Average velocity of falling mass						
Final velocity of falling mass						
Acceleration of falling mass						
Angular acceleration						
Tension in the string						
Torque acting on the rotating assembly						

Main Disk and Auxiliary Disk or Single Disk Alone

Apparatus of Fig. 11.1	Large-diameter drum		Medium-diameter drum		Small-diameter drum	
	50-g mass	100-g mass	50-g mass	100-g mass	50-g mass	100-g mass
Apparatus of Fig. 11.3	20-g mass	40-g mass	60-g mass	80-g mass	100-g mass	120-g mass
Time of fall of mass: 1.						
2.						
3.						
Average						
Average velocity of falling mass						
Final velocity of falling mass						
Acceleration of falling mass						
Angular acceleration						
Tension in the string						
Torque acting on the rotating assembly						

Main Disk and Steel Ring

Apparatus of Fig. 11.1	Large-diameter drum		Medium-diameter drum		Small-diameter drum	
	50-g mass	100-g mass	50-g mass	100-g mass	50-g mass	100-g mass
Apparatus of Fig. 11.3	20-g mass	40-g mass	60-g mass	80-g mass	100-g mass	120-g mass
Time of fall of mass: 1.						
2.						
3.						
Average						
Average velocity of falling mass						
Final velocity of falling mass						
Acceleration of falling mass						
Angular acceleration						
Tension in the string						
Torque acting on the rotating assembly						

Main Disk and Rectangular Plate (Apparatus of Fig. 11.1 only)

	Large-diameter drum		Medium-diameter drum		Small-diameter drum	
	50-g mass	100-g mass	50-g mass	100-g mass	50-g mass	100-g mass
Time of fall of mass: 1.						
2.						
3.						
Average						
Average velocity of falling mass						
Final velocity of falling mass						
Acceleration of falling mass						
Angular acceleration						
Tension in the string						
Torque acting on the rotating assembly						

CALCULATIONS

1. From the data of Procedures 4, 5, and 6, for the apparatus of Fig. 11.1, calculate the average time of descent. For the apparatus of Fig. 11.3 this data was obtained in Procedure 4.

2. Similarly, calculate the average time of descent for each set of data in Procedures 7, 8, and 9 for the apparatus of Fig. 11.1 This step does not apply to the apparatus of Fig. 11.3.

3. From the data of Procedure 2 for the apparatus of Fig. 11.1, compute the radius of each drum and the radius of the main disk. For the apparatus of Fig. 11.3 you need the radii of only the one drum and disk.

4. Using the average time of fall, compute the average velocity of the falling mass for each set of observations.

5. From the value of the average velocity, calculate the final velocity of the mass as it strikes the floor, for each set of observations. Notice that since the system starts from rest and experiences a uniform acceleration, the final velocity will be twice the average velocity.

6. From the average time of fall and the final velocity, calculate the acceleration of the falling mass for each set of observations.

7. The value of the acceleration found in Calculation 6 is also the tangential acceleration of a point on the circumference of the drum. Using Equation 11.3, find the angular acceleration α for each set of observations.

8. Using Equation 11.6, calculate the tension in the string for each set of observations from the values of the acceleration found in Calculation 6.

9. Calculate the applied torque for each set of observations by using Equation 11.7.

10. Plot the values of the angular acceleration α along the x axis and the corresponding values of the applied torque τ_α along the y axis. Draw the best straight line through the six points.

11. Determine the value of the moment of inertia I of the main disk and drum or of the rotating platform by computing the slope of the straight line drawn in Calculation 10.

12. Determine the value of the frictional torque τ_f by reading the value of the y intercept of the straight line drawn in Calculation 10.

13. Calculate the moment of inertia I of the main disk if you are using the apparatus of Fig. 11.1 or of the rotating platform if you are using the apparatus of Fig. 11.3. In the former case, use Equation 11.5 in which M is the disk's mass as measured in Procedure 2 for that apparatus and R is its radius as found in Calculation 3. In the latter case, use Equation 11.10 in which M is the platform's mass as measured in Procedure 3 for that apparatus, a is the platform's length from that same procedure, and b may be taken to be zero.

14. Find the percent difference between the measured and calculated moments of inertia found in Calculations 11 and 13.

15. Repeat Calculations 1–12 for the main disk and auxiliary disk using the data obtained in Procedures 10 and 11 if you are using the apparatus of Fig. 11.1. If you are using the apparatus of Fig. 11.3, repeat these calculations for your one disk using the data obtained in Procedure 5 for this apparatus.

16. Find the moment of inertia of the auxiliary disk by subtracting the moment of inertia of the main disk alone found in Calculation 11 from that of the complete rotating assembly (apparatus of Fig. 11.1 only).

17. Calculate the moment of inertia of the auxiliary disk from the formula $I = \frac{1}{2}MR^2$ using the mass of this disk measured in Procedure 10 and the radius obtained when you repeated Calculation 3. For the apparatus of Fig. 11.3, obtain the moment of inertia of your one disk using the mass and radius from Procedure 5 and Calculation 3.

18. Find the percent difference between the measured and calculated moments of inertia found in Calculations 16 and 17, or, in the case of the apparatus of Fig. 11.3, Calculations 15 and 17.

19. Repeat Calculations 1–12 for the main disk and steel ring using the data of Procedures 12 and 13. Notice that in repeating Calculation 3 for the steel ring, both an inside and an outside radius must be found. For the apparatus of Fig. 11.3 the pertinent data were obtained in Procedure 6 for that apparatus.

20. Repeat Calculations 16–18 to find a measured and calculated value for the moment of inertia of the steel ring and the percent difference between these results. The formula for the ring's moment of inertia is

$$I = \frac{1}{2}M(R_1^2 + R_2^2)$$ (11.9)

where M is the mass of the ring measured in Procedure 12 and R_1 and R_2 are the two radii obtained when you repeated Calculation 3. For the apparatus of Fig. 11.3, Calculations 15, 17, and 18 are applicable. The mass and radii of the ring were found in Procedure 6 for that apparatus.

21. Repeat Calculations 1–12 for the main disk and rectangular plate using the data of Procedures 14 and 15. Notice that in repeating Calculation 3 for the rectangular plate there is no radius to measure. The length and width were found in Procedure 14. This calculation applies to the apparatus of Fig. 11.1 only.

22. Repeat Calculations 16–18 to find a measured and calculated value for the moment of inertia of the rectangular plate and the percent difference between these results. The formula for the plate's moment of inertia is

$$I = \frac{1}{12}M(a^2 + b^2)$$

(11. 10)

where M is the mass of the plate and a and b are its length and width measured in Procedure 14. This calculation applies to the apparatus of Fig. 11.1 only.

QUESTIONS _____

1. The radius of gyration of a rotating body is the distance from the axis of rotation to the point at which the entire mass of the body may be considered concentrated without altering the moment of inertia. It is defined by the expression

$$I = MK^2$$

(11.11)

where I is the body's moment of inertia, M its mass, and K the radius of gyration in question. Use your measured values of the moments of inertia of the main disk, auxiliary disk, steel ring, and rectangular plate to find the radius of gyration of each of these objects. For the apparatus of Fig. 11.3, you can do this for the rotating platform, the disk, and the steel ring.

2. Using Equations 11.5, 11.9, 11.10, and 11.11, derive a theoretical expression for the radius of gyration of each of your rotating objects. Put the appropriate values for the masses and dimensions of these objects into these expressions to get numerical values of K and compare them with those found in Question 1. For the rotating platform in the apparatus of Fig. 11.3, use Equation 11.10 with a as the length of the bar and $b = 0$.

3. (a) If you used the apparatus of Fig. 11.3, estimate the width of the rotating platform and use this as the value of b in Equation 11.10 to obtain a new value of I. (b) Compare this new value with that obtained in Calculation 13 and see if you were justified in taking $b = 0$ in that calculation. (c) Repeat Question 2 for the rotating platform, treating it as a rectangular plate of length a and width b. Did neglecting b make a big difference in your computation of the radius of gyration for the platform?

4. Why is the radius of gyration of any wheel always smaller than the wheel's actual radius?

5. By referring to your graphs, determine if it is possible for the y intercept to be negative. What would this mean physically?

6. Consider the disk and drum of Fig. 11.2. Let the disk's mass be M and assume that the drum radius r is small enough so that its moment of inertia may be neglected. Also neglect the friction in the bearings; that is, take τ_f to be zero. Using Equations 11.3–11.7, show that the downward acceleration of the falling mass is

$$a = \frac{g}{1 + \dfrac{1}{2}\dfrac{M}{m}\left(\dfrac{R}{r}\right)^2}$$

7. In your work with the main disk in the apparatus of Fig. 11.1, you probably found that the value of the moment of inertia calculated from Equation 11.5 was larger than that obtained experimentally, that is, from the slope of your graph. Explain why this should be so. If you used the apparatus of Fig. 11.3, note that the shaft on which the rotating platform is mounted, the shaft coupling block, and the drum all form part of the rotating assembly, yet you did not include their moment of inertia in your calculations. Estimate their contribution to the total moment of inertia of the system and see if you were justified in neglecting it. Would your measured values of I from Calculations 11, 15, and 19 have been closer to the theoretical values if you had included this contribution?

8. How do the values of the frictional torque τ_f obtained from your four graphs compare? How should they compare?

9. Derive Equation 11.9. If you can, also derive Equations 11.5 and 11.10. *Hint:* The derivation of these last two requires the use of calculus, but Equation 11.9 may be derived without calculus if Equation 11.5 is assumed, for a ring may be considered to be a disk of radius R_2, with a hole of radius R_1 cut out of it.

10. A uniform disk 0.3 m in diameter and having a mass of 2 kg is free to rotate about its horizontal axis on frictionless bearings. An object with a mass of 0.05 kg is attached to a string wound around the rim of the disk. The object is released from rest and descends with constant linear acceleration. Calculate: (a) the moment of inertia of the disk; (b) the linear acceleration of the descending object; (c) the angular acceleration of the disk; and (d) the tension in the string.

Conservation of Angular Momentum 12

The principle of conservation of angular momentum is the rotational counterpart of the principle of conservation of linear momentum. According to this principle, the angular or rotational momentum of a rotating system remains constant provided no external torques act on it. In this experiment, the system will consist of a rotating platform with a catcher mounted on it into which a ball is fired. The system is designed this way so that the collision forces between the ball and the catcher are internal to the system. Thus, except for the friction in the bearing on which the platform is pivoted (which must be taken into account), no external torques act on the system, and its angular momentum should be conserved. The purpose of this experiment is to investigate the law of conservation of angular momentum in this arrangement.

THEORY

Just as the quantity mv is the linear momentum of a moving object of mass m and velocity v, so $I\omega$ is the angular momentum of a rotating body having moment of inertia I and angular velocity ω. And just as it follows from Newton's second law of motion that the linear momentum of a moving system remains constant in the absence of forces external to the system, so also do we find that the angular momentum of a rotating system is constant in the absence of external torques.

In this experiment, the rotating system consists of a heavy, bar-like platform mounted on a very good bearing (so that it is free to rotate about its center with very little friction), a catcher cup mounted on one end and balanced by a counterweight mounted on the other, and a metal ball that gets fired into the catcher by a spring gun or launcher, as shown in Fig. 12.1. After the ball has been fired but before it enters the catcher, it has a linear momentum mv, where m is its mass and v its velocity. However, any object moving in a straight line with a linear momentum can be considered to have an angular momentum about some selected pivot point, as illustrated in Fig. 12.2. This angular momentum is found by multiplying the linear momentum mv by the lever or moment arm l defined as the perpendicular distance from the pivot point in question to the line of action of the linear momentum vector. As a consequence of this definition, angular momentum is sometimes called the *moment of momentum*. Like linear momentum, it is a vector; its direction lies along the supposed axis of rotation, which is a line through the pivot point perpendicular to the plane defined by that point and the line of action of mv. This axis is a line perpendicular to the page in Fig. 12.2. The sense of the angular momentum vector along this axis is given by the right-hand rule: If the axis is gripped by the right hand so that the fingers curl in the direction of rotation, the thumb points along the axis in the direction of the vector. In the situation shown in Fig. 12.2, this is into the paper.

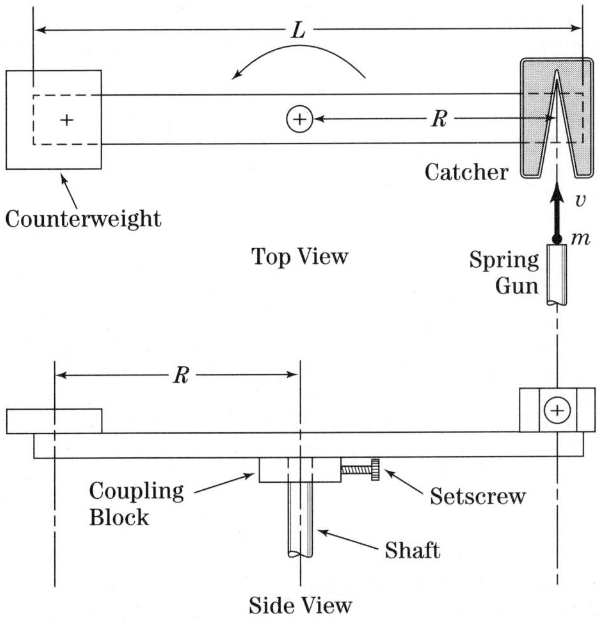

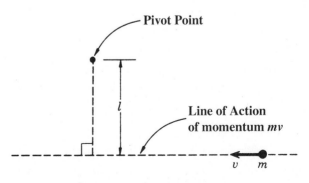

Figure 12.1 *Angular momentum apparatus*

Figure 12.2 *Moment of momentum*

After the ball is in the catcher, the system (consisting of the rotating platform, the catcher with the ball in it, and the counterweight) rotates with angular velocity ω. The moment of inertia of the system can be thought of as made up of three parts. The first is the moment of inertia of the platform alone and is given by

$$I_\mathrm{p} = \frac{1}{12}ML^2$$

where I_p is the moment of inertia of the platform in kilogram-meters^2, M is its mass in kilograms, and L is its overall length in meters.* The second is the moment of inertia of the catcher with the ball in it rotating in a circle of radius R, where R is the distance from the pivot to the center of mass of the catcher assembly (see Fig. 12.1). This moment of inertia is given by

$$I_\mathrm{cb} = m_\mathrm{cb}\,R^2$$

where m_cb is the total mass of the catcher and the ball. Similarly, the third part of the total moment of inertia is that of the counterweight. It moves in a circle of radius R just as the catcher does and has mass m_cw, which is close to but not necessarily the same as m_cb. Its moment of inertia is thus given by

$$I_\mathrm{cw} = m_\mathrm{cw}R^2$$

so that the total moment of inertia of the system is

$$I = I_\mathrm{p} + I_\mathrm{cb} + I_\mathrm{cw}$$
$$= \frac{1}{12}ML^2 + (m_\mathrm{cb} + m_\mathrm{cw})R^2 \qquad (12.1)$$

The angular momentum of the system right after the ball has been caught in the catcher is $I\omega$, where ω is the angular velocity of the platform right after the ball has been caught. The platform was at rest just before this, but the ball in flight from the spring gun had angular momentum mvl, where l is the perpendicular distance from

_____

*This formula treats the rotating platform as a simple bar of length L and negligible width.

the ball's line of flight to the platform's pivot point (the supporting bearing shaft). In Fig. 12.1, where the axis of the catcher cup is set perpendicular to the platform, $l = R$, but the catcher axis could be set at some smaller angle θ, in which case l would be equal to $R\sin\theta$ and the effective angular momentum reduced accordingly. Whatever the value of l, the angular momentum before the collision will be mvl and afterward $I\omega$, so that

$$mvl = \left[\frac{1}{12}ML^2 + (m_\mathrm{cb} + m_\mathrm{cw})R^2\right]\omega \qquad (12.2)$$

All the quantities in Equation 12.2 can be measured; hence, its validity and the principle of conservation of angular momentum can be checked.

Most of these quantities are easily found by direct measurement; but two—the speed v given the ball by the spring gun and the angular velocity ω of the rotating system just after the ball has been caught—have to be determined by special procedures. The initial velocity of the ball is easily found by firing it through a pair of light gates a fixed distance apart placed immediately in front of the spring gun as described in Experiment 5. The value needed for ω, however, is the one in effect immediately after the ball has been caught and the rotating assembly starts to turn. If there were no friction in the bearing supporting this assembly, rotation would continue at this angular velocity indefinitely. However, although the bearing is a very good one, it does exert a small frictional torque that will eventually bring the system to a stop. If the system rotates through a total angle ϕ before stopping and takes t_r seconds to do this, then the average value $\bar{\omega}$ of ω is ϕ/t_r and can be measured by simply timing the rotational motion. But to a very good approximation the decelerating torque due to friction in the bearing can be considered constant, so that the rules of constant negative angular acceleration apply. One of these is that the average angular velocity is half the sum of the initial and final velocities. Since the final angular velocity is zero, $\bar{\omega} = \frac{1}{2}\omega$ or $\omega = 2\bar{\omega}$. Thus, the desired value of ω is determined by dividing the total angle turned through by the time required for this motion and doubling the result.

APPARATUS

1. Rotating platform with catcher and counterweight mounted on a verticle shaft carried in a bearing in the base unit
2. Spring launcher with steel ball and ramrod
3. Mounting stand and clamp for launcher
4. Light-gate assembly with mounting bracket and timer

5. Meter stick
6. Stopwatch or stop clock
7. Triple-beam balance
8. Level
9. Draftsman's triangle
10. Backstop

PROCEDURE

1. Remove the catcher and counterweight from the rotating platform if you find them mounted on it. Then loosen the setscrew under the platform so that you can

take this latter off the shaft and unscrew the screws holding the shaft coupling block to the platform's bottom with the special tool provided. You can now weigh the platform alone. Also weigh the catcher with the ball from the launcher in it, weigh the ball separately, and weigh the counterweight. Enter these results in the data table.

2. Measure the overall length of the rotating platform with the meter stick. Notice that although there is a centimeter scale on the platform's upper surface, its zero is at the center to facilitate radius of rotation measurements and it does not extend to the platform's ends. Hence, the need for the meter stick to measure the *overall* length. Reassemble the coupling block to the platform and replace the unit on the shaft, being sure to tighten the setscrew against the shaft's flat.

3. Mount the launcher on its stand and clamp the stand to the bench in some convenient position from which you can fire the steel ball into the backstop, the rotating assembly being temporarily moved out of the way. Attach the mounting bracket with the two light gates to the launcher and determine the distance between them. See Procedures 1 and 3 of Experiment 5. Record this distance in the space provided in the data table.

4. Use the level to set the launcher exactly horizontal as described in Procedure 2 of Experiment 5. Then fire the launcher into the backstop as described in Procedure 6 of that experiment. If you have not used the launcher and light gates before, do this several times for practice, observing that the timer reads the time the ball takes to go between the light gates so that a direct calculation of the launcher's muzzle velocity can be made. Use the 0.1-millisecond range on the timer and don't forget to reset it to zero after each shot. Finally, make five shots and record the time for each in the data table.

5. You are now ready to start the conservation of angular momentum trials. Mount the catcher on the rotating platform at a radius of 20 cm by placing it so that the centimeter scale reads the same distance above and below 20 cm on each side. For example, if the catcher is 6 cm wide, it should be placed on the platform with one side at the $20 - 3 = 17$-cm mark and the other at the $20 + 3 = 23$-cm mark. The catcher has two tabs projecting from its underside that should be butted up against the front edge of the platform. This should ensure that the catcher's center line is perpendicular to the platform's long dimension, but check this with the draftsman's triangle and insert slips of paper behind the appropriate tab as needed. Then mount the counterweight on the platform at a 20-cm radius on the other side of center from the catcher. You will thus have $R = 20$ cm in Equations 12.1 and 12.2.

6. Remove the bracket carrying the light gates from the launcher and place the rotating assembly in front of the launcher, as shown in Fig. 12.3. The launcher should be aimed directly into the catcher, and at the same time its axis should be perpendicular to the rotating platform. This may necessitate moving both the rotating assembly and the launcher's stand. Note that aiming the launcher at the catcher is facilitated if the two are quite close, but space must be left so that the ends of the platform clear the launcher as they come around. Finally, level the platform by placing your level on its top surface and adjusting the screws in the base unit until the platform is seen to be level regardless of its orientation. It should then show no tendency to start turning out of any position in which it is left at rest.

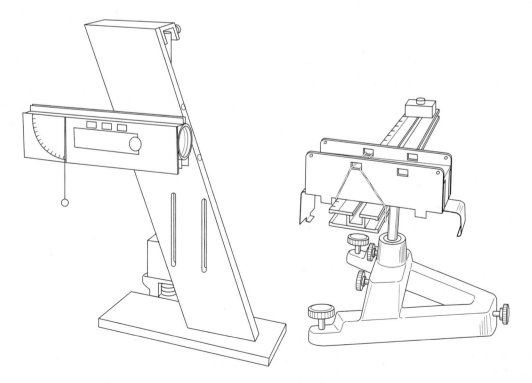

Figure 12.3 *Angular momentum apparatus and launcher*

7. Load the launcher and fire the ball into the catcher to see if there are any problems. You may want to make a few practice shots before taking data. When all is in order, set the rotating platform in the proper position, fire the launcher, start the stopwatch at the instant the ball enters the catcher, and stop it at the instant the rotating platform just stops turning. While the platform is in motion, count the number of turns completed before rotation stops, and thereafter estimate to 10° the angle through which the platform turned in the uncompleted final revolution. Record the number of turns and this angle.

8. Repeat Procedure 7 four more times to obtain a total of five sets of data on the angle turned through and the time required.

9. Reset the catcher and counterweight to radii of 10 cm as described in Procedure 5. The rotating assembly and the launcher will then have to be moved so that the launcher is once more properly aimed into the catcher. Repeat Procedures 6–8 with the new value of R.

DATA

Mass of the rotating platform _____

Mass of the counterweight _____

Mass of the catcher and ball _____

Calculated moment of inertia I of the rotating system

Length L of the rotating platform _____

Mass of the ball alone _____

$R = 20$ cm _____ $R = 10$ cm _____

Distance between light gates _____

Times from Procedure 4   1. _____   2. _____   3. _____   4. _____   5. _____

Launcher muzzle velocity   1. _____   2. _____   3. _____   4. _____   5. _____

Deviations   1. _____   2. _____   3. _____   4. _____   5. _____

Average value of muzzle velocity _____   Average deviation _____

R, meters	Rotation time, s	Number of turns completed	Angle of incomplete turn, degrees	Total angle turned, rad	Average angular velocity $\bar{\omega}$, rad/s	Initial angular velocity ω, rad/s	Deviations
0.20							
0.10							

$R = 20$ cm

Average value of the initial angular velocity ω _____

Average value of deviation _____

Angular momentum after the ball is caught _____

$R = 10$ cm

Average value of the initial angular velocity ω _____

Average value of deviation _____

Angular momentum after the ball is caught _____

Moment arm l of the ball in flight	_____	Moment arm l of the ball in flight	_____
Angular momentum before the ball is caught	_____	Angular momentum before the ball is caught	_____
Percent discrepancy	_____	Percent discrepancy	_____

CALCULATIONS _____

1. Calculate the moment of inertia of the rotating system using Equation 12.1 and the data of Procedures 1 and 2. Remember that m_{cb} is the mass of the catcher with the ball inside it.

2. From the data of Procedure 4, compute the average muzzle velocity of the ball, the deviation of each of these measurements from the average, and the average deviation (a.d.).

3. Calculate the total angle through which the rotating arm turned following each firing of the ball into the catcher. This is done by multiplying the number of complete turns by 360 and adding the number of degrees turned through in the incomplete final turn. The result is converted from degrees to radians by multiplying by $\pi/180$.

4. Calculate the average angular velocity $\bar{\omega}$ in each case by dividing the total angle turned through *in radians* by the corresponding time in seconds. Then double each value of $\bar{\omega}$, to obtain the corresponding initial angular velocity ω.

5. Find the average of your five values of ω for the case of the catcher set at $R = 20$ cm and multiply this result by the moment of inertia found in Calculation 1 to obtain the angular momentum of the system after the ball is caught.

6. Find the deviation of each of your values of ω obtained in Calculation 4 from the average value found in Calculation 5 and then find the average deviation (a.d.). Treating this as the error in your result for ω, find the error to be associated with the angular momentum determined in Calculation 5, assuming no error in the moment of inertia.

7. With the catcher set at 90° to the rotating platform, the moment arm l of the ball in flight is equal to the radius R of the circle in which the catcher moves. Multiply this radius (20 cm or 0.20 m for your first set of measurements), the mass of the ball, and the ball's velocity from Calculation 2 together to obtain the angular momentum mvl due to the ball before it is caught. Since the rotating system is at rest before the ball is caught, mvl is the total angular momentum of the complete system at that time and should be equal to the system's angular momentum afterward as found in Calculation 5. Attach the proper error to your value of mvl and compare your before and after values of the angular momentum by seeing if they fall within the limits of error established by your a.d.'s. Also calculate their difference and the percent discrepancy.

8. Repeat Calculations 5–7 for the case of the catcher set at $R = 10$ cm. Note that in this case, $l = R = 0.10$ m.

QUESTIONS _____

1. Explain in your own words how this experiment demonstrates the principle of conservation of angular momentum.

2. Where do you think errors came into your experiment? What could be done to reduce them?

3. The counterweight was considered to be a point mass traveling in a circle of radius R, but actually it is a square block about 4 cm on a side. Since it is mounted on the platform, it turns about its own center of mass with the same angular velocity as that of the rotating system. In other words, its total motion has two components: the motion of its center of mass in a circle of radius R and the rotational motion about its own center of mass. Both contribute to the moment of inertia of the complete rotating system. Why could you neglect the contribution due to the counterweight's own rotation?

4. You removed the shaft coupling block from the bottom of the rotating platform before weighing the platform, yet this block and the shaft that rotates with the assembly are both part of the complete rotating system. Estimate their contribution to the system's moment of inertia and state whether you were justified in neglecting it.

5. Calculate the energy $\frac{1}{2}mv^2$ of the system before the ball was caught in the catcher and the energy $\frac{1}{2}I\omega^2$ afterward. Which is greater? What happened to the difference? Was the difference greater when the catcher was at $R = 20$ cm or when it was at $R = 10$ cm? Why?

6. A wheel having a moment of inertia of 0.01 kg·m^2 is mounted in a horizontal plane on a frictionless shaft. The wheel has a diameter of 0.5 m and is initially at rest. A 40-g ball is fired in a direction tangent to the rim of the wheel with a velocity of 10 m/s and is caught in a cup mounted on the rim. What is the angular velocity of the wheel after the ball settles down in the cup?

7. (a) Calculate the conversion factor between angular velocity in radians per second and in revolutions per minute. (b) Express your answer to Question 6 in revolutions per minute.

Archimedes' Principle 13

Buoyancy is the ability of a fluid to sustain a body floating in it or to diminish the apparent weight of a body submerged in it. Archimedes' principle states that this apparent reduction in weight is equal to the weight of the fluid displaced. It is the purpose of this experiment to study Archimedes' principle and its application to the determination of density and specific gravity. In particular, the specific gravities of a solid heavier than water, a solid lighter than water, and a liquid other than water will be measured.

THEORY

The density of a body is defined as its mass per unit volume. It is usually expressed in grams per cubic centimeter.* The specific gravity of a body is the ratio of its density to the density of water at the same temperature. Since for a given location the weight of a body is taken as a measure of its mass, the specific gravity may be taken as the ratio of the weight of a given volume of a substance to the weight of an equal volume of water. Because the mass of 1 cm^3 of water at 4°C is 1 g, the specific gravity of a body at this temperature is also numerically equal to its density in grams per cubic centimeter.

If a body is totally immersed in a fluid, the volume of fluid displaced must be equal to the volume of the body, because if the body weren't there, its volume would be occupied by the fluid. Moreover, the hydrostatic pressure on the bottom of the body is greater than that on the top because hydrostatic pressure rises as depth increases. The situation is illustrated in Fig. 13.1, which shows a beaker of fluid with an object (which we have made rectangular in the interest of simplicity) submerged in it. Let us begin by imagining an area of 1 cm^2 on the bottom of the beaker as shown at the right in Fig. 13.1. The dotted lines indicate that we may think of this 1-cm^2 patch as supporting a column of fluid 1 cm^2 in cross section extending up to the surface a distance d above it, the depth of fluid being d cm. Thus, a volume of fluid $1 \times d$ cm^3 with a weight of $\rho g d$ dynes is sitting on that 1 cm^2 of the bottom. Here ρ is the fluid's density and g is the acceleration of gravity in centimeters per second squared. Hence, the pressure sustained by the bottom of the beaker is $\rho g d$ dynes weighing on every square centimeter, or $\rho g d$ dynes/cm^2. Note that this is the *gauge* pressure. The *absolute* pressure must include the pressure due to the atmosphere on the fluid surface—we could imagine the dotted lines in Fig. 13.1 going on up to the top of the atmosphere and indicating a column of air 1 cm^2 in cross section sitting on top of the column of liquid. Since atmospheric pressure does not affect the present experiment, we will not consider it further.

Turning now to the immersed solid, we note that it has height h, its bottom end being at depth h_2 and its top end at depth h_1 where $h_2 - h_1 = h$. The considerations of the last paragraph show that the pressure at any depth d in a fluid of density ρ is $\rho g d$; hence, the pressure on the bottom of the body is $\rho g h_2$ and the force exerted on the bottom, which will be an upward force, is $\rho g h_2 A$, where A is the body's cross-sectional area as shown in Fig. 13.1. Similarly, the downward force on the body's top surface is $\rho g h_1 A$, so that there is a net upward or so-called *buoyant force* $\rho g A (h_2 - h_1) = \rho g A h$. But Ah is just the body's volume V, so that the buoyant force may be written as simply $\rho g V$. Then, since ρ is the fluid's density, $\rho g V$ is identical with the weight of the displaced fluid. This result—namely, that the buoyant force (which, being an upward force, gives rise to an apparent loss of weight of the submerged body) is equal to the weight of the displaced fluid—is known as *Archimedes' principle*.

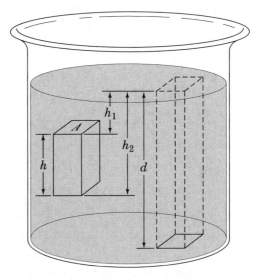

Figure 13.1 *A beaker of fluid with a submerged rectangular object*

---

*The CGS (centimeter-gram-second) system will be used in this experiment as it is both more usual and more convenient for this work.

The specific gravity of a solid heavier than water may be easily determined by the application of this principle. The body is weighed in air; then it is weighed in water, that is, suspended by a thread from the arm of a balance so as to be completely submerged (see Fig. 13.2). The loss of weight in water is $W - W_1$, where W is the weight in air and W_1 is the weight in water. But this loss of weight must be just the buoyant force, which, by Archimedes' principle, is equal to the weight of the water displaced, or the weight of an equal volume of water. Thus, the specific gravity S will be

$$S = \frac{W}{W - W_1} = \frac{Mg}{Mg - M_1 g} = \frac{M}{M - M_1} \quad (13.1)$$

where M is the body's mass in air and M_1 its apparent mass (from the apparent weight measurement) in water.

The specific gravity of a liquid may be found by measuring the loss of weight of a convenient solid body when immersed in that liquid and the loss of weight when immersed in water. The procedure is as follows: A heavy body is weighed in air; this weight is called W. Then it is weighed in water; this weight is called W_1. Finally, it is weighed in the liquid whose specific gravity is to be determined; this weight is called W_2. The specific gravity of the liquid will then be

$$S = \frac{W - W_2}{W - W_1} = \frac{M - M_2}{M - M_1} \quad (13.2)$$

since this expression represents the weight of a certain volume of the liquid divided by the weight of an equal volume of water. Here $W - W_1$ is the loss of weight in water, and $W - W_2$ is the loss of weight in the given liquid.

In order to find the specific gravity of a solid lighter than water, it is necessary to employ an auxiliary body, or sinker, of sufficient weight and density to hold the other body completely submerged. The specific gravity of a solid lighter than water, as obtained by the sinker method, is given by

$$S = \frac{W}{W_1 - W_2} = \frac{M}{M_1 - M_2} \quad (13.3)$$

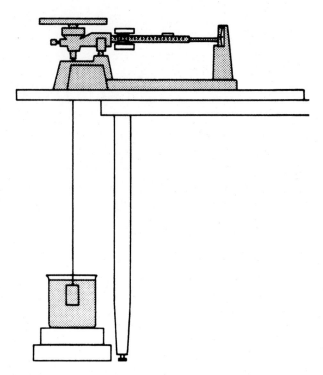

Figure 13.2 *Arrangement for weighing a body in water*

where W is the weight of the solid in air; W_1 is the weight of the solid and the sinker, with the sinker alone immersed; and W_2 is the weight when both solids are immersed in water.

The hydrometer is an instrument designed to indicate the specific gravity of a liquid by the depth to which it sinks in the liquid. To measure the specific gravity of a liquid by means of a hydrometer, it is only necessary to let the hydrometer float in the liquid and to read the specific gravity directly on the calibrated scale. The reading is taken, if possible, by placing the eye below the liquid surface and seeing where this surface cuts the hydrometer scale.

APPARATUS

1. Triple-beam balance with hook for suspending weights
2. Metal cylinder
3. Wooden cylinder
4. Lead sinker
5. Distilled water
6. Alcohol
7. Hydrometer
8. Hydrometer jar
9. 1000-cc Pyrex beaker
10. Fine thread

PROCEDURE

1. The triple-beam balance is set up with a fine thread attached to the underside of the pan carrier so that you can weigh bodies by hanging them on the thread rather than placing them in the pan. Make sure that the thread is of the proper length so that a body attached to its end will hang completely submerged in the fluid in the beaker (see Fig. 13.2). Also check that the beam balance balances with no body attached and adjust it accordingly. Then weigh the metal cylinder in air by suspending it on the end of the thread. Note that the balance is calibrated in grams, so that you will be recording the *mass* of the cylinder.

2. Fill the beaker with water and place it on the floor with the metal cylinder submerged in it as shown in Fig. 13.2. Be sure the cylinder is completely submerged and not touching the sides of the beaker. In this way, obtain the weight of the metal cylinder in water. Again, you will be reading in grams so that you will be observing an "apparent reduction in mass." When you have completed this measurement, remove and dry off the metal cylinder.

3. Empty the beaker, dry it thoroughly, and refill it with alcohol. Then repeat Procedure 2. When you are fin-

ished, pour the alcohol into the hydrometer jar and rinse out the beaker with distilled water.

4. Weigh the wooden cylinder in air as you did with the metal cylinder in Procedure 1.

5. Refill the beaker with distilled water. Attach the sinker to the wooden cylinder and weigh the combination with the sinker alone immersed in water.

6. Now weigh the two solids when they are both immersed in water.

7. Measure the specific gravity of the alcohol with the hydrometer. See that there is enough alcohol in the hydrometer jar and let the hydrometer float in the alcohol. Read the specific gravity directly.

DATA

Mass of metal cylinder in air	_____
Apparent mass of metal cylinder in water	_____
Apparent mass of metal cylinder in alcohol	_____
Specific gravity of metal cylinder	_____
Specific gravity of metal cylinder from Table IV at the end of the book	_____
Percent error	_____
Specific gravity of alcohol	_____

Specific gravity of alcohol from Table IV at the end of the book	_____
Percent error	_____
Mass of wooden cylinder in air	_____
Apparent mass of cylinder in air and sinker immersed in water	_____
Apparent mass of cylinder and sinker, both immersed in water	_____
Specific gravity of wooden cylinder	_____
Specific gravity of alcohol by using the hydrometer	_____

CALCULATIONS

1. From the data of Procedures 1 and 2, calculate the specific gravity of the metal cylinder.

2. From the data of Procedures 1–3, calculate the specific gravity of the alcohol.

3. From the data of Procedures 4–6, compute the specific gravity of the wooden cylinder.

4. Compute the percent error of your measurements by comparing your results for the specific gravity of the metal cylinder and of the alcohol with the accepted values.

5. Compare your measurement of the specific gravity of alcohol using the hydrometer with the accepted value and note whether this latter value falls within the limits of precision on the hydrometer reading.

QUESTIONS _____

1. (a) Explain how you can obtain the volume of an irregular solid insoluble in water. (b) How can you obtain the weight of an equal volume of water?

2. (a) When a block of wood is completely submerged in water, why does it apparently lose more than its entire weight in air? (b) Given the block of wood's weight and density, find an expression for the minimum weight of a sinker made of a material of density ρ_s (>1) that will completely submerge the combination.

3. Suppose there were a bubble of air on the bottom of the metal cylinder immersed in water. How would this affect the calculations of the density of the metal?

4. A piece of cork having a mass of 25 g in air and a specific gravity of 0.25 is attached to a lead sinker whose mass is 226 g in air. What will be the apparent mass of the two solids when they are both immersed in water?

5. The cork in Question 4 is allowed to float in water. What fraction of its volume is above the surface?

6. Derive Equation 13.3.

7. Look up Pascal's principle, state it in your own words, and tell how it was applied (without our saying so) in the derivation of Archimedes' principle given in the Theory section.

8. (a) Explain the apparent loss of weight suffered by an object when immersed in a liquid. (b) Suggest a modification of the apparatus shown in Fig. 13.2 that will demonstrate your answer to Part (a). (c) If your instructor suggests it and time permits, carry out the experiment you proposed in Part (b) and state your results below.

Simple Harmonic Motion 14

Simple harmonic motion is one of the most common types of motion found in nature, and its study is therefore very important. Examples of this type of motion are found in all kinds of vibrating systems, such as water waves, sound waves, the rolling of ships, the vibrations produced by musical instruments, and many others. In fact, any time a *linear restoring force* exists, simple harmonic motion results. A linear restoring force is a force that is proportional to the displacement of the body on which the force acts from an equilibrium position (where the force is zero) and is always directed back toward that equilibrium position. The object of this experiment is to study two important examples of such a force, the simple pendulum and the vibrating spring, and to determine the time of vibration (called the period) in each case.

THEORY

A linear restoring force as defined above may be expressed mathematically by the equation

$$F = -kx \qquad (14.1)$$

where F is the force, x represents the displacement from the equilibrium point $x = 0$, k is the proportionality constant, and the minus sign expresses the restoring nature of the force by indicating that the direction of F is always opposite to the displacement x from equilibrium.

If a body to which a linear restoring force is applied is displaced from equilibrium and released, the restoring force will bring the body back to its equilibrium position. But the body's inertia will then carry it beyond this position, and so a restoring force will build up in the opposite direction. This force first brings the body to rest and then accelerates it back toward the equilibrium position again. The action is repeated and the body executes a type of vibration known as simple harmonic motion. In this motion, the maximum displacement from the equilibrium point achieved by the body on either side is called the amplitude, one complete vibration is called a cycle, the time required for one cycle is called the period, and the number of cycles completed per second (the reciprocal of the period) is called the frequency. Clearly, the period is measured in seconds and the frequency in cycles per second (s⁻¹). The unit of cycles per second is now commonly called hertz in honor of Heinrich Hertz (1857–1894), who investigated electrical oscillations and first showed (1887) that high-frequency electrical oscillations could produce electromagnetic radiation. Further analysis of the vibrations resulting from a linear restoring force shows that the period is given by

$$T = 2\pi \sqrt{\frac{m}{k}} \qquad (14.2)$$

where T is the period, m is the mass of the vibrating body, and k is the proportionality constant from Equation 14.1. If m is expressed in kilograms and k in newtons per meter, T will come out in seconds.

Two physical examples of this type of force are considered in the present experiment. The first is the simple pendulum, which consists of a concentrated mass suspended at the end of a cord of negligible weight. A close approximation to this is a small metal sphere on a long, thin thread. Such an arrangement is diagrammed in Fig. 14.1, which shows a sphere of mass m hung on the end of

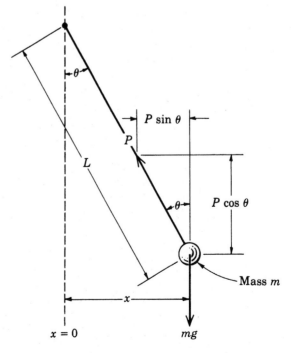

Figure 14.1 *The simple pendulum*

109

a thread whose length is such that the distance from the suspension point to the center of the sphere is L. The sphere has been drawn aside to a horizontal distance x, the thread consequently making an angle θ with the vertical. Note that the vertical line through the suspension point is at $x = 0$, for this is the equilibrium position, mass m being in equilibrium when it hangs straight down.

Inspection of Fig. 14.1 shows that the sum of the x components of all the forces acting on mass m is simply $-P \sin \theta$ and the sum of all the y components is $P \cos \theta - mg$, where P is the tension in the thread. If the pendulum is never swung by more than a very small amount so that θ is always a small angle even at the extremes of the swing, then two simplifying assumptions may be made. The first of these is that for small angles the cosine remains approximately equal to 1, so that $P \cos \theta - mg$ may be approximated by $P - mg$. Secondly, in a small swing there is very little vertical motion of mass m, so that zero vertical acceleration may be assumed. Hence,

$$P \cos \theta - mg \approx P - mg \approx 0$$

or

$$P \approx mg$$

Further inspection of Fig. 14.1 shows that, by definition, $\sin \theta = x/L$, and therefore by substitution the x-directed force $-P \sin \theta$ becomes $-mgx/L$. This qualifies as a linear restoring force with the proportionality constant k equal to mg/L, and Equation 14.2 shows that the period of the simple pendulum should then be

$$T = 2\pi \sqrt{\frac{mL}{mg}} = 2\pi \sqrt{\frac{L}{g}} \qquad (14.3)$$

Again the period will be in seconds if the length L is in meters and g, the acceleration of gravity, is in meters per second per second (9.8 m/s²).

The second example of a linear restoring force presented in this experiment is that of a stretched spring. Ideally, the elongation of a spring is proportional to the stretching force. Therefore, if a mass is hung on a spring, the spring will be stretched by an amount proportional to the weight of the mass. Furthermore, an equilibrium position will be reached at which the spring pulls up on the mass with a force just equal to the mass's weight. The situation is illustrated in Fig. 14.2, where initially the spring hangs relaxed, with no mass attached, and has a length y_1. When a weight of mass m is attached, the spring elongates to length y_0, and if the elongation is proportional to the stretching force (the weight of the mass), $mg = k(y_0 - y_1)$.

When the weight is at any arbitrary distance y below the spring support, the elongation of the spring is $y - y_1$, the upward pull of the spring on the weight is $k(y - y_1)$, and the downward force of gravity is mg. Thus, the net force on the weight is

$$F = mg - k (y - y_1) \qquad (14.4)$$

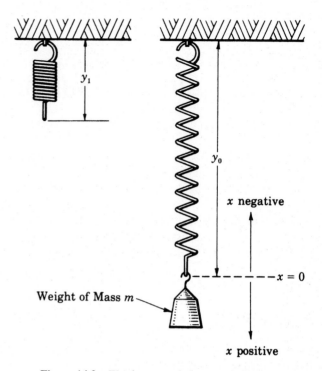

Figure 14.2 *Weight suspended on an ideal spring*

If a new coordinate x, measured in the same direction as y but having the equilibrium point as its origin, is introduced as shown on the right in Fig. 14.2, then $x + y_0 = y$, for the value of y will always be y_0 greater than the value of x. Substitution in Equation 14.4 yields

$$F = mg - k(x + y_0 - y_1) \qquad (14.5)$$

But since $mg = k(y_0 - y_1)$, Equation 14.5 becomes

$$F = k(y_0 - y_1) - k(x + y_0 - y_1) = -kx$$

Thus, again, we have a linear restoring force, the constant k here being a characteristic of the spring called its force constant. The force constant can be determined by measuring the elongation in meters suffered by the spring when a stretching force of a given number of newtons is applied. A part of this experiment will be devoted to making such a determination.

If the weight in Fig. 14.2 is now pulled down below its equilibrium position and released, it will execute simple harmonic motion with a period given by Equation 14.2. Note, however, that m in this equation will be the mass of the weight only if the spring has negligible mass. In the general case where the spring's mass cannot be neglected, m is the effective mass of the vibrating system, which is made up of the mass of the suspended weight plus a part of the mass of the spring, since the spring itself is also vibrating. It is found by analysis that one third of the mass of the spring must be added to the mass of the suspended weight in calculating the value of m to be used in Equation 14.2.

APPARATUS _____

1. Metal sphere on a long string (the simple pendulum)
2. Supporting rod with right-angle clamp, short rod, and hook for suspending spring
3. 2-m stick with caliper jaws
4. Vernier caliper
5. Spring (cylindrical type, coils not touching when relaxed)

6. Set of hooked weights (100-g, 200-g, 200-g, 500-g)
7. Stopwatch or stop clock, *or* (optional) light gate and timer
8. Triple-beam balance

PROCEDURE _____

1. Measure the diameter of the metal sphere with the vernier caliper.

2. Attach the string so that the center of the sphere is about 0.6 m below the point of support. Carefully measure the distance from this point to the upper surface of the sphere. Record this distance as the length of string used. The length of the pendulum will be this distance plus the radius of the metal sphere.

3. If you are using the light gate and timer, position this assembly and orient the gate so that the sphere will swing through it, breaking the light beam.

4. Displace the sphere to one side through an angle of no more than about 5° and let the pendulum oscillate. Record the time it takes the pendulum to make 50 vibrations. In counting vibrations, be sure to start the stopwatch on the count of zero and not on the count of one. If you are using the light gate and timer, switch the timer to its pendulum mode and observe its reading as the sphere swings through the gate. When the timer is in the pendulum mode, it is started when the sphere breaks the beam but is *not* stopped the next time the sphere comes through. It *is* stopped on the second time through, thus measuring the time for a complete swing of the simple pendulum. The reading is therefore the actual period of the pendulum and may be entered as such in the data table without the bother of counting and timing 50 swings and then dividing by 50. However, the timer must be reset to zero by pushing the reset button after each reading. Practice doing this and take the average of several readings as the period. Use the 0.1-millisecond scale on the timer. *Note:* On some timers of this type, the seconds digit will not go beyond 1 when this scale is in use, although the count for fractions of a second remains correct. In this case, use the 1-millisecond scale to get the number of seconds and then switch to the 0.1-millisecond scale to get the requisite precision.

5. Repeat Procedures 2 and 4, making the length of the pendulum successively 0.8, 1.0, 1.2, 1.4, and 1.6 m. In the last three cases, record the time for 25 vibrations instead of 50 if you are not using the light gate and timer.

6. Using a length of about 0.5 m, determine the time for 50 vibrations when the sphere is displaced about 5°, 30°, and 45° from the equilibrium position. Record the length of the pendulum and the time for 50 vibrations at each displacement. If you are using the light gate and timer you can measure the period at each displacement directly without counting vibrations as described in Procedure 4.

7. Weigh the spring and record its mass in kilograms.

8. Remove the pendulum from the support and suspend the spring in its place. Using the 2-m stick and a caliper jaw, observe the position of the lower end of the spring and record the reading.

9. Suspend 0.1 kg from the spring and again record the position of its lower end.

10. Repeat Procedure 9 with loads of 0.2, 0.3, 0.4, and 0.5 kg suspended from the spring. Record the position of the spring's lower end in each case.

11. Suspend a mass of 0.2 kg from the spring. If you are using the light gate and timer, position this assembly and orient the gate so that the mass passes through it, breaking the beam. Note, however, that you should arrange the gate so as to have the mass pass through it somewhat to one side rather than through the center, in order to keep the spring from interrupting the beam. You want only the mass to trigger the timer as it passes through. Now displace the mass about 5 cm downward from its position of equilibrium, and set it into oscillation. Time the period as described in Procedure 4, or, if you are *not* using a light gate and timer, measure the time for 50 complete oscillations with the stopwatch and record it.

12. Repeat Procedure 11, again using a mass of 0.2 kg but displacing it about 10 cm from its position of equilibrium and setting it into vibration. Again record the time for 50 complete oscillations or use the timer and light gate as already described.

13. Suspend a mass of 0.5 kg from the spring, displace it about 5 cm from its position of equilibrium, and set it into vibration. Record the time for 50 complete oscillations or the period as measured by the light gate and timer.

DATA

Diameter of sphere _____   Radius of sphere _____

Length of string used	Length of pendulum	Number of vibrations	Time	Period	Square of period

Value of g from slope _____   Percent error _____

Length of string used _____   Length of pendulum _____

Initial displacement of sphere	Number of vibrations	Time	Period	Square of period
5°				
30°				
45°				

Mass suspended from the spring	Force stretching the spring	Scale reading	Elongation
0 kg			
0.1 kg			
0.2 kg			
0.3 kg			
0.4 kg			
0.5 kg			

Mass of the spring _____   Force constant of the spring _____

Mass suspended from the spring	Mass of the vibrating system	Amplitude of vibration	Time for 50 vibrations	Period Experimental value	Period Calculated value	Percent discrepancy
0.2 kg		5 cm				
0.2 kg		10 cm				
0.5 kg		5 cm				

CALCULATIONS _____

1. Calculate the period of the pendulum for each observation if you timed a number of vibrations. You already have the period if you used the light gate and timer.

2. Calculate the square of the pendulum's period for each observation. Use only three significant figures in recording the square of the period, rounding off the third figure.

3. Plot a curve using the values of the square of the period as ordinates and the lengths of the pendulum as abscissas. Use the entire sheet of graph paper. Read the instructions on plotting graphs in the Introduction.

4. Determine your measured value of g from the slope of the curve just plotted and compare it with the known value, $g = 9.8$ m/s^2, by finding the percent error. The acceleration of gravity varies slightly with latitude and elevation. The value given is approximately correct for sea level and 45° latitude.

5. From the data of Procedures 8–10, determine the elongation of the spring produced by each load by subtracting the zero reading from the reading corresponding to each load. Also multiply each suspended mass load by 9.8 m/s² to obtain the force stretching the spring in each case.

6. Plot a curve using the values of the elongation as ordinates and the forces due to the corresponding loads as abscissas.

7. Obtain the force constant of the spring from the slope of the curve plotted in Calculation 6. See the instructions in the Introduction.

8. From the data of Procedures 11–13, calculate the experimental value of the period of the oscillating weight for each set of observations.

9. Using Equation 14.2, calculate the value of the period for each of the two masses used. The effective mass of the oscillating system will be the mass suspended from the spring plus one third of the mass of the spring.

10. Calculate the percent discrepancy between the experimental values of the period and the calculated values.

QUESTIONS _____

1. Show that if *m* is in kilograms and *k* in newtons per meter, *T* will in fact come out in seconds in Equation 14.2.

2. State what you conclude from your work with the simple pendulum.

3. Why was it unnecessary to weigh the metal sphere?

4. Explain in your own words why the angle through which the pendulum swings must be no more than about 10°.

5. (a) At which point on the path of its vibration does the weight suspended on the spring have the greatest acceleration? (b) Where does it have the greatest velocity? (c) Where does it have its least acceleration? (d) Where does it have its least velocity?

6. Is the motion of the piston in a one-cylinder steam engine simple harmonic? Explain.

7. What does your curve show about how the elongation of the spring depends upon the applied force?

8. (a) If it takes a force of 0.5 N to stretch a spring 2 cm (0.02 m), what force is needed to stretch the spring 5 cm? (b) How much work is done in stretching the spring 5 cm?

9. What percent error is introduced in the calculated value of the period of vibration, for the 0.2 kg load and the 0.5 kg load, respectively, if the mass of the spring is neglected?

10. If adding a certain weight in Procedure 10 would double the period, what would the mass of the added weight be?

11. Rotational simple harmonic motion occurs when there is a linear restoring *torque* given by $\tau_r = -K\theta$, where θ is the *angular* displacement from equilibrium. The period of the resulting rotational oscillation is given in analogy with Equation 14.2 by $T = 2\pi\sqrt{I/K}$. Derive Equation 14.3 by treating the simple pendulum as a rotating system with the suspension point as the pivot. Note that for small angles, $\sin\theta \approx \theta$.

12. (a) Discuss what would happen if there were no minus sign in Equation 14.1. In doing so, distinguish between *stable* and *unstable* equilibrium. (b) *(Optional)* If you can, derive the equation of motion, that is, the equation that describes the motion called simple harmonic, and hence Equation 14.2, from the force equation, Equation 14.1. What would happen to these equations if the minus sign were replaced by a plus sign in Equation 14.1?

Standing Waves on Strings 15

The general appearance of waves is well illustrated by standing waves (so called because they do not seem to move along the medium) on a string. This type of wave is very important because most of the vibrations of bodies, such as piano or violin strings or the air column in an organ pipe, are standing waves. The purpose of this experiment is to observe standing waves on a stretched string, to determine the natural vibrational frequencies of that string, and to study the relation between those frequencies, their associated wavelengths, and the stretching force.

THEORY

Standing or stationary waves are produced by the interference of two wave trains of the same wavelength, velocity of propagation, and amplitude traveling in opposite directions through the same medium. The necessary condition for the production of standing waves on a stretched string can be met by fixing one end of the string and sending a train of waves down from a vibrating body at the other end. These waves reflect from the fixed end and travel back toward the source, thus interfering with the oncoming waves.

A stretched string has many modes of vibration, all determined by the fact that the two ends must remain fixed and non-vibrating. A non-vibrating point is called a *node*, and we therefore see that we have imposed on our string the constraint that the two ends must be nodes. Because the vibrating string assumes a sinusoidal shape (see, for example, Fig. 15.1), a node or zero point occurs every half wavelength, and so our constraint requires that the string's length between its two fixed ends be an integer number of half wavelengths. The string may thus vibrate as a single segment, in which case its total length is equal to one half wavelength, or in two segments with a node in the middle so that the string's total length is two half wavelengths or one whole wavelength. Similarly, vibration may occur with the string's length being equal to three half wavelengths with two intermediate nodes as shown in Fig. 15.1, or in four half wavelengths (two whole wavelengths) with three intermediate nodes, or in five, six, etc. half wavelengths. In short, because in a standing wave the nodes occur every half wavelength, the fact that the ends of the string must correspond to nodes requires that the whole length of the string accommodate an integer number of half wavelengths.

Now, since the wavelengths of the various modes are determined by the above considerations, the frequencies at which the string wants to vibrate are also determined, the equation

$$V = f\lambda \tag{15.1}$$

expressing the relationship between the wavelength λ, the frequency f, and the velocity V with which waves propagate on the string. This velocity is given by the equation

$$V = \sqrt{\frac{F}{d}} \tag{15.2}$$

where F is the string tension in newtons and d is the string's mass per unit length in kilograms per meter. Thus, if we define L as the distance between adjacent nodes as shown in Fig. 15.1, then for any given mode $\lambda = 2L$ and

$$f = \frac{V}{\lambda} = \frac{1}{2L}\sqrt{\frac{F}{d}} \tag{15.3}$$

We conclude that the frequencies at which the string wants to vibrate are determined by the allowed values of λ, and hence L, and also by the tension and mass per unit length of the string. Consequently, if the frequency of a vibrating body driving one end of the string is set to one of the frequencies given by Equation 15.3, the string will vibrate in the corresponding mode with a relatively large amplitude, whereas it will hardly vibrate at all when driven at other frequencies. Setting the driving frequency equal to one of the mode frequencies establishes a condition called *resonance* between the vibrating body and the string.

In this experiment, standing waves are set up on a stretched string by a string vibrator, which is basically the voice-coil section of a loudspeaker but which has, instead of a speaker cone, a projecting rod that bears on the string so as to move it up and down. The vibrator is driven by a signal generator that provides a sinusoidal output at the frequency selected by the tuning dial. The arrangement of the apparatus is shown in Fig. 15.1. The tension in the string is fixed by the weights suspended

119

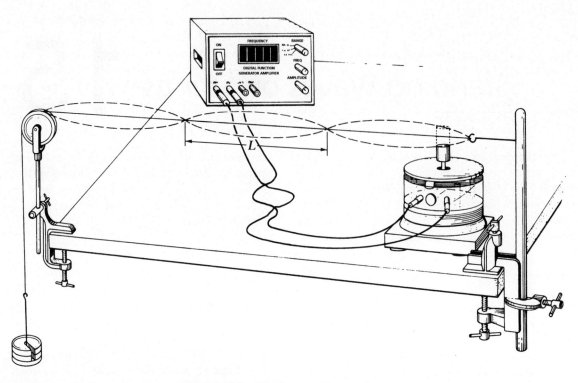

Figure 15.1 *Standing waves on a string*

over the pulley by means of a weight hanger and can be altered by changing these weights. Thus, different sets of modes may be established by varying F in Equation 15.3,

and within a particular set, the resonant frequencies corresponding to the various allowed values of L may be determined and measured.

APPARATUS

1. String vibrator
2. Signal generator
3. Rod-mounted pulley and clamp
4. Support rod and clamp for string
5. String

6. Triple-beam balance
7. 2-meter stick
8. Weight hanger
9. Set of slotted weights

PROCEDURE

1. Measure and cut off a piece of string exactly 4 m long. Weigh this length of string and record its mass. This length is chosen for convenience and also to provide enough string for an accurate weighing. It is enough for two setups.

2. Tie one end of the string to the vertical support rod, attach the weight hanger at the other end, and pass the string over the pulley mounted as shown in Fig. 15.1 at the other corner of the laboratory bench about a meter away.

3. Place the vibrator under the string as near to the end opposite the pulley as you can get it and adjust the string's height so that it just rests on top of the vibrator's driving rod. Adjust the height of the pulley so that the string is level.

4. Connect the vibrator to the signal generator, plug the latter in, and turn it on. Start with a mass of 50 g on the string (usually the mass of the weight hanger alone) and set the generator frequency to a low value, say

10 Hz. Increase the generator's output amplitude until you can see the vibrator post oscillating up and down, and search for resonance at the lowest frequency mode by slowly increasing the generator frequency until you see the string vibrating in one segment. Adjust the generator frequency to maximize the vibration amplitude and note this frequency.

5. Search for resonance in the second mode, that is, with the string vibrating in two segments with a node in the middle. As this mode has a wavelength half of that associated with the first mode, the frequency, according to Equation 15.1, should be doubled, so that your search will be facilitated by looking for frequencies in the neighborhood of twice the value noted in Procedure 4.

6. Tune the signal generator to exact resonance with the second mode by maximizing the vibration amplitude. Record this frequency. Then measure the distance from the point where the string contacts the pulley to the first node beyond the vibrator (see Fig. 15.1). With the string vibrating in only two segments, this node is the one in the middle. The measurement may be made by simply holding the 2-m stick alongside the vibrating string. However, a more accurate result is usually obtainable by grasping the string between your thumb and forefinger with your thumbnail right at the node and measuring the distance from your nail to the pulley. In doing this, be sure not to pull on the string in such a direction as to raise the weight and thus increase the length being measured. If you pull at all, make sure you pull against the support rod. Record the value of L so obtained.

7. Search for resonance in the third mode, which should be at three times the frequency noted in Procedure

4. Determine and record the frequency for this mode as you did for the second mode in Procedure 6. Also measure the distance from the point where the string contacts the pulley to the first node beyond the vibrator as in Procedure 6, but this time observe that your measured length is $2L$ (see Fig. 15.1). Record your results.

8. Repeat Procedure 7 for the fourth and fifth modes, noting that the frequencies for these modes should be close to four and five times that noted in Procedure 4 and that the measured string lengths will be $3L$ and $4L$, respectively.

9. Repeat Procedures 4–8 with weights of 100, 150, 200, 250, 300, 350, and 400 g attached to the string. Record all your results.

DATA _____

Mass of 4 meters of string _____ Mass per unit length _____

Measurements with 50-Gram Weight

Tension _____ Square root of tension _____

Number of segments	Frequency	Length to first node beyond vibrator	L
2			
3			
4			
5			

Measurements with 100-Gram Weight

Tension _____ Square root of tension _____

Number of segments	Frequency	Length to first node beyond vibrator	L
2			
3			
4			
5			

Measurements with 150-Gram Weight

Tension _____ Square root of tension _____

Number of segments	Frequency	Length to first node beyond vibrator	L
2			
3			
4			
5			

Measurements with 200-Gram Weight

Tension _____   Square root of tension   _____

Number of segments	Frequency	Length to first node beyond vibrator	L
2			
3			
4			
5			

Measurements with 250-Gram Weight

Tension _____   Square root of tension   _____

Number of segments	Frequency	Length to first node beyond vibrator	L
2			
3			
4			
5			

Measurements with 300-Gram Weight

Tension _____   Square root of tension   _____

Number of segments	Frequency	Length to first node beyond vibrator	L
2			
3			
4			
5			

Measurements with 350-Gram Weight

Tension _____   Square root of tension   _____

Number of segments	Frequency	Length to first node beyond vibrator	L
2			
3			
4			
5			

Measurements with 400-Gram Weight

Tension _____   Square root of tension   _____

Number of segments	Frequency	Length to first node beyond vibrator	L
2			
3			
4			
5			

Number of segments	Slope	Average value of L	$1/(2L\sqrt{d})$	Percent error
2				
3				
4				
5				

CALCULATIONS _____

1. Determine the mass per unit length of the string. Express it in kilograms per meter.

2. Calculate the tension in newtons and the square root of this tension for each weight used on the string.

3. For each observation, calculate the value of L, the length of a single vibrating segment.

4. Plot the square root of the tension against frequency for the 2-segment mode using your values of $\sqrt{F}$ as abscissas and those of f for that mode as ordinates. Use the entire sheet of graph paper. Draw the best straight line through your plotted points, remembering that the origin should be a point on your line.

5. Repeat Calculation 4 using the frequencies for the 3-segment mode, the 4-segment mode, and the 5-segment mode. Use the same sheet of graph paper so that you have four lines of different slope on the one sheet.

6. According to Equation 15.3, f and $\sqrt{F}$ are proportional with proportionality constant $1/(2L\sqrt{d})$. This constant (with the appropriate value of L) should therefore be the slope of each line drawn in Calculations 4 and 5. Obtain these slopes and record their values in the spaces provided in the last data table.

7. Calculate and record an average value of the segment length L from the values of L obtained each time you had the string vibrating in 2 segments. Similarly, find average values of L from the values obtained with the string vibrating in 3, 4, and 5 segments.

8. Using the average values of L found in Calculation 7, calculate the value of the proportionality constant $1(2L\sqrt{d})$ in each case and compare with the corresponding values found in Calculation 6 by finding the percent discrepancy.

QUESTIONS _____

1. In this experiment, why did we consider the length of the string between the first node beyond the vibrator and the pulley rather than the entire length between the vibrator and the pulley? What error would have been introduced if the entire length and total number of vibrating segments had been used? With your experimental setup, is this error likely to have been large?

2. Why should string that doesn't stretch easily be used in this experiment?

3. Why should the curves drawn in Calculations 4 and 5 pass through the origin? What does this mean physically?

4. What is meant by resonance?

5. A copper wire 1 m long and weighing 0.061 kg/m vibrates in two segments when under a tension produced by a load of 0.75 kg. What is the frequency of this mode of vibration?

6. Upon what physical properties do (a) the loudness, (b) the pitch, and (c) the quality of a musical note depend?

7. Compute the velocity of the wave on the string for each of the weights used and note how your results compare with the speed of sound in air, which, for air at room temperature, is 343.4 m/s.

The Sonometer 16

A stretched string that is set into vibration will emit a musical note, the frequency of which is determined by the length of the string, its mass per unit length, and the tension. As any violinist, guitarist, or mandolin player knows, the frequency (which, to the musician, means the note in the musical scale) may be changed by varying any one of these quantities. Thus, the violinist tunes his instrument by twisting the knobs that set the tension in the violin strings, the different strings have frequency ranges because they differ in mass per unit length, and

our performer plays different notes by "stopping" a string with his finger and thereby setting that string's free length. The purpose of this experiment is to determine (1) how the frequency of a vibrating string varies with its length when the tension is kept constant, (2) how the frequency of a vibrating string varies with the tension in the string when its length is kept constant, and (3) how the frequencies of two strings of the same length and material and under the same tension depend on their cross-sectional areas.

THEORY

The frequency, or number of vibrations per second, of the fundamental of a stretched string depends upon the length of the string, the tension, and the mass per unit length. The relation between these quantities is

$$f = \frac{1}{2L} \sqrt{\frac{F}{d}} \qquad (16.1)$$

where f is the frequency of the note emitted in vibrations per second or hertz, L is the length of the vibrating segment in meters, F is the tension in newtons, and d is the mass per unit length of the string expressed in kilograms per meter. Note that if we limit ourselves to the *fundamental* or lowest resonant frequency of the string, L is also the string's length between its fixed ends or the bridges that fix two points on it.

In this experiment, the above relation will be tested by the use of a sonometer. This device, which is rather like a guitar except for the method of tensioning the strings and setting them vibrating, consists of a resonant case provided with wires that may be stretched. The tension on the wires, or the stretching force, can be easily measured. Adjustable bridges are provided so that the lengths of the vibrating strings may be varied and

measured. The general appearance of a sonometer having only one string is shown in Fig. 16.1.

The experimental work consists of tuning the sonometer to several known frequencies produced by an audio signal generator so that the law expressed by Equation 16.1 may be tested. Thus, the variation of the frequency of a vibrating string with stretching force may be studied, as well as the variation of frequency with length. Finally, the effect of the mass per unit length may be shown by using wires of different diameters as stretched strings.

To set the wire vibrating and find the length at which it vibrates with maximum amplitude in one segment at a particular frequency, a driver coil is provided, which exerts a magnetic force on the steel wires used in this experiment. The driver coil is connected to the signal generator so that its magnetic field and hence the force on the wire varies with the generator's output current. Resonance—that is, maximum vibration amplitude at the driving frequency—may be detected visually, by careful listening to the tone produced by the vibrating wire, or by a detector coil placed under the wire near its center and connected to an instrument suitable for observing the

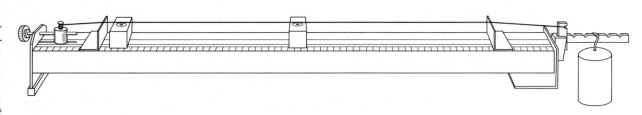

Figure 16.1 *A typical sonometer*

amplitude of the signal so obtained. This instrument may be an oscilloscope, a small amplifier and loudspeaker that allow you to hear an amplified version of the tone, or a sensitive ac voltmeter. *Note:* The signal generator's output is *alternating* current—that is, it flows first one way and then the other—and a cycle consists of one complete back-and-forth of this current. But the steel wire is attracted to the driver coil no matter which way the current is flowing (the wire, being unmagnetized, is attracted by either a north or a south magnetic pole); hence the wire

gets *two* pulses for each cycle of the signal generator's output. In short, the frequency driving the wire is really twice that shown on the generator's dial, and account must be taken of this when analyzing the data. A column for entering this actual driving frequency is provided in the data table. Using a signal generator as the source of the known frequency has the advantage that its tuning dial can serve to make fine adjustments so that the generator and string frequencies agree exactly.

APPARATUS

1. Sonometer with two steel wires of different diameters and two bridges
2. Audio signal generator covering the range 10–200 Hz
3. Wire driver and detector coil set
4. Instrument to measure detector coil output
5. Micrometer caliper
6. Card giving mass per unit length of the wires (optional)
7. Hooked weight having a mass of 1 kg

PROCEDURE

1. When you come into the laboratory, study the sonometer and get your instructor to answer any questions you may have about how it works. Note that this form of sonometer uses just one steel wire as the stretched string, but changing wires is easy. You are provided with two such wires: one of small diameter and a larger one. Begin by mounting the smaller one on the sonometer by hooking the lug at one end over the post on the movable carrier and slipping the ball at the other end into the slot in the tensioning lever. Observe that this lever has five notches into which a known mass may be hooked. When the mass is placed in the first notch (nearest the sonometer proper), the tension in the wire is equal to the weight of this mass. When the mass is placed in the second notch, the tension in the wire is twice the mass's weight; in the third notch, three times; in the fourth notch, four times; and in the fifth notch, five times. Start the experiment by hooking the 1-kg mass in the second notch so that the tension in the wire is $1 \times 2 \times 9.8 = 19.6$ newtons. Adjust the screw that moves the carrier until the lever is pulling straight on the wire, that is, so the horizontal part of the lever is indeed horizontal. Your setup should now look like that shown in Fig. 16.1.

2. Select a 60-cm length for the vibrating segment of the wire by placing the two bridges under the wire 60 cm apart. The bridges slide along the top of the sonometer box and can be positioned quite precisely by means of the scale that's on top of the box. Then place the driver coil near one end of the 60-cm segment and the detector coil in the segment's center. *Note:* We will be studying only the lowest, or fundamental, frequency at which the wire vibrates. This mode has nodes only at the points fixed by the bridges, therefore the wire vibrates with maximum amplitude at the center of the segment. The

detector coil is placed there so as to give maximum possible output.

3. Connect the driver coil to the low-impedance output of the signal generator, and the detector coil to the oscilloscope or other instrument for observing the detected signal. Turn on the signal generator and set its output level control about halfway up. Start with a low frequency and search for *resonance*, that is, the frequency at which the wire vibrates with maximum amplitude. This will be very critical, so that you should search carefully, changing the frequency quite slowly. Be careful to select the fundamental rather than the harmonic. The fundamental is the *lowest* resonant frequency and can be verified by looking at the wire. You should be able to observe maximum vibration amplitude at the center of the segment, this amplitude decreasing to zero at the bridges. Adjust the signal generator output to give a good vibration amplitude but not one so large that the vibrating wire strikes the metal core of the detector coil. In the data table record the generator frequency at which maximum vibration occurs.

4. Repeat Procedure 3 with segment lengths of 50, 40, and 30 cm, these segments being selected by moving the bridges appropriately. In each case, place the driver coil near one end and the detector coil at the center. Record the resonant frequencies so determined.

5. Change the tension in the small-diameter wire to $1 \times 5 \times 9.8 = 49.0$ newtons by hooking the 1-kg mass in the fifth notch. Move the bridges to give you a vibrating segment 50 cm long and search for the resonant frequency as in Procedure 3.

6. Repeat Procedure 5 with the 1-kg mass in the fourth, third, and first notch. Note that you already have the frequency for a 50-cm segment and the mass in the second notch.

7. Replace the wire on your sonometer with the other, larger-diameter wire, place the 1-kg mass in the fifth notch, and adjust the carrier so that the horizontal part of the lever is truly horizontal, as described in Procedure 1. Set the signal generator to the same frequency as that found in Procedure 5 and move the bridges so that the new wire resonates at this frequency. To do this, place one of the bridges near the carrier, locate the driver coil a short distance in front of it, and search for resonance by moving the other bridge, simultaneously moving the detector coil so that it is always near the center of the selected segment. What you are looking for is the length of the large wire for which the resonant frequency is the same as that for 50 cm of the small wire under the same tension. Record this length along with the generator frequency and the 49.0-newton tension from Procedure 5.

8. Using the micrometer caliper, measure the diameter of each wire and record these values.

DATA _____

Smaller wire

Diameter _____

Mass per unit length: Given   _____

From diameter and steel density   _____

From frequency vs. length curve   _____

From frequency vs. tension curve   _____

Percent error:

Value of d from frequency vs. length curve   _____

Value of d from frequency vs. tension curve   _____

Larger wire

Diameter _____

Mass per unit length: Given   _____

From diameter and steel density   _____

Signal generator frequency	Driving frequency f	Length L of the string	Tension F	Square root of tension	Reciprocal $1/L$ of length of the string

CALCULATIONS

1. Calculate the mass per unit length of each wire by treating a one-meter length as a steel cylinder one meter long having the diameter measured in Procedure 8 and the density listed for steel in Table IV at the end of this book. Be careful of your units. If you have been given a mass per unit length for each of your wires, enter these values in the appropriate spaces in the data table and compare them with those you just found.

2. From the data of Procedures 2, 3, and 4, calculate the reciprocals of the length of the string. Plot a curve using these reciprocals as abscissas and the corresponding frequencies as ordinates. What does this curve show about the way the frequency of a string under constant tension depends on the length of the string?

3. Referring to Equation 16.1, write the expression for the slope of the curve plotted in Calculation 2. Then find a numerical value for the slope of your curve and, from this, determine a value for the mass per unit length d for the small wire. Calculate the percent error in the value of d thus found as compared to the given or directly measured value noted in Calculation 1.

4. From the data of Procedures 4, 5, and 6, calculate the square roots of the tension. Plot a curve using the square roots of the tension as abscissas and the corresponding frequencies as ordinates. What does this curve show about the dependence of the frequency of a string of constant length on the tension in the string?

5. Again referring to Equation 16.1, write the expression for the slope of the curve plotted in Calculation 4. Then find a numerical value for the slope of this curve and again determine d for the small wire. See how well it agrees with the previously obtained values by again calculating a percent error.

6. From the data of Procedures 5 and 7, show how the lengths of two strings compare with their cross-sectional areas if the strings vibrate at the same frequency, are under the same tension, and are made of the same material.

QUESTIONS _____

1. Upon what physical properties do (a) the loudness, (b) the pitch, and (c) the quality of a musical note depend?

2. (a) Find the slope of the curve plotted in Calculation 2 and compare it with the theoretical value obtained from Equation 16.1 using the pertinent values of the tension and mass per unit length of the wire. (b) Repeat (a) for the curve of Calculation 4, comparing the slope with the theoretical value obtained from the values of the length and mass per unit length.

3. A wire of mass 0.0003 kg/m and 0.5 m long is vibrating 200 times per second. What must be the tension in newtons? What mass hung on the wire would produce this tension?

4. The first overtone of a stretched string 0.75 m long has a frequency of 200 Hz when the string is stretched by a weight of 4 kg. What is the mass per unit length of the vibrating string?

5. Two wires equal in length and made of the same material are subjected to tensions in the proportion of 1 to 4; the one under the greater tension is the thicker wire. If they vibrate at the same frequency, how do their diameters compare?

6. What is meant by "harmonics" and "overtones"? Distinguish between these two terms.

Resonance of Air Columns 17

The velocity with which a sound wave travels in a substance may be determined if the frequency of the vibration and the length of the wave are known. In this experiment, the velocity of sound in air is found by using a sound wave of known frequency whose wavelength can be measured by means of a resonating air column. The purpose of this experiment is thus to determine the wavelengths in air of sound waves of different frequencies by the method of resonance in closed pipes and to calculate the velocity of sound in air using these measurements.

THEORY

If a sound wave source is held over a tube open at the top and closed at the bottom, it will send disturbances made up of alternate compressions and rarefactions—that is, a sound wave—down the tube. These disturbances will be reflected at the tube's closed end, thus creating a situation in which identical waves are propagating in opposite directions in the same region. This is the condition for a standing wave, and so we have the possibility of a standing wave being set up in the air column in the tube. However, closing one end of the tube requires that there be a zero-amplitude point or node at that end, while at the other end, which is open, we must have a loop, or maximum-amplitude point. These requirements will be satisfied if the length of the tube is one quarter of the sound wave's wavelength in air, for this is the distance from a node to an adjacent maximum, or loop. The requirements will also be satisfied if the tube's length is a quarter wavelength plus any integer number of half wavelengths, because half a wavelength is the distance from one node to the next. Whenever the tube has one of these lengths, the tube and the sound source are said to be in resonance, or, put another way, the tube resonates at the source's frequency. The condition of resonance is indicated by an increase in the loudness of the sound heard when the air column has the resonant length.

When a sound source such as a loudspeaker or telephone receiver is held over a tube closed at one end, resonance will occur whenever the tube length is an odd number of quarter wavelengths, that is, when $L = \frac{1}{4}\lambda$, $\frac{3}{4}\lambda$, $\frac{5}{4}\lambda$, etc., where L is the length of the tube and λ is the wavelength of the sound waves in air. The relation between this wavelength and the frequency of the sound source is

$$V = f\lambda \qquad (17.1)$$

where V is the velocity of sound in air in meters per second, f is the frequency (the number of vibrations or cycles per second), and λ is the wavelength in meters. From this relation, the velocity can be calculated from the known source frequency and the wavelength obtained from measurements of L at resonance. It should be noted, however, that the center of the loop at the tube's open end does not fall right at the end but outside it by a small distance that depends on the wavelength and the tube diameter. However, the distance between successive tube lengths at which resonance is obtained gives the exact value of a half wavelength.

In this experiment, a closed pipe of variable length is obtained by changing the level of the water contained in a glass tube. The length of the tube above the water level is the length of the air column in use. The apparatus is shown in Fig. 17.1 and consists of a glass tube 1.2 m long closed at the bottom with a cup and mounted on a

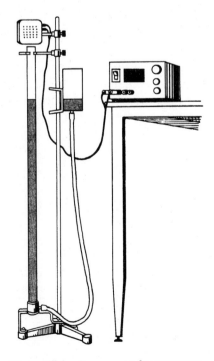

Figure 17.1 *Resonance tube apparatus*

heavy tripod base. A supply tank, which is connected to the cup by a rubber hose, is attached to the support rod by means of a clamp, which permits easy adjustment. The sound source used in this experiment consists of a small loudspeaker driven by a signal generator, and a clamp on the support rod holds this loudspeaker directly above the tube. If the sound level is to be detected by a microphone, it, too, is held just above the tube and next to the loudspeaker by means of a second clamp.

The velocity of sound in air is 331.4 m/s at 0°C. At higher temperatures, the velocity is slightly greater than this and is given by

$$V = 331.4 + 0.6t \qquad (17.2)$$

where V is the velocity in meters per second and t is the temperature in degrees Celsius. Equation 17.2 can be used to get a more accurate value for the speed of sound at the laboratory room temperature.

APPARATUS

1. Resonance tube apparatus
2. Signal generator
3. Loudspeaker or telephone receiver
4. Meter stick
5. Clamp to mount loudspeaker
6. Celsius thermometer
7. Microphone and mounting clamp (optional)
8. Sound level indicator (optional)

PROCEDURE

1. Using the clamp provided, mount the loudspeaker or telephone receiver so that it faces into the open end of the long glass tube and is about 2 cm above it. If a microphone is to be used to detect the sound level, it, too, should be mounted facing the top of the tube. In this case, the loudspeaker may be mounted a little to one side and slanted to face toward the tube end, thus making room for the microphone beside it.

2. Fill the supply tank with water and raise it to adjust the water level in the long tube to a point near the top. Do not overfill the supply tank or it will overflow when you lower it.

3. Turn on the signal generator and set it to a frequency of 500 cycles/s (or Hz). Connect the loudspeaker or telephone receiver to the generator output and adjust the level to give a low but audible volume of the 500-Hz tone.

4. Determine the shortest tube length for which resonance is heard by slowly lowering the water level while listening for resonance to occur. At resonance, that is, when the air column is adjusted to the proper length, there will be a sudden increase in the intensity of the

sound that you can detect either by listening, or, if you have the equipment, by noting the rise in the reading of the sound level indicator connected to the microphone. When the approximate length for resonance has been found, run the water level up and down near this point to determine the position for which the sound level is maximum. Measure and record the length of the resonating air column to the nearest millimeter. Make two additional determinations of this length by changing the water level and relocating the position of the maximum. Record these two readings also.

5. Lower the water level until the second position at which resonance occurs is found and repeat Procedure 4 to determine the air-column length for this case. As in Procedure 4, make three independent determinations of this length and record each to the nearest millimeter. *Do not change either the frequency or the sound level from the settings used in Procedure 4.*

6. Repeat Procedures 3–5 with signal-generator frequencies of 400 and 300 Hz. Record your three frequencies in the spaces provided in the data table.

7. Record the room temperature of your laboratory.

DATA

Room temperature   _____

Calculated value of the velocity of
    sound in air (average)   _____

Known value of the velocity of
    sound at room temperature   _____

Percent error   _____

Source frequency	First position of resonance				Second position of resonance				Wavelength	Velocity of sound in air
	1	2	3	Average	1	2	3	Average		

CALCULATIONS _____

1. For each length of resonating air column, calculate the average of the three readings taken.

2. For each frequency, determine the value of the wavelength in air from the lengths of the resonating air column.

3. For each frequency, calculate the velocity of sound in air from the wavelength found in Calculation 2. Find the average of these three values.

4. Calculate the velocity of sound in air at room temperature from Equation 17.2.

5. Compare the average value of the velocity of sound as found in Calculation 3 with the known value as found in Calculation 4 by finding the percent error.

QUESTIONS _____

1. Suppose that in this experiment the temperature of the room had been lower. What effect would this have had on the length of the resonating air column for each reading? Explain.

2. How would an atmosphere of hydrogen affect the pitch of an organ pipe?

3. A 128-Hz sound source is held over a resonance tube. What are the two shortest distances at which resonance will occur at a temperature of 20°C?

4. A sound source is held over a resonance tube, and resonance occurs when the surface of the water in the tube is 10 cm below the source. Resonance occurs again when the water is 26 cm below the source. If the temperature of the air is 20°C, calculate the source frequency.

5. An observer measured an interval of 10 s between seeing a lightning flash and hearing the thunder. If the temperature of the air was 20°C, how far away was the source of sound?

6. From the spread in your data for this experiment, estimate limits of error to be attached to your measured value of the velocity of sound in air. Does the accepted value fall within these limits? If it does not, can you offer an explanation for the discrepancy?

Coefficient of Linear Expansion

18

Most substances expand with an increase of temperature. It is found that the change in length of a solid is proportional to the original length and to the change in temperature. The factor of proportionality, which is called the coefficient of linear expansion, depends on the material of which the solid is made. The purpose of this experiment is to determine the coefficient of linear expansion of several metals.

THEORY

The coefficient of linear expansion of a substance is the change in length of each unit of length when the temperature is changed one degree. The coefficient α may thus be expressed by the relation

$$\alpha = \frac{L_2 - L_1}{L_1(t_2 - t_1)} = \frac{\Delta L}{L_1(t_2 - t_1)} \qquad (18.1)$$

where L_1 is the original length in meters at the temperature t_1 degrees Celsius, L_2 is the length at the temperature t_2, and $\Delta L \equiv L_2 - L_1$ is the observed change in length.

The apparatus shown in Fig. 18.1 consists of a metal rod that is placed in a brass jacket through which steam may be passed. The temperature of the rod is measured by a thermometer inserted in the jacket. One end of the rod is fixed and the other end is allowed to move. A micrometer screw, which may be adjusted to make contact with the movable end, is used to measure the change in length.

In determining the coefficient of linear expansion of a metal rod, the length of the rod is measured at room temperature. The rod is then placed in the brass tube of the apparatus and the initial reading of the attached micrometer screw is taken. Steam is then run through the tube, and the increase in length is measured by means of the micrometer screw. The base of the apparatus carries binding posts for connecting an electric buzzer or light bulb to indicate the exact point when the screw touches the rod. The coefficient of linear expansion is finally calculated from Equation 18.1.

APPARATUS

1. Coefficient of linear expansion apparatus
2. Boiler and tripod stand or electrically heated steam generator
3. Bunsen burner (if needed)
4. 0–100°C thermometer
5. Rubber tubing
6. Meter stick
7. 6-V dry cell or power supply
8. Electric buzzer or 6-V light bulb (if not built into apparatus)
9. Three metal rods: aluminum, copper, and steel
10. Paper towels or wipers

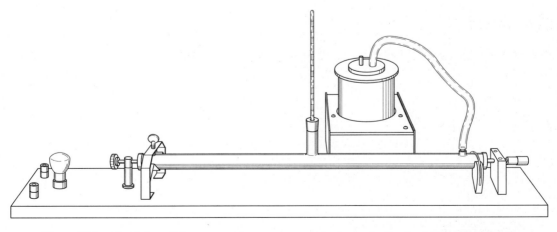

Figure 18.1 *Coefficient of linear expansion apparatus with micrometer head and light bulb indicator*

PROCEDURE

1. Measure the length of the aluminum rod with the meter stick to 0.1 mm at room temperature. Record the length and the temperature. Avoid handling the rod with your bare hands so as not to raise its temperature above that of the room. Use dry paper towels or wipers.

2. Mount the rod in the brass tube and record the setting of the micrometer screw for the rod at room temperature. Make sure that one end of the rod is touching the fixed end of the apparatus and the other end is facing the micrometer screw, but not touching it. Carefully turn the micrometer screw until the buzzer sounds or the light lights, showing that the screw has made contact with the rod. Record the setting of the micrometer screw.

3. Turn the micrometer screw back several millimeters to allow room for the expansion of the rod. CAUTION: Be sure to do this, as the apparatus will be badly damaged if the expanding rod pushes on the screw. Insert the thermometer into the brass tube through the port provided for it and pass steam through the tube. Let the steam continue to flow until the temperature has stabilized at a value at or near 100°C. Note that the temperature of the steam at normal atmospheric pressure will be 100°C, but it may take some time before the rod is brought up to this temperature, and in fact, heat losses from the apparatus may prevent the rod from ever getting to the full 100°C. In any case, record the final temperature reached. Then, with the steam still flowing, carefully turn the micrometer screw until the buzzer sounds or the light lights, showing that the screw has made contact with the rod. Record the setting of the micrometer screw. Remove the thermometer from the apparatus and let it return to room temperature when you have finished your measurement.

4. Repeat Procedures 1–3 for the copper rod.

5. Repeat Procedures 1–3 for the steel rod.

DATA

	Aluminum	Copper	Steel
Room temperature			
Length of rod at room temperature			
Setting of micrometer screw at room temperature			
Final temperature of rod			
Setting of micrometer screw at final temperature			
Change of length of the rod			
Coefficient of linear expansion			
Value of the coefficient of linear expansion from Table VI, at end of book			
Percent error			

CALCULATIONS

1. Compute the change in length of each rod in centimeters from the two readings of the micrometer screw.

2. Calculate the coefficient of linear expansion of aluminum, copper, and steel from your data.

3. Compare your results with those given in Table VI at the end of the book, and find the percent error.

QUESTIONS _____

1. Why must the increase in length of the rod be measured so carefully when the length itself can be determined by using an ordinary meter stick?

2. A compound bar made of an aluminum strip and an iron strip fastened firmly together is heated. Explain what happens to the shape of the bar.

3. When an ordinary mercury thermometer is placed in hot water, its reading drops at first but quickly rises again. Explain why.

4. Why is the numerical value of the coefficient of linear expansion when expressed in Fahrenheit degrees different from what it is when expressed in Celsius degrees?

5. An aluminum hoop whose inside diameter is 40 cm at room temperature (20°C) is to be "shrunk" on to a disk to form a wheel with an aluminum tire or rim. The process involves heating the hoop to make it expand, placing it on the disk, and letting it shrink down on the disk as it cools. The available furnace can heat the hoop to 300°C. To what diameter should the disk be machined so that the hoop will just fit on when so heated?

6. The Golden Gate Bridge over San Francisco Bay is 4200 ft long. Compute the total expansion of the bridge as the temperature changes from –20°C to +40°C. The coefficient of linear expansion for steel is given in Table VI at the end of the book.

7. If a solid material is in the form of a block rather than a rod, its volume will grow larger when it is heated, and a *coefficient of volume expansion β* defined by

$$\beta = \frac{V_2 - V_1}{V_1(t_2 - t_1)}$$

may be quoted. Here V_1 and V_2 are the initial and final volumes of the block, and t_1 and t_2 are the initial and final temperatures. Find the relation between the coefficients α and β. *Hint:* Suppose the block is a rectangular parallelepiped whose volume is its length times its width times its height.

Specific Heat and Calorimetry 19

To raise the temperature of a body from a given initial temperature t_1 to a final temperature t_2 requires a total quantity of heat that depends on the mass of the body, the specific heat of the body's material, and the temperature difference. The process of measuring quantities of heat is called calorimetry. The purpose of this experiment is to determine the specific heat of one or more metals by a calorimetric procedure called the method of mixtures.

THEORY

The measurement of heat quantities by the method of mixtures makes use of the principle that when a heat interchange takes place between two bodies initially at different temperatures, the quantity of heat lost by the warmer body is equal to that gained by the cooler body, and some intermediate equilibrium temperature is finally reached. This is true provided no heat is gained from or lost to the surroundings.

Any given body is characterized by a *heat capacity,* which is the number of calories required to raise the temperature of that particular body 1°C. Clearly, the heat capacity depends both on the mass of the body and the nature of its material. To eliminate the dependence on mass and obtain a characteristic depending only on the material involved, the *specific heat* is defined as the heat capacity of a particular amount of the material in question. Any unit of amount may be used, but it must be explicitly stated in the units of specific heat. Because it is common practice in calorimetry, we shall use the gram of mass as our unit of quantity in this experiment. Our units of specific heat will then be calories per gram per degree Celsius, in contrast to the units of heat capacity, which are just calories per degree. Note that the calorie is defined as the heat required to raise 1 g of water (at 4°C) 1°C. Thus, by definition the specific heat of water is 1 cal/g-°C. The heat capacity of m grams of water is thus numerically equal to m, whereas the heat capacity of the same mass of a substance whose specific heat is c cal/g-°C is mc. Since $mc/m = c$, the specific heat of any substance may be thought of as the ratio of the heat capacity of a body made of that substance to the heat capacity of an equal mass of water.

In this experiment, the specific heat of two different metals will be measured by studying the heat interchange between a sample of each metal and a mass of water. The vessel in which this interchange takes place is called a calorimeter. Any cup may be used as a calorimeter provided account is taken of the heat given up or absorbed by the cup and care is exercised to prevent heat exchange with the cup's surroundings. A normal way to do this is to use a thin metal cup whose heat capacity is small and easily measured and to mount it in an insulating jacket that prevents external heat exchange. A simpler and quite satisfactory procedure is to use a styrofoam cup, which both provides the needed insulation and has a negligibly small heat capacity.

To insure that an equilibrium temperature is reached shortly after the metal sample is placed in the water in the calorimeter, the water should be stirred. The thermometer used to measure the equilibrium temperature may also serve as a stirrer. The thermometer itself will absorb some heat in coming to the equilibrium temperature, but its heat capacity is usually so small that negligible error is made by ignoring it.

In the experimental determination of the specific heat of a metal by the method of mixtures, a metal sample of known mass that has been heated to a known high temperature is dropped into a known mass of water at a known low temperature. After equilibrium has been established, the new temperature of the water-metal-calorimeter combination is measured. In arriving at this equilibrium temperature, the water and calorimeter must have gained the heat lost by the metal sample. This situation is expressed by the relation

$$Mc(t_1 - t_2) = (mc_w + m_1c_1)(t_2 - t_3) \qquad (19.1)$$

where M is the mass of the metal sample in grams, c is the specific heat of the metal, t_1 is the initial temperature of the metal sample, t_2 is the final equilibrium temperature, t_3 is the initial temperature of the water and calorimeter, m is the mass of water, m_1 is the mass of the calorimeter cup, c_1 is the specific heat of the material of which this cup is made, and c_w is the specific heat of water, which, in our present set of units, is 1 cal/g-°C. Note that m_1c_1 is the heat capacity of the calorimeter cup. This quantity is sometimes called the *water equivalent* of the calorimeter because it is numerically equal to the mass of water having that heat capacity. Equation 19.1 may be used to determine c if all the other quantities in it have been measured.

145

APPARATUS

1. Calorimeter
2. Boiler and tripod stand or electrically heated steam generator
3. Bunsen burner (if needed)
4. Triple-beam balance
5. 0–100°C thermometer
6. 0–50°C thermometer
7. Sample cylinder no. 1 (metal to be specified by the instructor)
8. Sample cylinder no. 2 (metal to be specified by the instructor)

PROCEDURE

1. Fill the boiler or steam generator with enough water to cover either sample cylinder with a couple of inches to spare and place it on the tripod. Put the Bunsen burner under it or turn on the electric heater and bring the water to a boil.

2. Weigh the empty calorimeter cup. Note that you will be determining its *mass* in *grams*.

3. Weigh sample cylinder no. 1 and then lower it into the boiling water by means of a thread. Be sure the cylinder is completely immersed in the water but do not allow it to touch the bottom or sides of the boiler.

4. Pour cold water (about 3° below room temperature) into the calorimeter cup until the cup is about half full. Then weigh the cup with the water in it. Note that the temperature of the cold water should be about as much below room temperature as the equilibrium temperature will be above it so as to balance out errors due to transfer of heat by radiation to or from the surroundings.

5. Replace the calorimeter in its insulating jacket and measure the temperature of the water with the 0–50°C thermometer. Record this temperature.

6. Quickly transfer the sample cylinder from the boiler to the calorimeter without splashing any water. Stir the water and record the equilibrium temperature as indicated on the 0–50°C thermometer.

7. Measure and record the temperature of the boiling water using the 0–100°C thermometer.

8. Repeat Procedures 3–7 for sample cylinder no. 2.

DATA

Mass of calorimeter cup _____   Specific heat of calorimeter cup material _____

	Data for cylinder no. 1	Data for cylinder no. 2
Mass of cylinder		
Mass of calorimeter and cold water		
Mass of cold water		
Initial temperature of cold water		
Temperature of boiling water		
Equilibrium temperature		
Calculated specific heat		
Specific heat from Table VII		
Percent error		

CALCULATIONS _____

1. Calculate the specific heat of the sample no. 1 material from the data of Procedures 2–7.

2. Calculate the specific heat of the sample no. 2 material from the data of Procedure 8.

3. Compare your results for the specific heats with those given in Table VII at the end of the book, and find the percent error.

QUESTIONS _____

1. What is the purpose of starting with the temperature of the water lower than room temperature and ending with it about the same amount above room temperature?

2. How would the computed value of the specific heat be affected if some boiling water were carried over with the metal?

3. What will be the biggest source of error if too much water is used?

4. What is meant by the water equivalent of a body?

5. Check that Equation 19.1 is dimensionally correct. The factor c_w (the specific heat of water) is sometimes omitted because its numerical value is 1. What is wrong with doing this?

6. Atomic scientists often speak of the *molar* specific heat of a substance. What is meant by this term?

7. A platinum ball weighing 100 g is removed from a furnace and dropped into 400 g of water at 0°C. If the equilibrium temperature is 10°C and the specific heat of platinum is 0.04 cal/g-°C, what must have been the temperature of the furnace? Neglect the effect of the mass of the calorimeter.

The present experiment deals with *changes of phase,* that is, the changing of a substance from the solid to the liquid state and from the liquid to the gaseous state. These three states, or phases, are normally exhibited by all substances, but the temperatures at which changes from one state to another take place and the heat involved vary widely. It is the purpose of this experiment to study the heats involved in the change of water from a solid to a liquid and from a liquid to a gas by the method of mixtures. This method has been described in detail in Experiment 19 on calorimetry, and a review of the Theory section of that experiment will be helpful preparation for the present work.

THEORY

When a solid body is heated, its temperature rises at a rate determined by the rate at which heat is being supplied and the heat capacity of the body. However, when a certain temperature called the melting point is reached, the body starts to melt, and as melting proceeds the temperature remains constant even though heat is being continually supplied. Increasing the supply rate merely increases the rate at which the solid changes to a liquid. No change in temperature occurs until the body has completely melted, after which further heating will raise the temperature of the resulting liquid at a rate determined by the liquid's heat capacity (its mass multiplied by its specific heat) and the rate at which heat is being added.

Eventually, a temperature called the boiling point will be reached. Here the liquid changes rapidly to a vapor, and again there is no change in temperature while this is going on even though the addition of heat continues. When all the liquid has been evaporated, continued addition of heat to the resulting gas (which must now be contained in a suitably sealed vessel) will once more raise the substance's temperature—this time at a rate determined by the heat supply rate and the heat capacity (mass times specific heat) of the gas.

Upon cooling, the gas begins to condense when the boiling point is reached, and again there is no change in temperature while condensation takes place. The same amount of heat is removed during this condensation process as was added during vaporization; this heat represents the difference in the energy of the molecules in the gaseous state and their energy when bound (although loosely) in the liquid. Similarly, once the temperature has been reduced to the freezing point (which is the same as the melting point), further removal of heat causes the liquid to solidify. No change of temperature occurs during this process, and the heat removed (which is the same as the heat added during melting) represents the difference in the molecular energy when the molecules are loosely bound in the liquid and when they are tightly bound in the solid.

You should note that vaporization takes place at all temperatures, because there are always a few molecules in the liquid with enough energy to break the bonds at the surface and emerge into the space above. One can thus think of a *vapor pressure* due to these escaped molecules above any liquid surface. Naturally, the higher the temperature, the more molecules will have enough energy to break through the surface and the higher the vapor pressure will be. If the liquid surface is exposed to the atmosphere, atmospheric pressure will prevent molecules entering the gaseous state near this surface from forming bubbles until the vapor pressure and atmospheric pressure become equal. At this point, the vapor can form bubbles in opposition to the pressure of the atmosphere and the action we call boiling occurs. Molecules can then break from the liquid in large numbers, energy in the form of heat must be added to the liquid to replace the energy of the departing molecules, and a high rate of heat supply merely permits more molecules to evaporate per second. Note, however, that the boiling point is now seen to be that temperature at which the liquid's vapor pressure becomes equal to atmospheric pressure and is therefore highly dependent on the atmospheric pressure's value. This is why water boils at a lower temperature on top of a mountain, where the atmospheric pressure is less than at sea level. Thus, when the boiling point of a liquid is given without further information, normal sea-level atmospheric pressure is assumed. Because there is usually a small volume change when a solid changes to a liquid or vice versa, the melting point may also be affected by the external pressure, but this effect is usually so small that it can be ignored.

The amount of heat required to change a unit mass of a substance from the solid to the liquid state without a change in temperature is called the *latent heat of fusion* of the substance, or simply its heat of fusion. Similarly,

149

the amount of heat required to change a unit mass of a substance from the liquid to the vapor state without a change in temperature is called its *latent heat of vaporization,* or simply the heat of vaporization. Since evaporation can go on at any temperature and the heat of vaporization is different at different temperatures, the temperature should be quoted when a heat of vaporization is given. However, heats of vaporization are usually given without this information, it being assumed that the temperature of the boiling point is meant. Since this temperature is dependent on the atmospheric pressure, normal sea-level pressure is also assumed.

In doing heat experiments where the quantity of heat added to or taken from a substance is to be determined, the range of temperature should extend equally above and below room temperature so that the amount of heat absorbed from the surroundings during the course of the experiment will be approximately equal to the amount of heat radiated to the surroundings. Best results will thus be obtained by starting with a calorimeter of warm water when the heat of fusion of ice is to be measured and starting with a calorimeter of cold water for the measurement of the heat of vaporization of steam. This is easily accomplished by measuring first one and then the other of the two latent heats of water. It is the purpose of this experiment to carry out these two measurements.

The experimental determination of the heat of fusion of ice and the heat of vaporization of water is made by the method of mixtures. This makes use of the principle that when a heat interchange takes place between two bodies initially at different temperatures, the quantity of heat lost by the warmer body is equal to that gained by the cooler body, and some intermediate equilibrium temperature is finally reached. This is true provided no heat is gained from or lost to the surroundings.

In determining the heat of fusion of ice, a few small pieces of ice are placed, one by one, into a calorimeter containing water. As the ice melts, heat is absorbed from the water and calorimeter until the mixture comes to a final equilibrium temperature. The heat absorbed by the

ice in melting plus the heat absorbed by the ice water thus produced is equal to the heat lost by the warm water and calorimeter. The working equation is

$$ML_f + Mc_w(t_2 - 0) = (mc_w + m_1c_1)(t_1 - t_2) \quad (20.1)$$

where M is the mass of ice in grams, L_f is the latent heat of fusion of ice, t_2 is the equilibrium temperature, t_1 is the initial temperature of the water and calorimeter, m is the mass of warm water, c_w is the specific heat of water, m_1 is the mass of the calorimeter, and c_1 is the specific heat of the material of which the calorimeter is made. Note that if all masses are in grams and specific heats in calories per gram per degree Celsius, c_w has the numerical value 1 and m_1c_1 is the water equivalent of the calorimeter. Note also that the zero in Equation 20.1 is the temperature of the ice water produced by the melting ice and that we have further assumed the original pieces of ice to be at 0°C when they were put in the water. Thus one shouldn't take these pieces directly out of some deep freeze for this experiment unless we include in Equation 20.1 a term giving the heat lost by the ice in coming from its initial temperature (whatever that may be) up to 0°C.

The heat of vaporization of water is determined in a similar manner by the method of mixtures. A quantity of steam is passed into a known mass of cold water in a calorimeter, where it condenses and raises the temperature of the water. The heat lost by the steam in condensing plus the heat lost by the condensed steam is equal to the heat absorbed by the cold water and calorimeter. The working equation is

$$ML_v + Mc_w(100 - t_2) = (mc_w + m_1c_1)(t_2 - t_1) \quad (20.2)$$

where M is the mass of steam in grams, L_v is the latent heat of vaporization of water, t_2 is the equilibrium temperature, t_1 is the initial temperature of the water and calorimeter, m is the mass of cold water, c_w is the specific heat of water, m_1 is the mass of the calorimeter, and c_1 is the specific heat of the calorimeter material as before. Note here also that the steam is assumed to be at 100°C when it enters the water.

APPARATUS

1. Calorimeter
2. Boiler and tripod stand or electrically heated steam generator
3. Water trap
4. Rubber tubing
5. Bunsen burner (if needed)

6. Triple-beam balance
7. 0–100°C thermometer
8. 0–50°C thermometer
9. Ice
10. Paper towels

PROCEDURE

1. Weigh the empty calorimeter. Note that you are finding the calorimeter's *mass* in *grams.*

2. Fill the calorimeter half full of water about 10° above room temperature and weigh it. Replace it in the

outer calorimeter jacket and record the temperature of the water.

3. Dry some small pieces of ice on a paper towel and add them to the water without touching the ice with your fingers, so as not to melt it. Add the ice until the temperature is about 10° below room temperature, keeping the mixture well stirred with the 0–50°C thermometer. Record the equilibrium temperature when the ice is entirely melted. Weigh the calorimeter with its contents again, but without the thermometer.

4. Fill the calorimeter three-quarters full of water about 15° below room temperature and again weigh it. Replace it in the outer calorimeter jacket.

5. Set up the steam generator as shown in Fig. 20.1 and adjust it so that the steam passes through the water trap and flows freely from the steam tube. Record the temperature of the cold water and quickly immerse the steam tube. Allow the steam to pass in and condense until the temperature of the water is about 15° above room temperature, stirring continuously with the 0–100°C thermometer. Remove the steam tube and record the equilibrium temperature. Weigh the calorimeter with its contents again, but without the thermometer.

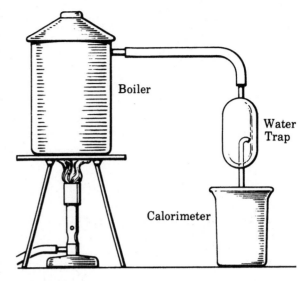

Figure 20.1 *Heat of vaporization apparatus*

DATA _____

Mass of calorimeter _____   Specific heat of calorimeter _____

Heat of Fusion

Mass of calorimeter and water _____   Mass of ice _____

Mass of water _____   Calculated heat of fusion of ice _____

Initial temperature of water _____   Accepted value of heat of fusion of ice _____

Equilibrium temperature _____   Percent error _____

Mass of calorimeter, water, and
    melted ice _____

Heat of Vaporization

Mass of calorimeter and water _____   Mass of steam _____

Mass of water _____   Calculated heat of vaporization
    of water _____

Initial temperature of water _____

Equilibrium temperature _____   Accepted value of heat of
    vaporization of water _____

Mass of calorimeter, water, and
    condensed steam _____   Percent error _____

CALCULATIONS _____

1. From the data of Procedures 1–3, compute the heat of fusion of ice.

2. From the data of Procedures 4 and 5, compute the heat of vaporization of water.

3. Compare your results for the heat of fusion and the heat of vaporization with the known values by finding the percent error.

QUESTIONS _____

1. Discuss the principal sources of error in this experiment.

2. Why is it necessary to stir the mixtures?

3. What error would have been introduced if the ice had not been dry?

4. Suppose the steam generator produced steam so fast that some of it bubbled up through the water and escaped from the calorimeter without condensing. Would this cause an error in your results? Explain.

5. In the Theory section of this experiment it was pointed out that Equation 20.1 assumed the ice was at 0°C before it was put in the water and that, similarly, Equation 20.2 assumed the entering steam to be at 100°C. (a) How good do you think these assumptions are? (b) What should you have done to make sure they're very good? (c) How should Equations 20.1 and 20.2 be modified if the above assumptions are *not* good—that is, if the ice is much colder than 0°C and the steam much hotter than 100°C when they are put in the water? (d) Could you run the experiment on this basis?

6. (a) What is the purpose of the water trap in determining the heat of vaporization? (b) How would its absence have affected the results?

7. The Theory section of this experiment suggests that when a liquid starts to boil, the formation of bubbles should begin near the surface. Why do the bubbles actually form first at the bottom of the containing vessel?

8. A glass tumbler of 200 g mass and 5 cm inside diameter is filled to a depth of 10 cm with tap water whose temperature is 10°C, and 50 g of ice at 0°C are then added. At the end of an hour, it is noted that the ice has just finished melting. What was the average rate, in calories per second, at which heat flowed into the glass from its surroundings?

When a substance is in the gaseous state, its molecules fly about at random, since there are only very small forces of interaction between them (sometimes called Van der Waals forces); and because the molecules are small compared to their average separations, there is only a very small chance of any two of them colliding. A so-called ideal gas is defined as the idealization of the situation just described, that is, a gas in which the molecules are mass points that have *no* chance of colliding with each other and that also have *no* interaction forces, not even weak ones. Actual gases like nitrogen, oxygen, and hydrogen satisfy this definition to a good approximation at ordinary temperatures and pressures; thus, the study of the ideal gas has a practical application. Only at extreme temperatures and pressures do the molecules of real gases get close enough together for their interaction forces to have an appreciable effect and their finite size to become significant.

In order to specify fully the condition of any substance, its pressure, volume, and temperature must be known. These quantities are always interrelated through a relationship characteristic of the substance called its *equation of state*, so that if two of them are known, the equation of state gives the third one. The equation of state for an ideal gas (often called the ideal gas law) is a particularly simple one, derivable theoretically on the basis of the assumptions noted above with regard to no collisions and no interaction forces among the ideal gas molecules. It applies to real gases at ordinary temperatures and pressures and can thus be checked experimentally. The purpose of this experiment is to carry out such a check and incidentally to determine the so-called absolute zero of temperature.

THEORY

In studying the behavior of a gas under different conditions of pressure, temperature, and volume, it is convenient to keep one of these constant while varying the other two. Thus, if the temperature is kept constant, a relation is obtained between the pressure and the volume; if the volume is kept constant, a relation between the pressure and temperature can be found; and if the pressure is kept constant, there is a relation between temperature and volume. Such studies were carried out independently by three different investigators whose names are now associated with the results they obtained. The three possible relations, collectively referred to as the gas laws, are as follows: pressure and volume with temperature constant (Boyle's law); volume and temperature with pressure constant (Charles's law); and pressure and temperature with volume constant (Gay-Lussac's law).

Boyle's Law   If a gas is kept at constant temperature, we would expect an increase in pressure to "squeeze the gas together," that is, to cause a decrease in volume. Conversely, reducing the pressure allows the gas to expand, that is, to occupy a larger volume. This relation was studied quantitatively by Robert Boyle (1627–1691), who found that for a fixed amount of gas at constant temperature the pressure and volume are inversely proportional over any ordinary pressure range. This result may be stated mathematically as

$$P = \frac{C_1}{V}$$

or
$$PV = C_1 \qquad (21.1)$$

where P is the pressure, V is the volume, and C_1 is a proportionality constant whose value depends on the (constant) temperature and the amount of gas involved. The property of gases expressed by Equation 21.1 is called *Boyle's law*.

Charles's Law   If a gas is heated and the pressure remains constant, it expands so that it occupies an increased volume. The French physicist Jacques Alexandre César Charles (1746–1823) found that under these circumstances a linear relation exists between the volume and the temperature, expressible by the relation

$$V = V_0(1 + \beta t) \qquad (21.2)$$

where V is the volume at temperature t, t is the temperature in degrees Celsius, V_0 is the volume at 0°C, and β is a constant called the coefficient of expansion of the gas at constant pressure. Since the zero of temperature on the Celsius scale is arbitrarily chosen as the freezing point of water at standard atmospheric pressure, negative temperatures on this scale are perfectly meaningful, and it is interesting to note the temperature at which an extrapolation of Equation 21.2 indicates the volume would go to zero. This temperature is clearly $-1/\beta$, and since β is very nearly $\frac{1}{273}$ or 0.00366 per degree Celsius for all gases, Equation 21.2 predicts zero volume at a temperature of $-273°C$. If a new temperature scale is now defined as having degrees the same size as Celsius degrees but a zero at $-273°C$, we have

$$T = t + 273 \qquad (21.3)$$

where T is the temperature on the new scale, called the absolute or Kelvin scale, and is expressed in degrees Kelvin (K). Substitution of Equation 21.3 in Equation 21.2 with $\beta = \frac{1}{273}$ yields

$$V = \frac{V_0}{273} T = C_2 T \qquad (21.4)$$

where C_2 is a constant that is seen to depend on V_0 and hence on the quantity of and pressure on the gas. Thus, if the temperature is given on the absolute scale, volume and temperature are not just linearly related but are proportional. The linear dependence of volume on temperature at constant pressure is called *Charles's law*.

Gay-Lussac's Law   If a gas is heated but not allowed to expand (that is, maintained at constant volume by being sealed in a container of fixed dimensions), the pressure will rise. The investigations of the French scientist Joseph Louis Gay-Lussac (1778–1850) showed that, like the volume-temperature relation at constant pressure, the pressure-temperature relation at constant volume is linear and expressible by the relation

$$P = P_0 (1 + \beta t) \qquad (21.5)$$

where P is the pressure at $t°C$, P_0 the pressure at $0°C$, and β the pressure coefficient of the gas at constant volume. Note that the same symbol is used for the pressure coefficient in Equation 21.5 and the coefficient of expansion in Equation 21.2. This is because the same value, $\frac{1}{273}$ per degree Celsius, is found for both and both have the same significance—namely, that the volume and therefore the pressure of an ideal gas should become zero at $-273°C$ or $0°K$. Fig. 21.1 is a plot of pressure versus temperature for a typical gas and shows this extrapolation to zero pressure. If Equation 21.3 is substituted in Equation 21.5 in the same manner as in Equation 21.2, we get

$$P = \frac{P_0}{273} T = C_3 T \qquad (21.6)$$

where C_3 is a constant that depends on P_0 and hence on the quantity of gas involved and the volume in which it is confined. Again, use of the absolute temperature scale makes pressure and temperature proportional. As a result of Gay-Lussac's work, the linear dependence of pressure on temperature at constant volume is usually known by his name.

The Ideal Gas Law   Equations 21.1, 21.4, and 21.6 can be combined in the single relation

$$PV = C_4 T \qquad (21.7)$$

where C_4 is a new constant whose value depends only on how much gas is involved. In view of Avogadro's law (which states that equal volumes of all gases at the same pressure and temperature contain the same number of molecules), stating how much gas we have by giving the actual number of molecules present allows us to write Equation 21.7 as

$$PV = NkT \qquad (21.8)$$

where $C_4 = Nk$, N is the number of molecules in question, and k is a *universal* constant applying to all gases with no dependence on temperature, pressure, volume, or quantity. It is called *Boltzmann's constant* and has the value 1.38×10^{-23} joule per molecule-degree absolute. The student should note carefully that the product PV has the dimensions of energy so that the units for Boltzmann's constant are those of energy per molecule-degree. The value of this constant thus depends on the units of energy used. Expressing the amount of gas by the number of molecules present requires inconveniently large numbers, so the number n of moles is more often given. Since by definition a mole contains Avogadro's number N_0 of molecules, the total number of molecules N is equal to nN_0, and Equation 21.8 may be written

$$PV = nRT \qquad (21.9)$$

where $R = N_0 k$ and is called the *universal gas constant*. It serves the same purpose as k but has the dimensions of energy per *mole*-degree rather than energy per *molecule*-degree. Equation 21.9 (or 21.8) is a relation between the pressure, volume, and temperature of an ideal gas and is therefore the equation of state for such a gas. It is called the *ideal gas law*.

In the present experiment, we will check the laws of Boyle and Gay-Lussac by seeking experimental verification of Equations 21.1 and 21.5. Moreover, if the linear relation predicted by Equation 21.5 is obtained, simple extrapolation to $P = 0$ will give the location of the absolute zero of temperature on the Celsius scale and hence the value of β. In this way, Equation 21.6 will also be verified. Then, if pressure is found to be proportional to the absolute temperature at constant volume and inversely proportional to volume at constant temperature, the proportionality of volume and temperature at constant pressure (Charles's law), and hence the ideal gas law, is implied. The experimental test of Boyle's law consists of observing a series of different volumes, measuring the corresponding pressures, and plotting P against $1/V$ to see if a straight line is obtained. The apparatus is made up of two glass tubes of uniform bore connected by a piece of flexible rubber tubing and mounted on a supporting frame to

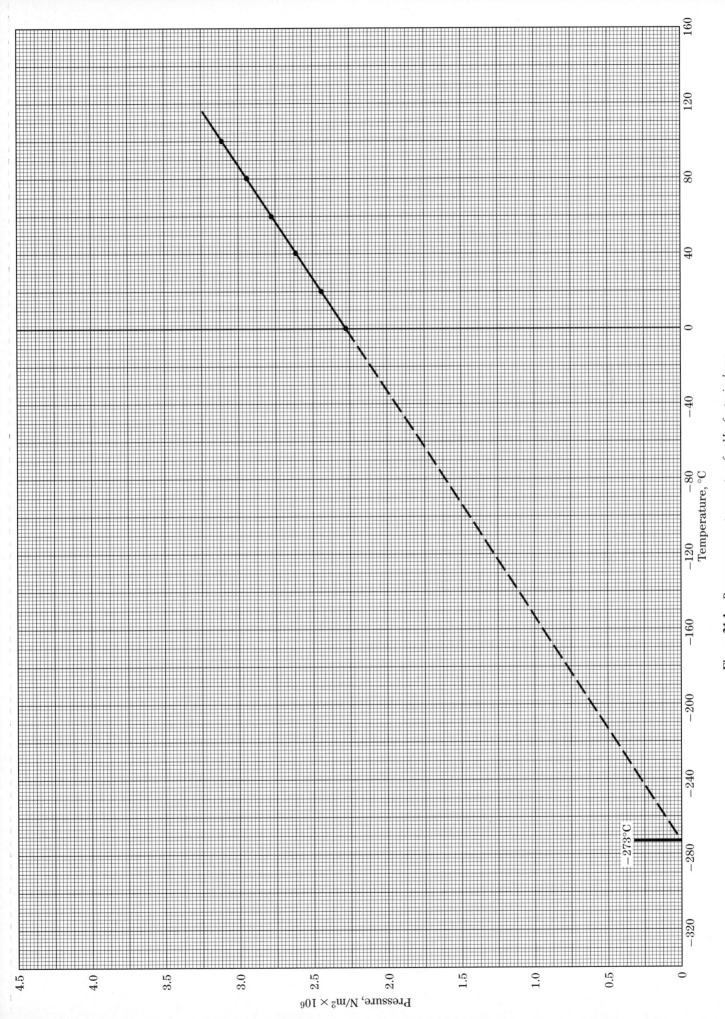

Figure 21.1 *Pressure versus temperature for 1L of a typical gas*

which a scale is attached (see Fig. 21.2). The apparatus is partially filled with mercury. One of the glass tubes is provided with a stopcock; for this part of the experiment the air upon which observations are to be made is confined in this tube between the stopcock and the mercury column. The other tube is open to the atmosphere and can be raised or lowered, thus changing the height of the mercury column that confines the air. The difference in height of the mercury in the two tubes causes a pressure to be exerted on the gas trapped in the closed tube. This is called the gauge pressure because it is the pressure that would be measured by a gauge that reads zero when exposed to atmospheric pressure if it were connected to the confined air. The total or absolute pressure on the confined air is the sum of the gauge pressure and the atmospheric pressure that is applied on top of the mercury column in the open

tube. Thus, the total pressure P acting on the confined air is given by

$$P = A + (h_2 - h_1) \qquad (21.10)$$

where A is the atmospheric pressure as read on the barometer; h_2 is the level of the mercury in the open tube; and h_1 is the mercury level in the tube closed by the stopcock. Notice that Equation 21.10 is consistent if h_1 and h_2 are in centimeters and all pressures are expressed in centimeters of mercury. Note also that if h_1 is greater than h_2 the quantity $(h_2 - h_1)$ is negative and subtracts from atmospheric pressure to give the total pressure on the confined air.

For the test of Gay-Lussac's law, a bulb is connected to the stopcock by a short length of rubber tubing as shown in Fig. 21.3. The stopcock is opened so that the volume of confined air includes the volume in the glass

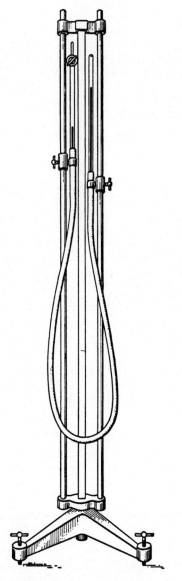

Figure 21.2 *Boyle's law apparatus*

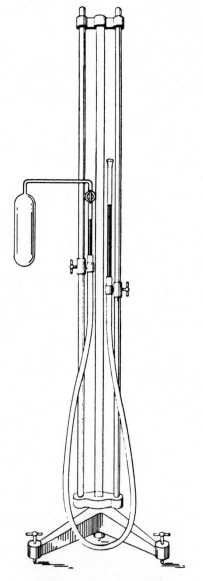

Figure 21.3 *Air thermometer*

tube above the mercury column, the volume of the rubber hose, and the volume of the bulb. The total volume is thus very much larger than that used for the test of Boyle's law, and the bulb provides for convenient heating to a known temperature. This is done by immersing the bulb in water contained in a Pyrex beaker, the temperature of the confined air being varied by heating or cooling the water bath. The total volume of confined air is kept constant by adjusting the height of the open glass tube so that the mercury level in the tube connected to the bulb is brought to a particular scale division before each set of readings is taken. Pressure readings are taken by using Equation 21.10 as in the test of Boyle's law and temperature readings by means of a thermometer in the water bath. Note that in the present experiment the rela-

tion in Equation 21.5 is being checked and β determined by measuring the gas temperature with an external thermometer and obtaining the pressure from the difference in height of the mercury columns. However, if the linear relation postulated by Gay-Lussac is correct and β known to high accuracy, the arrangement of Fig. 21.3 could serve as a thermometer, the difference in height of the mercury columns being a direct measure of the temperature of the gas in the bulb. In fact, the linear dependence of gas pressure on temperature at constant volume is so good and the value of β has been determined so accurately that the constant-volume hydrogen thermometer— an instrument similar to the present setup but using hydrogen instead of air—is the standard thermometer used to calibrate other types of thermometers.

APPARATUS _____

1. Boyle's law apparatus
2. Bulb with connecting tube
3. Barometer
4. 0–50°C thermometer
5. 0–100°C thermometer

6. 1000-cc Pyrex beaker
7. Ring stand and clamps
8. Bunsen burner
9. Ice

PROCEDURE _____

1. Record the reading of the barometer and the room temperature as read on the 0–50°C thermometer. CAUTION: Suspend the thermometer in the room for several minutes to allow it to come to equilibrium and do not touch it during this period.

2. Check that the apparatus has the proper amount of mercury in it. To do this, open the stopcock and see if each tube is about half full. The mercury level will, of course, be the same in both tubes when the stopcock is open. Adjust the tube support clamps so that the top of each tube is at about the same height and the clamps are about midway up their support bars, so that the open tube may be both raised above and lowered below the closed one through as great a range as possible.

3. Close the stopcock and thereafter do not handle the closed tube or allow the enclosed air to be heated above or cooled below room temperature in any way. Record the level of the top of the closed tube (just below the stopcock) and the level of the mercury columns in the two tubes. Note that atmospheric pressure should continue to exist in the closed tube after the closing of the stopcock, so that the mercury level should remain the same in both tubes.

4. Decrease the pressure on the gas trapped in the closed tube by lowering the open tube as far as the instrument will permit. Allow it to remain in this position for a few minutes to test for leaks. If there is any leak-

age in the apparatus, consult the instructor. Record the level of the mercury column in the closed tube and in the open tube.

5. Vary the pressure in nine approximately equal steps from the smallest to the largest value that can be obtained so as to get ten different observations. Record the level of the mercury column in each tube for each setting. After changing the level, allow a minute or two before taking readings to permit the confined air to reach equilibrium at room temperature. Remember not to handle the closed tube or allow spurious heating or cooling of the enclosed air to take place.

6. When the Boyle's law measurements have been taken, return the open tube to the position at which the mercury levels are equal and open the stopcock. Connect the bulb to the stopcock nipple by means of the rubber tube.

7. Mount the Pyrex beaker on the ring stand with the Bunsen burner under it. Support the bulb inside the beaker by means of a clamp, not allowing the bulb to touch the sides or bottom.

8. Fill the beaker with ice and water until the bulb is entirely covered. Watch the motion of the mercury in the closed tube to make sure it does not get up to the stopcock. If the mercury appears to be rising too high, lower the open tube, being careful not to let mercury spill out of it.

9. Pick a scale division a few centimeters below the bottom of the stopcock as your constant volume mark, and adjust the open tube so as to bring the mercury column in the closed tube to this level. Wait for temperature equilibrium to be reached, keeping the closed-tube mercury level at your constant volume mark until equilibrium is established. This will be indicated by the pressure stabilizing so that no further adjustment of the open tube is required. Stir the water well and measure its temperature with the 0–50°C thermometer. Note that if you still have a mixture of ice and water in the beaker as called for in Procedure 8, this temperature should be 0°C. Record it and the height of the mercury level in the open tube.

10. Remove the ice and add enough water to cover the bulb. Light the Bunsen burner and heat the water slowly until its temperature is about 10°C. Remove the Bunsen burner and stir the water well with the 0–50°C thermometer. Keep the mercury level in the closed tube at your constant volume mark by raising the open tube as required. When equilibrium has been reached, record the temperature of the water and the height of the mercury in the open tube.

11. Repeat Procedure 10, raising the temperature of the water in steps of about 10° until the boiling point is reached. Use the 0–100°C thermometer in place of the 0–50°C one as soon as the temperature goes above 40°C.

12. When the experiment is completed, remove the bulb from the water and lower the open tube progressively as the bulb cools so as to keep the mercury from rising up to the stopcock. Leave the apparatus in the condition it was in at the end of Procedure 6.

13. Check the barometer reading to be sure there has been no appreciable change since it was read in Procedure 1. If there is, use the earlier reading for the Boyle's law work and the later one for the test of Gay-Lussac's law.

DATA

Barometer reading, beginning _____   Level of the top of the
                    end _____   closed tube, h_3 _____

Room temperature _____

| Mercury level | | Gauge pressure $h_2 - h_1$ | Volume of air $h_3 - h_1$ | Total pressure | PV | 1/V |
Closed tube h_1	Open tube h_2					

Height of chosen constant volume
  mark   _____

Accepted value of the
  absolute zero   _____

Measured value of the absolute
  zero of temperature on the
  Celsius scale   _____

Percent discrepancy   _____

Mercury level in the open tube	Gauge pressure	Total pressure	Temperature of the bulb

CALCULATIONS _____

1. Compute the value of $h_2 - h_1$ for each setting of the apparatus in Procedures 4 and 5.

2. Compute the value of P, the total pressure on the confined air, for each observation taken in Procedures 4 and 5. Express the results in centimeters of mercury. *Note:* For these and the following calculations, use three significant figures in the numbers representing the pressure, the volume, and the reciprocal of the volume. In calculating the product, use four significant figures.

3. Calculate the volume of the enclosed air for each of the observations of Procedures 4 and 5. *Note:* Since the tube is of uniform bore, the tube's volume is proportional to its length. The length $h_3 - h_1$ of the column of air confined in the closed tube may therefore be taken as a measure of this air's volume.

4. Calculate the product *PV* for each corresponding pressure and volume found in Calculations 2 and 3.

5. Calculate the value of 1/*V* for each volume found in Calculation 3.

6. Plot a curve using the values of the pressure as abscissas and the corresponding values of 1/*V* as ordinates. Use an entire sheet of graph paper for the curve. Read the instructions on plotting graphs in the Introduction.

7. For each temperature measured in Procedures 9–11, calculate the gauge pressure exerted on the air in the bulb by subtracting the height of the mercury in the closed tube (height of the constant volume mark) from the height of the mercury in the open tube. Note that this pressure will be negative in many cases. Then add the atmospheric pressure to obtain the total pressure in centimeters of mercury.

8. Plot a curve using the values of the temperature as abscissas and the values of the total pressure obtained in Calculation 7 as ordinates. For more accurate results, construct the graph of the data using as large a scale as possible; that is, utilize the whole sheet of graph paper and extend the line only to 0°C.

9. From the graph of Calculation 8 determine the position of the absolute zero of temperature on the Celsius scale; that is, calculate the temperature at which this graph predicts that the pressure would be zero. A simple proportion may be used to do this since, according to Equation 21.5, the curve should be a straight line. Compare your result with the accepted value by calculating the percent discrepancy.

QUESTIONS _____

1. Show how your results and the graph from Calculations 4–6 verify Boyle's law. Discuss in particular your results in Calculation 4.

2. Explain how an error is introduced into the Boyle's law experiment when the room temperature changes.

3. How does the difference in height of the mercury in the two tubes cause a pressure to be exerted on the enclosed air?

4. Does Boyle's law hold accurately for a real gas at any temperature or pressure, even a very high or a very low one? Explain.

5. Show that Equation 21.7 is in fact a combination of Equations 21.1, 21.4, and 21.6, and in the process show how the constants C_1, C_2, and C_3 are related to C_4.

6. In the apparatus used to test Gay-Lussac's law, is an error introduced by the fact that the air in the rubber tubing between the bulb and the stopcock and in the glass tube between the stopcock and the mercury column is not in the water bath? Estimate how big and in which direction such an error might be.

7. A spherical basketball 10 in. in diameter is pumped up to a gauge pressure of 44.1 lb/in.^{2} If the normal atmospheric pressure is 14.7 lb/in.^{2}, what volume of air at this pressure has been pumped into the ball?

8. In an air thermometer apparatus like that used to test Gay-Lussac's law, the level of the mercury in the open tube is 20 cm higher than that in the closed tube when the bulb is at 0°C. At what temperature will the level in the open tube be 40 cm higher than that in the closed tube? Assume standard atmospheric pressure.

9. Use the graph you drew in Calculation 8 to determine P_0 and β in Equation 21.5 as applied to your setup.

10. Calculate the volume occupied by 1 mole of an ideal gas at $0°C$ and 1 atmosphere pressure. Will your result be true for real gases such as nitrogen or oxygen? What difference will the choice of gas make?

11. A constant-volume gas thermometer indicates an absolute pressure of 50.0 cm of mercury at $0°C$ and 68.3 cm of mercury at $100°C$. At what temperature will it read an absolute pressure of 56.1 cm of mercury?

The Mechanical Equivalent of Heat

22

Before the work of Benjamin Thompson (Count Rumford, 1753–1814) and James Prescott Joule (1818–1889, after whom the SI unit of energy is named), mechanical energy and heat were considered to be two quite different things. It was Thompson who first noted, while boring cannon for the Elector of Bavaria, that while the large amount of work put into turning the drill seemed to disappear rather than become stored as kinetic or potential energy, the drill and cannon got hot with no apparent source of heat. He suggested that the work was the source of the heat, that mechanical work was transformed into heat by friction and that, therefore, heat was not something separate and apart but just another form of

energy. Joule put this idea on a quantitative footing by transforming a known amount of work into a measurable quantity of heat. In this way, he not only showed that heat was a form of energy, so that calories and joules were really different size units of the same thing, but also measured the conversion factor between these units. This conversion factor is called the *mechanical equivalent of heat*. The purpose of the present experiment is to determine the mechanical equivalent of heat by doing a known amount of work against friction and measuring the amount of heat thereby produced. The apparatus used is a highly improved version of Joule's original setup.

THEORY

When two surfaces slide one over another, a frictional force arises parallel to the surfaces and must be opposed by an equal parallel applied force if motion at constant speed is to be maintained. The applied force will then do a work equal to its magnitude multiplied by the distance through which it moves in pushing one surface over the other. But this work is not stored as potential energy, nor does it appear as an increase in the kinetic energy of the moving body. The sliding surfaces do get hot, however, and Joule's and Thompson's idea was that the friction transforms the mechanical work done into heat, which thus appears as a different form of energy. Evidence in favor of this concept is obtained by showing that the amount of heat that appears is always proportional to the amount of work done against friction. Joule's experiment, like the present one, consisted of a setup by which a predetermined amount of work could be done against friction (in his case, against the viscous friction encountered by a paddle turning in a fluid) and the resulting heat could be measured. The conclusions to be drawn from all this are simply that heat is just another form of energy, that how much heat you have can be measured in joules as well as in calories, that we don't really need these two separate units, which came into being separately only because at the time heat and energy were considered distinct entities, and that "mechanical equivalent of heat" is a somewhat grandiose term for a simple conversion factor between two different-size units measuring the same thing. Saying that there are 4.186 joules in a calorie is no more profound than saying that there are four quarts in a gallon.

Fig. 22.1 shows the experimental arrangement to be used. A copper band is wrapped around a cylindrical copper calorimeter drum and carries a weight on its lower end. The other end is not attached to the drum but is allowed to go slack. Thus, when the drum is turned (by hand cranking in such a direction as to hoist the weight as if the drum were a windlass), the band slips on the drum surface, and the weight, while supported clear of the floor, is not raised. The value of this weight is thus equal to the frictional force between the band and the drum surface on which it slides, and the distance through which the force acts (the distance the sliding surfaces move with respect to each other) is simply the distance moved by any point on the drum surface. This is just the drum's circumference multiplied by the number of revolutions through which it is turned. The work done is thus

$$W = \pi DNMg \qquad (22.1)$$

where W is the work in joules, D is the diameter of the drum in meters, N is the number of revolutions through which the crank is turned, M is the mass in kilograms of the weight suspended on the copper band, and g is the acceleration of gravity in meters per second squared. Notice that the units of the expression on the right in Equation 22.1 are newton-meters, or joules, a joule being a newton-meter by definition.

The copper drum serves as a calorimeter. It is filled with water and insulated from the base of the apparatus by the plastic holder and shaft on which it is mounted. The copper band is in thermal contact with the drum and must be included with it. In operation, a thermometer is

167

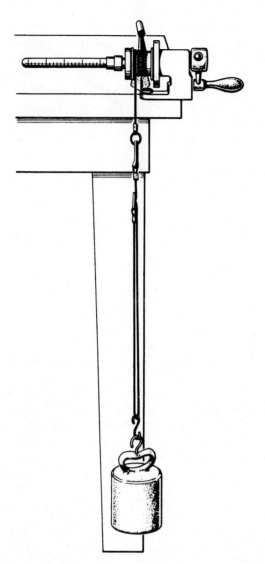

Figure 22.1 *Apparatus for measuring the mechanical equivalent of heat*

inserted through a leakproof seal so that the temperature of the water in the drum may be measured. The relation between the heat delivered to the calorimeter assembly and the observed temperature rise is

$$Q = [m_\text{w}c_\text{w} + (m_1 + m_2)c + C_\text{t}] \, (t - t_0) \qquad (22.2)$$

where Q is the heat delivered, m_w is the mass of the water, m_1 the mass of the calorimeter drum, and m_2 the mass of the copper band. The specific heats of water and copper are c_w and c, respectively, C_t is the effective *heat capacity* of the thermometer, t is the temperature measured after turning the drum, and t_0 is the initial temperature. In Equation 22.2, considerable attention must be paid to the units. As the mechanical equivalent of heat is a relation between calories and joules, we will want Q to be in calories, and as masses in this experiment will most conveniently be measured in grams and temperatures in degrees Celsius, the specific heats must be in calories per gram-degree Celsius and the heat capacity C_t in calories per degree Celsius. In these units, the specific heat of water is 1, but the symbol c_w has been included in Equation 22.2 to make it dimensionally consistent.

If, in fact, the work done against friction is converted into heat, then W should be proportional to Q and we may write

$$\pi DNMg = J[m_\text{w}c_\text{w} + (m_1 + m_2)c + C_\text{t}](t - t_0) \qquad (22.3)$$

where J is the proportionality constant and has the dimensions of joules per calorie. J is called the mechanical equivalent of heat and can be computed from Equation 22.3 if all other quantities are known. In the present experiment, the proportionality of W and Q will be checked and the value of J measured.

APPARATUS

1. Mechanical-equivalent-of-heat apparatus, complete with 5-kg weight and special thermometer
2. Vernier caliper
3. Triple-beam balance

PROCEDURE

1. The apparatus will be set up when you come to the laboratory. Inspect it carefully, noting in particular how the copper calorimeter drum is attached to the plastic mounting disk, so that you will know how to remove it and reattach it. A special thermometer is provided that does not have a very large range but whose expanded scale is arranged to allow small temperature changes near room temperature to be easily read. This thermometer is extremely fragile. Be particularly careful when it is inserted in the calorimeter drum, as a careless knock will break it off. Also note that it is larger in diameter than most thermometers and rolls easily. Take care not to let it roll off the bench.

2. Weigh the empty copper calorimeter drum with the thermometer sealing ferule but without the thermometer. Record the mass m_1 thus measured.

3. Weigh the copper band; record its mass as m_2.

4. Measure the diameter of the calorimeter drum with the vernier caliper.

5. Fill the drum nearly full of water (about 50 or 60 g) and again weigh it. Then push the thermometer gently through the seal, insert it into the drum, and tighten the sealing ferule. Do not overtighten. Normal tightening with the thumb and forefinger should be sufficient to prevent leaks. Record the thermometer's effective heat capacity C_t. This is 0.80 cal/°C for the thermometer supplied by Leybold-Heraeus but may be different for other brands. Your instructor will give you C_t for the particular thermometer you are using.

6. Attach the drum to its mounting disk, being careful not to hit the protruding thermometer by accident. Wrap the copper band around the drum until almost its full length is wound up (four or five turns). Attach the 5-kg weight to one end so that it hangs freely and attach the other end to a spring hooked to a peg in the apparatus base. The peg may be placed in any of a number of holes in the base; its position should be chosen so that approximately the same length of copper band extends away from the drum at either end. The cord by which the weight is suspended from the band has a length adjustment in it; this should be set so that the weight clears the floor when the crank is being turned in the direction tending to raise the weight. The band should slide smoothly on the drum surface when the crank is turned at a steady rate, and the end attached to the spring should be slack so as not to detract from the force applied to the band by the suspended weight. Check that there are no leaks around the thermometer.

7. The experiment should be started with the temperature of the calorimeter drum and its contents two or three degrees below room temperature. The calorimeter temperature will then be a few degrees above room temperature at the end of the run. In this way, the heat exchange between the calorimeter and its surroundings will just about balance out to zero over the running time. If the initial reading of the thermometer is more than three or four degrees below room temperature due to the calorimeter drum having been filled with cold water, turn the crank until the temperature reaches a better initial value. Notice that a lock is provided on the base that will hold the crank in a fixed starting position and that the thermometer may be turned in its seal so that the scale faces upward for easy reading when the crank is in the starting position just noted. If the thermometer is not adjusted in this way, the apparatus also has a mirror that may be swung out under the thermometer so that its scale may be read when it is facing down. Record your initial temperature value t_0 as the temperature corresponding to zero turns of the crank.

8. Start your experimental run immediately after recording t_0 and work rapidly so as to minimize the opportunity for heat exchange between the calorimeter drum and its surroundings. Turn the crank through 50 turns, then observe and record the temperature. Repeat for another 50 turns, and continue, recording the temperature after each 50-turn interval, until a total of 250 turns have been completed.

9. At the end of the experiment, detach the weight and unwind the copper band from the calorimeter drum. Remove the drum from the plastic mounting disk, loosen the sealing ferule, and carefully take out the thermometer. Replace the thermometer in its case and put it away safely. *Do not* leave it lying on the laboratory bench. Empty the calorimeter drum and leave it with the rest of the apparatus.

DATA _____

Mass of empty calorimeter drum
   and sealing ferule, m_1 _____

Mass of copper band, m_2 _____

Mass of calorimeter drum, sealing
   ferule, and water   _____

Mass of water, m_w _____

Diameter of calorimeter drum, D _____

Heat capacity of thermometer, C_t _____

Number of turns	0	50	100	150	200	250
Temperature						
Temperature rise $t - t_0$	✕					

Mechanical equivalent of heat J,
   from the graph   _____

Percent error   _____

Value of the mechanical equivalent
   of heat from Table I at
   end of book   _____

CALCULATIONS _____

1. Subtract the mass of the empty calorimeter from its mass when full to get the mass of the water.

2. Subtract the initial temperature t_0 (corresponding to zero turns of the crank) from the temperature observed after 50 turns to get the temperature rise $t - t_0$ due to 50 turns. Repeat using the observed temperatures at 100, 150, 200, and 250 turns.

3. Plot a curve using the values of the temperature rise $t - t_0$ as abscissas and the numbers of turns as ordinates. Choose your scales so as to use an entire sheet of graph paper. Note that the origin should be a point on your curve.

4. Check that your curve is a straight line and find its slope. According to Equation 22.3, this slope should be $J[m_w c_w + (m_1 + m_2)c + C_t]/\pi DMg$; hence J can be calculated if the other quantities are known. All have been recorded except the specific heat of copper, which is given in Table VII at the end of the book. Enter your calculated value of J on your data sheet.

5. Look up the accepted value of *J* and compute your percent error.

QUESTIONS _____

1. Explain how your work in this experiment and, in particular, your graph indicate that heat is a form of energy.

2. What problem would be encountered if a standard 0–50°C thermometer were used in place of the special one provided?

3. A clever feature of the apparatus used in this experiment is the fact that the frictional force is developed between the copper band and a *rotating* surface. Why is this a great advantage?

4. A 2000-lb car is traveling at 60 mi/h when the engine is disengaged and the brakes are applied, bringing the car to a stop. How many calories of heat are produced in the stopping process?

5. A Prony brake is a device for measuring the power output of an engine. It consists of a brake applied to a drum attached to the engine shaft so that the engine does work against the frictional force provided by the brake. The system is arranged so that the heat produced can be measured. In some Prony brakes, this is done by cooling the brake with water whose rate of flow and temperature rise is measured. If a 50°C rise is noted for the water cooling a Prony brake while the engine under test is developing 200 hp, how many gallons per minute must be flowing through the brake?

Coulomb's Law and the Force Between Charged Plates

23

A fundamental experiment in physical science is the demonstration of Coulomb's law, which states that two electric charges exert a force, one upon the other, that is proportional to the magnitude of each charge and inversely proportional to the square of the distance separating them. Charles Augustin Coulomb (1736–1806) established this law experimentally with the so-called Coulomb balance (Fig. 23.1). Similar to the Cavendish balance for measuring the inverse-square gravitational force, the Coulomb balance consists of two balls on the ends of a stick suspended by a quartz fiber so that this assembly can rotate under the influence of forces due to charges placed on the balls interacting with appropriately placed fixed charges. Quantitatively, this experiment is difficult to carry out, largely because of the complications in measuring the charges involved. In fact, Coulomb worked for years with this apparently simple apparatus before obtaining reliable results. Therefore, we do not attempt to duplicate his work here but instead perform a related experiment that is easier to carry out. Nevertheless, this experiment checks a direct consequence of Coulomb's law and thus constitutes a demonstration of the law itself.

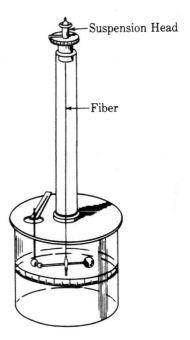

Suspension Head

Fiber

Figure 23.1 *The original Coulomb torsion balance*

THEORY

The mathematical expression for the force between two point charges q_1 and q_2 separated by a distance r is

$$F = k\frac{q_1 q_2}{r^2} = \frac{q_1 q_2}{4\pi\epsilon_0 r^2} \qquad (23.1)$$

where F is the force in newtons, q_1 and q_2 are in coulombs, r is in meters, and k is a proportionality constant having very nearly the value $9 \times 10^9 \ \mathrm{N \cdot m^2/C^2}$. The quantity ϵ_0 is called the *permittivity of free space*. We will shortly see why it is convenient to write the constant k as $1/4\pi\epsilon_0$, but for the moment note only that $\epsilon_0 = 1/4\pi k = 1/(36\pi \times 10^9) = 8.85 \times 10^{-12} \ \mathrm{C^2/N \cdot m^2}$.

Equation 23.1 gives the force between two *point* charges, but in this experiment we deal with two flat, parallel metal plates that are charged—one positively and one negatively—by the simple expedient of connecting a power supply (voltage source) between them so as to "pump" charge from one plate to the other. As a result, the upper surface of the lower plate and the lower surface of the upper plate become sheets of charge (one positive and one negative), each having a uniform areal charge

density $\sigma \ \mathrm{C/m^2}$. We are interested in the force of attraction between these two oppositely charged sheets rather than the force between positive and negative point charges.

It is a direct consequence of Coulomb's law that the electric field near a sheet of charge has the value $\sigma/2\epsilon_0$. Note the simplification that has occurred in this expression as a result of using ϵ_0 instead of k. Also, the electric field is by definition the force that a point charge of +1 C would feel if placed at the location in question. In this case, the expression $\sigma/2\epsilon_0$, which does not depend on position, shows that the field is constant and uniform. Remember, however, that it is derived on the assumption that the sheet of charge is infinite in extent so that its edges do not enter the picture. Actually, our plates are not infinitely large, so that $\sigma/2\epsilon_0$ is an approximation valid only when our test charge is separated from a plate by a distance that is small compared to the plate dimensions. In other words, in this experiment we keep the spacing between the plates small compared to their length and width.

173

We may now calculate the force on one plate due to the field of the other. Because this field is constant and is defined as the force per unit charge, the force in question is $q\sigma/2\epsilon_0$, where q is the charge on the plate on which the force is being measured. But because this charge has been obtained by "pumping" it from the other plate, the plates have equal and opposite charge, so that $\sigma = q/A$, where A is the plate area. Hence, the force with which one plate attracts the other is

$$F = \frac{q^2}{2A\epsilon_0} \qquad (23.2)$$

The question now arises as to how to determine q. The advantage of the present experiment is that q is simply related to the voltage or potential difference V imposed between the plates by the power supply. Imagine a point test charge of $+1$ C between the plates. The total field here (due to both plates) is $\sigma/\epsilon_0 = q/A\epsilon_0$, and therefore the work required to carry this test charge across the gap between the plates is $qd/A\epsilon_0$, where d is the plate spacing. But this is just the work per unit charge required to go from one plate to the other, which is by definition the potential difference between the plates. Thus,

$$V = \frac{qd}{A\epsilon_0}$$

or

$$q = \frac{VA\epsilon_0}{d} \qquad (23.3)$$

Substituting Equation 23.3 in 23.2 yields

$$F = \frac{A\epsilon_0 V^2}{2d^2} \qquad (23.4)$$

This is the result we propose to check. We will do so by measuring corresponding values of F and V and plotting F against V^2. The test consists of noting whether this plot is a straight line. Moreover, the slope of this line should be $A\epsilon_0/2d^2$, and, because A and d are easily measured, a value of ϵ_0 can then be found. It should agree with the value already noted in connection with Coulomb's law.

In the present experiment, a sensitive balance is used to measure the force of attraction F between two parallel plates of area A. High sensitivity is needed because, although V may be made fairly large, ϵ_0 is quite small; thus the forces to be measured are small. The apparatus, which is shown in Fig. 23.2, consists of a heavy base on which one plate is mounted by means of insulating posts. The second or movable plate forms part of a frame that constitutes one arm of a beam balance. The other arm consists of a threaded rod on which counterweights may be screwed back and forth to balance the beam assembly. This assembly rotates on knife edges, and a small mirror is attached to the central section to allow very small rotations to be observed. This is done by means of a so-called optical lever, as shown in Fig. 23.2. A laser shines a light ray on the mirror, which reflects it to a wall several feet away. This reflected ray is the "lever" because the displacement of the light spot on the wall is twice the angle through which the mirror rotates multiplied by the ray's length (mirror-to-wall distance). As we will see, an actual measurement of the mirror's rotation will be unnecessary, but the mirror-to-wall distance, although it need not be known, should be sufficient to insure detection of even small rotations of the beam balance. In addition, the beam assembly is provided with a damping system consisting of a conducting blade moving in the field of a permanent magnet and a beam lift that raises the assembly off the knife edges to prevent damage during transportation and to position it correctly on the knife-edge supports. Finally, the electrostatic force between the plates can be measured by placing weights in the center of the movable plate. This process is described in the Procedure section.

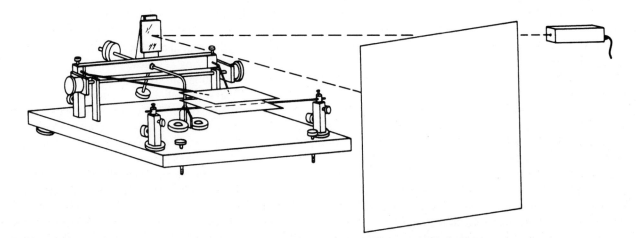

Figure 23.2 *Electrostatic balance arranged to measure the force between charged plates*

APPARATUS

1. Electrostatic balance with calibrated weights
2. Laser with support stand and necessary clamps
3. Level
4. Micrometer caliper
5. $\frac{1}{8}$-in. and $\frac{1}{4}$-in. spacer blocks
6. Vernier caliper
7. Variable power supply capable of delivering 0–1000 V dc at low current (less than 1 mA)
8. Voltmeter, 0–1000 V dc
9. Resistor, 10,000 Ω, 10 W
10. Roll of white calculator paper
11. Masking tape

PROCEDURE

1. Set up the electrostatic balance on the laboratory bench and carefully lift off the balance beam. Using the level provided, level the base by means of the two thumbscrews at its front.

2. Note the two tapered screws on the lifting mechanism and the matching tapered holes in the bottom of the beam assembly near the knife edges. Rotate the lifting shaft so that the tapered screw points project upward, and carefully place the beam assembly on these screws so that the points enter the tapered holes. Then gently lower the assembly onto the knife-edge bearing surfaces. Bring the movable plate into contact with the fixed plate and insure that these two plates touch over the full extent of their facing surfaces; that is, make sure they are exactly parallel. If they are not, squeeze them together *gently* with the thumb and forefinger of one hand while loosening the set screws retaining the fixed plate to its mounting posts with the other. Note that there are four such set screws, one pair allowing rotation of the plate about a longitudinal axis and the other permitting it to be tipped transversely. Be sure to tighten the set screws after making this adjustment. Lift the beam assembly off the knife edges with the beam lift, noting that an important function of the lifting mechanism is to position the beam assembly and hence the movable plate correctly. Lower the beam onto the knife edges and check the position of this plate. It should cover the lower or fixed plate exactly, the edges matching all the way around. If it does not, loosen the set screws that retain the movable plate support wires to the balance beam and adjust the movable plate position until it is correct. Consult your instructor if the plates are bent or if you have some other difficulty in making these adjustments. Be sure all set screws are tight after these adjustments have been completed.

3. Use the lifting mechanism to position the balance assembly correctly, and adjust the horizontal counterweights until the assembly is balanced and pivots freely on its knife edges. *Note:* The counterweights that hang vertically downward have already been set and should not be touched. Check that the damping blade rides in the middle of the hole cut for it in the base and does not rub on any of the damping magnet poles. If the blade comes too close to one side of the hole, lift the beam assembly off its supports and *gently* press the blade right or left as required. Then replace the assembly, positioning it prop-

erly with the lifting mechanism. If either of the damping magnets is too close to the blade, loosen that magnet's retaining screw and adjust its position as necessary.

4. Take the beam assembly out of the apparatus and measure the length and width of either plate with the vernier caliper. Record these measurements in meters. The plates used in this apparatus are usually square, so that you may get the same result for both length and width, but measure and record each independently. Then reposition the beam assembly on its supports.

5. Wire the apparatus as shown in Fig. 23.3. Notice that there is a connection terminal at each support post holding the fixed plate. The two terminals are needed when the apparatus is used as a current balance (Experiment 25), but in the present case either one may be used for connecting to the fixed plate. A similar consideration holds for the terminals at each knife-edge support, through which connection is made to the movable plate. The 10,000-Ω resistor serves to protect the power supply from being directly short-circuited should the plates come together while high voltage is applied. If this happens during the course of the experiment, however, you should be on the alert to reduce the power supply voltage to zero immediately. Fig. 23.3 shows the voltmeter connected directly across the power supply, although it is supposed to measure the voltage between the plates. However, because the resistance between the plates (when they are not touching) is almost infinite, there should be no current through and therefore no voltage drop across the protecting resistor. We may therefore assume that the voltmeter reading is the same as the actual plate-to-plate voltage.

6. Measure the thickness of the $\frac{1}{8}$-in. spacer block with the micrometer caliper. Note that the $\frac{1}{8}$ in. is a nominal value only. You need much greater precision than this; hence you should make three independent measurements with the micrometer and record the results *in meters* to four significant figures.

7. Place the $\frac{1}{8}$-in. spacer between the plates and put some weights on top of the movable plate (at its center) to bring this plate down solidly on the spacer. Lift the balance assembly momentarily with the beam lift to assure that it is correctly positioned. Then check that the plates are truly parallel. If you suspect that they are not, squeeze them together on each side of the spacer with the

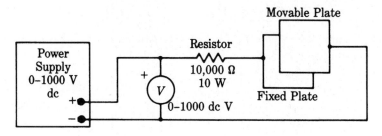

Figure 23.3   *Circuit for the electrostatic balance*

thumb and forefinger of one hand and loosen the four set screws that hold the fixed plate in position with the other. Then retighten these screws.

8. Mount the laser, turn it on, and position it so that its emitted ray strikes the mirror on the apparatus and is reflected back to a wall or screen at least 2 m away. Tape a strip of white paper vertically on the wall so that the light spot falls at about its center. Check that the movable plate is weighted down on top of the spacer and mark the position of the light spot on the paper. CAUTION: The low-power laser used in this experiment cannot hurt you even if its light falls directly on your skin, *but do not allow anyone, including yourself, to look directly into the emitted light ray.*

9. Now remove the weights on the movable plate, remove the spacer from between the plates, and place weights totaling a mass of 70 mg on the top center of the movable plate. Adjust the horizontal counterweights until the beam assembly balances with the light spot at the marked position on the paper strip. This is a critical adjustment that you should make with great care. Lift the balance assembly and return it to its supports to assure correct positioning several times and check that the spot returns to the marked position. Note that this procedure assures that the plates are spaced by exactly the thickness of the spacer block.

10. Remove the 70-mg mass from the top of the movable plate and observe that the balance assembly now balances with the light spot some distance above the position marked on the paper strip. Make sure the power supply's voltage control is turned all the way down and switch the supply on. Gradually increase its voltage while observing the motion of the spot. Note from Equation 23.4 that the force between the plates varies inversely as the square of their separation, so that if you allow the spot to overshoot the mark the plates are likely to be drawn together until they touch. If this happens, reduce the voltage to zero at once, reposition the balance assembly with the beam lift, and try again. With a little practice, you will be able to find a voltage for which the beam assembly balances with the spot right on the marked position. Do this three times, starting from zero voltage each time, and record the voltages obtained.

11. Repeat Procedures 9 and 10 using a mass of 50 mg instead of 70 and again with a mass of 20 mg. Record the voltages obtained in three independent measurements with each mass. Satisfy yourself that in each case, when the light spot has been returned to the marked position, the electrostatic force between the plates is exactly equal to the weight of the mass just removed.

12. Repeat Procedures 6–11 using the $\frac{1}{4}$-in. spacer and masses of 100, 70, and 50 mg.

DATA

Length of plate _____   Width _____   Plate area A _____

Thickness of $\frac{1}{8}$-in. block:

    Zero reading _____   Trial 1 _____

    Thickness d _____   Trial 2 _____

Trial 3 _____

Average _____

Mass removed from plate	Force on plate	Voltage measurement				
		Run 1	**Run 2**	**Run 3**	**Average**	V^2

Thickness of $\frac{1}{4}$-in. block: Trial 1   _____
    Zero reading   _____   Trial 2   _____
    Thickness d _____   Trial 3   _____
                                        Average   _____

Mass removed from plate	Force on plate	Voltage measurement				
		Run 1	Run 2	Run 3	Average	V^2

Slope with $\frac{1}{8}$-in. separation   _____   Value of ϵ_0 _____
Slope with $\frac{1}{4}$-in. separation   _____   Value of ϵ_0 _____
Average value of ϵ_0 _____   Percent error   _____

CALCULATIONS _____

1. Multiply the length of your movable plate by its width to obtain the area. Note that this experiment must be done strictly in MKS (SI) units, so that you should express your result in square meters.

2. Take the average of your three measurements of the thickness of the $\frac{1}{8}$-in. spacer block. Subtract the zero reading of the micrometer if it differs from zero. Write your result in meters using as many significant figures as you can justify. Repeat for the $\frac{1}{4}$-in. spacer block.

3. Calculate the electrostatic force on the movable plate for each case considered in Procedures 10–12 by finding the weight of the mass removed in each case.

4. Take the average of the voltages found in the three runs made for each value of removed mass and plate separation. Then square these averages, entering your results in the proper boxes in the V^2 columns on your data sheet.

5. Using your data for the $\frac{1}{8}$-in. spacer, plot the forces found in Calculation 3 as ordinates against the values of V^2 found in Calculation 4 as abscissas. Choose reasonable scales so as to cover as much of your sheet of graph paper as possible. Draw the best straight line through your plotted points, remembering that the origin (zero force, zero voltage) should be a point on this line.

6. Find the slope of the line plotted in Calculation 5, and, by reference to Equation 23.4, calculate your measured value of ϵ_0.

7. Repeat Calculations 5 and 6 using your data for $\frac{1}{4}$-in. spacing. This will give you a second straight-line graph from which you can again obtain a value of ϵ_0.

8. Calculate the average of your two values of ϵ_0 and compare this result with the accepted value by finding the percent error.

QUESTIONS _____

1. Explain in your own words how your graphs are a test of the theory proposed in this experiment. How well do your plotted points actually support this theory?

2. (a) Starting with Equation 23.3, derive an expression for the capacity of the parallel-plate capacitor formed by your two plates. (b) Compute the value of this capacity in picofarads for each of your two plate spacings.

3. Check that Equation 23.4 is dimensionally correct. What must be the dimensions of F, A, and d for this to be true?

4. Explain the statement in the Theory section that the displacement of the light spot on the wall is *twice* the angle through which the mirror rotates multiplied by the mirror-to-wall distance. Estimate the smallest angle of rotation of the balance that you could detect by observing the spot.

5. While you are bringing the plates to the right spacing as determined by the light spot on the wall, the movable plate may swing up and down a few times so that the plate spacing d varies with time. Suppose that this plate moves so that the spacing changes from d_1 to d_2 in a short time interval Δt while the power supply remains at a fixed voltage V. Derive an expression for the current that flows in the connecting wires during Δt.

6. What is the purpose of the damping blade and magnet arrangement? Even if you have not studied magnetic fields yet, find out and explain briefly how this arrangement works.

7. Assume that the plates used in your apparatus are made of aluminum sheet about 0.08 cm thick. Suppose one plate, instead of being mounted by supporting wires, is lying on a table and the other is held just above it. Find the plate-to-plate voltage required to just lift the lower plate off the table when the initial spacing between the plates is (a) the spacing you had with the $\frac{1}{8}$-in. spacer and (b) the spacing you had with the $\frac{1}{4}$-in. spacer.

8. A parallel-plate capacitor whose plates measure 20×20 cm and are spaced by 0.5 cm is connected to a power supply delivering 1000 V dc. (a) How much charge does the power supply transfer from one plate to the other? (b) How much work does the power supply do in charging this capacitor?

The Electric Field 24

The electrostatic force between electric charges is a so-called action-at-a-distance force, that is, a force between bodies that are not in contact. As a result, the question arises of how one body knows that the other is there so that it feels a force. For our present purposes, we must assume that the presence of a charged body creates a condition in the surrounding space such that another charged body located in that space feels a force due to it. Qualitatively, this condition is what we mean by an electric field—we introduce the more precise definition of what a physicist means by a "field" in the Theory section. It is the purpose of this experiment to study the electric field and some aspects of the related problem of electric potential. The latter arises because we are dealing with conservative forces on charges, and motion of those charges will result in work being done and hence the storing of potential energy. We will be interested in drawing two kinds of lines, namely, lines representing the electric field and a second set joining points at the same potential. A result of this experiment will be the determination of these lines for some simple but important configurations of electric charge.

THEORY

In physics, a field is a region of space in which every point is characterized by the value assumed at that point by some quantity that varies throughout the space. Thus, we could take the space inside the laboratory room, measure the temperature at a large number of points in the room, and draw a three-dimensional map showing our results. We would be able to state the temperature as a function of position in the room and would say that we thus had a *temperature field*. In the electrostatic case, if we have a given arrangement of charged bodies in a certain space and we bring in a test charge (that is, a charged body that we will place at various points in order to measure the force experienced there), we will be able to determine *a force field*, which is a value of the force felt by the test charge at every point in the space. Because the force is proportional to the value of the test charge, we standardize to a unit test charge, by which we mean a charge of +1 C. We then define the *electric field* at any given point to be the value of the force exerted on a +1-C test charge placed at that point. Notice that this will be a *vector* field because a force has a direction; thus both a direction and a magnitude must be given to specify the field at any point. In contrast, the *scalar* temperature field mentioned earlier needs only one quantity to describe it.

Now if the test charge is moved through a distance $\Delta\mathbf{r}$, an amount of work $\mathbf{E} \cdot \Delta\mathbf{r}$ will be done because the electric field vector $\mathbf{E}$ is the force on +1 C, the value of the test charge; hence, $\mathbf{E} \cdot \Delta\mathbf{r}$ is simply the usual expression for the work done when a force moves through a distance.* The potential energy difference between the two ends of $\Delta\mathbf{r}$ is then $-\mathbf{E} \cdot \Delta\mathbf{r}$, and as this is a potential energy *per unit charge*, it is called the *electric potential difference* or *voltage V* between those two points. Its units are clearly joules per coulomb or volts, named in honor of Alessandro Volta (1745–1827), who first produced batteries that would sustain a potential difference between their terminals. Note that as with any example of potential energy, it is the potential *difference* between two points that is of interest. Thus we cannot give the electric potential *at a point* without first stating what point is to be considered at zero potential; as we know, this choice is arbitrary. Because electric forces go to zero when we go infinitely far away from the charge configuration giving rise to them, infinity is often chosen as the zero potential point. In this case, every point of space may be labeled with a potential value giving the work required to bring +1 C to that point from infinity, that is, the potential difference or voltage between infinity and the point in question. Such a value is called the *absolute potential*. Obviously, the work required to take +1 C from one point to another is simply the difference between the absolute potentials at the two points; that is, the voltage between them, and the work required to take *any* charge from one to the other is just the value of that charge times the voltage. We can write

$$W_{AB} = qV_{AB} \tag{24.1}$$

where W_{AB} is the work required to take charge q from A to B and V_{AB} is the voltage between those two points.

We can now construct a surface made up entirely of points that are at the same potential. In the language of

---

*We assume here that $\Delta\mathbf{r}$ is a small enough distance so that $\mathbf{E}$ is constant over it. In the general case, those who know some calculus will recognize that a $\Delta\mathbf{r}$ that is "small enough" becomes the differential $d\mathbf{r}$ and that the work done in moving the test charge from a point A to a point B may be represented by the integral $\int_A^B \mathbf{E} \cdot d\mathbf{r}$. The electric potential difference (or potential energy per unit charge) is then $V = -\int_A^B \mathbf{E} \cdot d\mathbf{r}$.

geometry, such a surface would be called the *locus* of points characterized by a particular potential. It is called an *equipotential,* and we should note that no work is required to move a charge around on an equipotential surface because by definition there is no potential difference between any two points on it. This implies that there is no component of the electric field lying along such a surface, for if there were, a charge moving along the surface would feel a force parallel to its direction of motion so that the motion would result in work being done and hence a potential difference between points on the surface contrary to our definition. Therefore, *the electric field direction must be perpendicular to an equipotential surface at every point.*

Equipotentials are important in physics, and the simple example of the contour map may help us to understand them. A contour map is a plan view of some area of terrain. Thus, distances measured along the map correspond to horizontal distances over the ground. Elevations are shown by the contour lines, each of which is a line drawn through points all having the same height. A contour line is thus the locus of points having a particular height. Such a line may also be thought of as a gravitational equipotential because gravitational potential energy near the earth's surface is *mgh* and is thus proportional to *h,* the height above some point chosen as the zero of potential energy (which for our map is usually taken to be sea level).

An example of a contour map is shown in Fig. 24.1. Each contour line is marked with a number giving the height above sea level of every point on that line. Clearly, the map shows a hill slightly over 600 ft high, and we note that the right (east) side of the hill has quite a gentle slope whereas the left (west) side is very steep. We know this because close spacing of the contour lines means that the difference in height of the adjacent lines is achieved over a short horizontal distance, whereas wide spacing means a relatively long horizontal distance from one height to the next. We also note that it takes no work to walk around the hill on a contour line because every point of such a line is at the same height, so that as we walk along it we never ascend or descend. Conversely, the steepest path (one that might be followed by a skier in "schussing" the hill) is at right angles to the contour lines. This is the shortest distance between adjacent lines and therefore the path over which there is the greatest change in height and hence the greatest change in gravitational potential energy per unit distance traversed. This maximum change in potential energy per unit distance is called the *gradient.* It is a vector pointing straight downhill, that is, perpendicular to the contour lines, and its magnitude is a measure of the steepness of the hill and

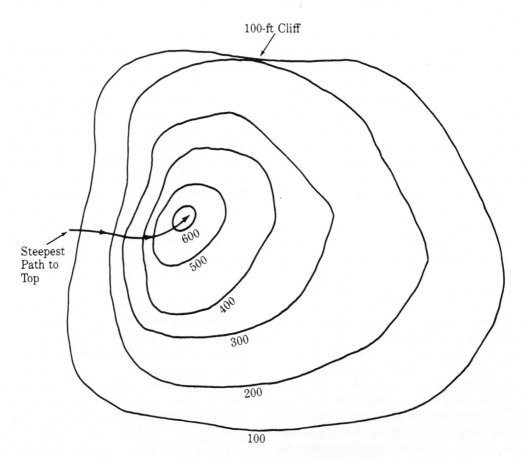

Figure 24.1 *A contour map of a hill*

hence of the force urging you downward and against which you must work to climb upward. You need not, of course, climb or descend by the steepest path, that is, along the gradient vector. If you choose a zigzag route, such as a switchback road in the mountains, you will be traveling in a direction having one component along a contour line, with which no force and hence no work is associated, and another along the gradient, which will be responsible for the work done in climbing. Thus, when you reach the top, you will have done the same amount of work as you would have done by climbing along the short-est path (the gradient), but you will have traveled a lot far-ther. You will not, however, have had to climb so steeply.

All of these ideas are immediately applicable to the electrical case. The equipotentials are surfaces of con-stant potential, so that, as already noted, no work is re-quired to move a charge along any one of them. When two equipotentials are close together, there is only the small spacing between them over which the potential must change from its value on one equipotential to its value on the other. Thus, suppose that in the expression $-\mathbf{E} \cdot \Delta \mathbf{r}$ for the work done in traversing $\Delta \mathbf{r}$, this latter were the perpendicular distance between adjacent equipotentials. Then $\Delta \mathbf{r}$ would be in the direction of the gradient and hence of $\mathbf{E}$, and $-\mathbf{E} \cdot \Delta \mathbf{r} = -E \Delta r$. Then if the equipotentials are close together so that Δr is small, E must be large. Notice that $\mathbf{E}$ is the force urging a unit test charge from one equipotential to the next lower one. It thus represents the "steepness" of the electric potential hill, and in fact, as we just noted, it is the gradient of the potential. Thus, if we could make a map of the equipo-tential surfaces similar to the contour map of Fig. 24.1, we would have a very good idea of the direction and magnitude of the electric field throughout the mapped re-gion. We could then present this knowledge by drawing lines representing the direction of $\mathbf{E}$ at every point, these lines being constructed perpendicular to the equipoten-tial surfaces. An example of such a construction is given in Fig. 24.2, which shows the field of a point charge q. Coulomb's law tells us that the force on $+1$ C in the re-gion surrounding q will be $q/4\pi\epsilon_0 r^2$, where r is the dis-tance from the center of q, and as it will be a repulsive force, its direction will be radially outward. The solid lines thus indicate the direction of $\mathbf{E}$. As the amount of work required to bring a unit test charge in from infinity to a certain distance r from q is quite independent of the direction from which the charge is brought in, all points a distance r from q will be at the same potential; hence a spherical surface of radius r, being by definition the lo-cus of points all at a distance r from the center, is an equipotential. The dashed circles in Fig. 24.2 represent several such surfaces.

Fig. 24.2 gives us a good idea of what the field of a point charge looks like and what we would expect a test charge placed anywhere in the region to feel. Note that although the field lines give us a good picture of the field, they are construction lines drawn by us as a con-venience in understanding the effects we will observe on test charges placed at various points in the region. Of course there are no actual lines in the space surrounding q. Our drawing merely represents the condition in space set up by q and resulting in the action-at-a-distance forces felt by other charges. Moreover, we have only an indirect presentation of the *strength* of the electric field at any point, for the lines of $\mathbf{E}$ simply show its direction. In Fig. 24.2, the equipotentials have been drawn with a constant potential difference between each, and we see that they get closer and closer together as we get nearer q. The conclusion is that the field gets stronger as we get nearer q, although it is not clear from the drawing that it varies as $1/r^2$. Note, however, that the radial field lines are everywhere perpendicular to the equipotential sur-faces, as they should be, and that as a result we can think of a "density of lines," that is, the number of lines cross-ing unit area of the normal surface. In Fig. 24.2, we have arbitrarily chosen to draw 16 lines from charge q, and these lines must all traverse each of the equipotential sur-faces at right angles. But any one of these surfaces has area $4\pi r^2$ where r is the radius of the particular surface. Therefore, the line density is $16/4\pi r^2$. Comparing this result with Coulomb's law for the field at distance r from q, we see that if q/ϵ_0 is represented by the total number of lines we have selected, the two expressions are identi-cal. Thus, the line density is a direct measure of the field strength, so that in looking at a map of electric field lines we can conclude that the magnitude of $\mathbf{E}$ will be large in regions where these lines are close together.

In the present experiment, we will determine some equipotentials arising from three simple charge configu-rations and plot some electric field lines using our knowledge that they must be perpendicular to the equipotentials at every point. We will do this in two di-mensions using sheets of carbon-impregnated paper that will conduct electricity over its surface, although not very well; that is, it will put up a resistance to the pas-sage of electric current. We will set up fixed-charge con-figurations by painting the appropriate shapes on the paper with a conducting silver paint that makes the painted area a very good conductor. Two shapes or elec-trodes will be used in each case with a battery or power supply connected between them. The battery or power supply pumps charge from one electrode to the next, and as the electrodes are conductors, their edges become equipotentials, the potential difference between them be-ing the supply voltage. But because the paper is a resis-tance, current will flow from one electrode to the other and there will be a fall in potential from the value at the positive electrode to the value at the negative one, which we can call zero to avoid working with negative num-bers. Moreover, the current must surely flow in the direc-tion of the electric field at each point so that the flow lines will be in the same direction as the field lines. We cannot see either one, of course, but we can measure the potential at various points by touching the positive probe

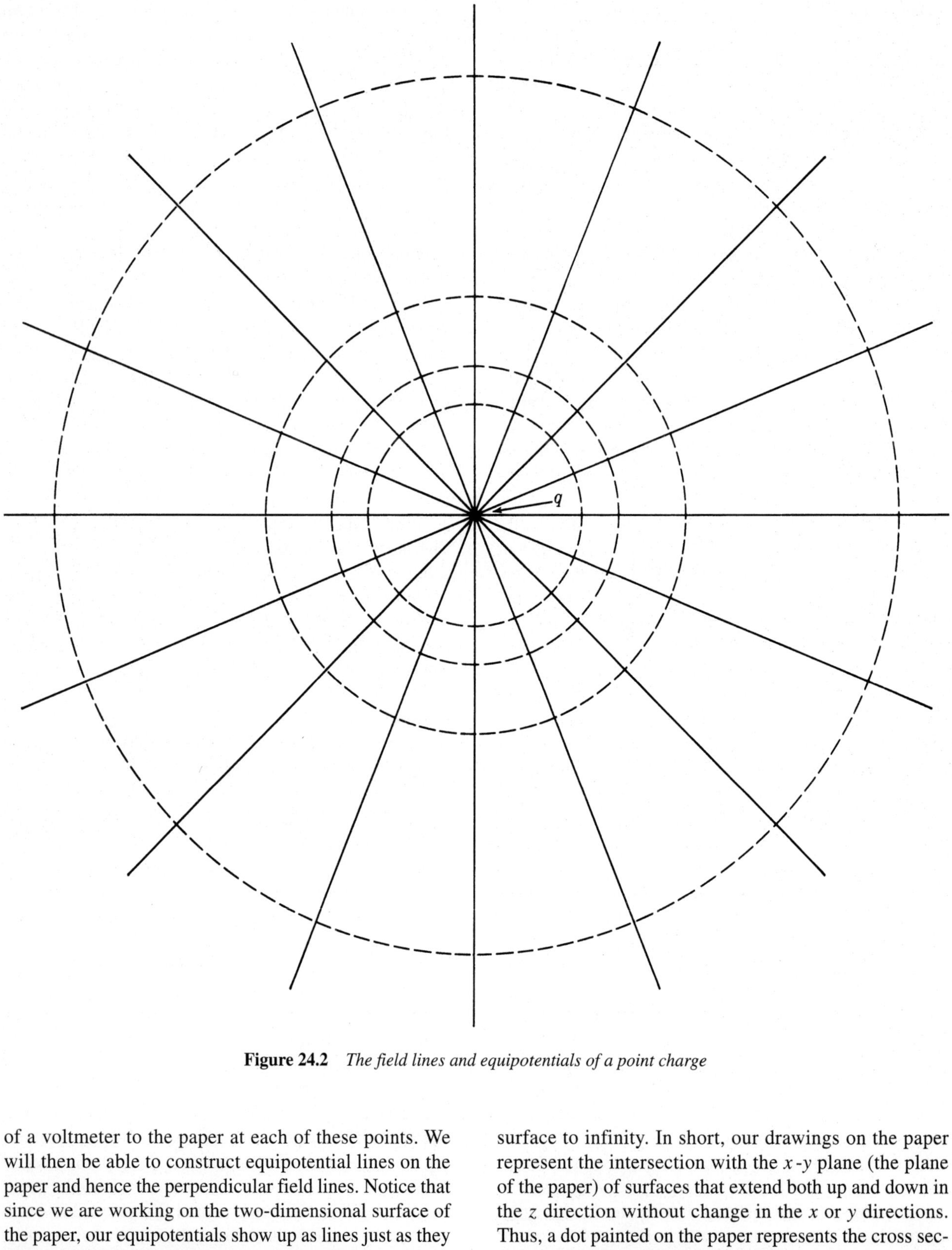

Figure 24.2 *The field lines and equipotentials of a point charge*

of a voltmeter to the paper at each of these points. We will then be able to construct equipotential lines on the paper and hence the perpendicular field lines. Notice that since we are working on the two-dimensional surface of the paper, our equipotentials show up as lines just as they do in Fig. 24.2. In that figure, the circles represent cross sections of spherical surfaces. Our experimental work will be applicable to the three-dimensional world if we think of the equipotential lines as being cross sections of cylindrical surfaces that project normally from the paper

surface to infinity. In short, our drawings on the paper represent the intersection with the x-y plane (the plane of the paper) of surfaces that extend both up and down in the z direction without change in the x or y directions. Thus, a dot painted on the paper represents the cross section of an infinitely long vertical rod; a circle, an infinitely long vertical cylinder; etc. Our results will thus apply to three-dimensional models that can be placed in a three-dimensional *cylindrical* coordinate system with no variation along the z-axis.

APPARATUS

1. Sheets of resistance paper with centimeter grid
2. Conductive ink in dispenser
3. Corkboard
4. Metal pushpins
5. 15-V dc power supply
6. Electronic voltmeter
7. Connecting leads
8. Masking tape
9. Straightedge and drawing compass

PROCEDURE

1. This experiment uses sheets of a carbon-impregnated paper that has a high but finite resistance across its surface. Electrodes may be drawn on this paper with silver conductive ink that makes the covered areas highly conducting. Such an area then assumes a single potential and may serve as an electrode when connected to the power supply. We will prepare three sheets, each with a pair of electrodes on it, between which we will connect the power supply so that a potential difference of 15 V will be established between them. Fig. 24.3 shows these three configurations. To prepare the sheets, tape each one down flat on the laboratory bench with masking tape. Draw the outline of each electrode using the centimeter grid to establish the dimensions shown in Fig. 24.3, a straightedge to draw the straight lines, and a drawing compass or circle template to draw the circles. Do this with an ordinary pencil and then fill in the outlined areas with the conductive ink. This ink is supplied in a dispenser resembling a small oil can and must be mixed thoroughly before using. Shake the can vigorously until you hear the stirring ball rattling around inside and then continue shaking for at least 30 s thereafter. Remove the cap and practice drawing with the dispenser on a piece of scrap paper. The ink should flow reasonably freely from the nozzle. If it doesn't, the nozzle can be cleaned out with a pin. If you have trouble getting a proper flow of ink, consult your instructor. Note that the silver pigment tends to settle out whenever you put the dispenser down for a minute, so it is good practice to shake it again before each use.

When you are adept at using the dispenser, fill in your electrode outlines as shown in Fig. 24.3. For the parallel-line configuration of Fig. 24.3(a), fill in the areas between the lines spaced 0.5 cm apart; be careful to bring the ink right up to your penciled lines but do not run over them. You should then have two parallel 0.5-cm-wide bands of conductive ink spaced by 6 cm between their facing edges. For the concentric circles, Fig. 24.3(b), fill in the 1-cm-radius central circle so that you have a circular spot of ink 2 cm in diameter there. The outer circle must have a smooth, accurate 7-cm inner radius, but the outer radius doesn't matter and need not be smooth. Paint the ink on the paper approaching your 7-cm penciled circle from the outside and coming carefully up to the line so as to produce the necessary exact 7-cm radius, but do not worry about the smoothness of the outer edge of the circular band you are making. Finally, for the configuration of Fig. 24.3(c), fill in the two circles to produce two circular conductive spots in the same way you did for the central spot in Fig. 24.3(b).

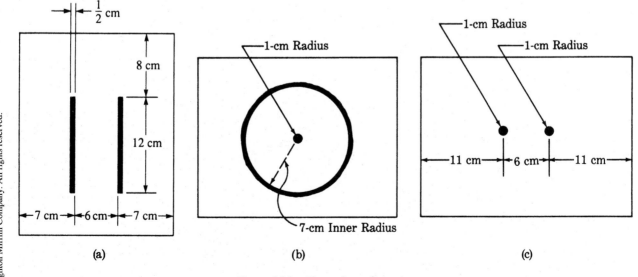

Figure 24.3 *Electrode configurations*

When you have finished using the ink, cap the dispenser and leave the three sheets taped down on the bench to dry. This will take 20–30 min, depending on how thick a coat you applied. Go to another bench and start on Procedure 2.

2. Copy the three electrode configurations of Fig. 24.3 on the first three sheets of graph paper found at the end of this experiment. This graph paper has centimeter-scale main divisions like the resistive paper grid, so that you can make an exact copy of your electrodes. However, the overall size of the graph paper is not the same as that of the resistive paper sheets, so the dimensions given in Fig. 24.3 for the spacing of the electrodes from the paper edges will not apply. This is of no consequence, however. Simply follow the dimensions for the electrode sizes and spacings, and arrange to center each complete configuration on the graph sheet as closely as you can.

3. When you are sure the ink is dry, remove the parallel-electrode configuration (a) from the laboratory bench and tack it down flat on the corkboard with a push-pin at each corner. Insert pushpins into the two conducting bands to make electrical contact with them and wire your setup as indicated in Fig. 24.4. Notice that the negative lead from the electronic voltmeter is connected to the same pushpin as the negative lead from the power supply. The electrode contacted by this pin will be considered at zero potential. Potentials at various points on the conductive paper may be determined by touching the point in question with the voltmeter probe.

4. When you are ready to make measurements, have your instructor check your setup and then turn on both the power supply and the electronic voltmeter. Allow both to warm up, after which you should set the voltmeter to a scale on which voltages up to +15 V dc are conveniently measured. Check its zero by touching the probe to the zero-potential pin (the one to which you connected the negative voltmeter and power supply leads) and adjusting the zero control until the meter needle is exactly on zero. Then touch the probe to the other pin (the one connected to the other, or positive, electrode) and set the power supply to exactly 15 V.

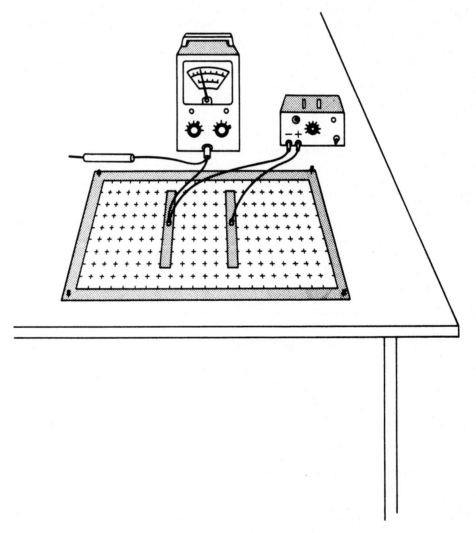

Figure 24.4 *The experimental setup*

5. You are now ready to investigate the equipotential lines between your two electrodes. Begin by determining a number of points on your resistive paper, all of which are at +12 V. A good way to do this is to look at the horizontal line 4 cm below the bottom ends of your two electrode strips. Run the voltmeter probe along this line, lightly touching the resistive paper until you locate a point where the meter reading is +12 V. Mark this point on the accompanying graph sheet prepared for the parallel-electrode configuration (a) in Procedure 2. Be careful not to try to measure the potential right at a grid mark on the resistive paper, as these marks prevent the voltmeter probe from making good contact. Now look at the horizontal line 3 cm below the bottom ends of your electrode strips and again run the voltmeter probe along this line until you find a point where the potential is +12 V. Plot this point on your graph sheet as before. Repeat this procedure along each horizontal line in the centimeter grid until you reach 4 cm above your electrodes. On your graph sheet, draw a curve through your plotted points and label it "12 V." This is your 12-V equipotential. Note that in the region between the electrodes it should be a straight line, and you may want to use a straightedge to draw the best straight line through these points, but note carefully where your equipotential begins to curve as the ends of the electrodes are approached.

6. Repeat Procedure 5 for voltmeter readings of 9, 6, and 3 V. When you have completed this work, your parallel-electrode graph sheet should have four equipotentials plotted on it for 12-, 9-, 6-, and 3-V potentials, respectively.

7. Remove the parallel-electrode resistive paper from the corkboard and replace it with the circular electrode configuration of Fig. 24.3(b). Use the metallic push-pins to make contact with the 7-cm circular band and the central circular spot, and connect the negative leads of the power supply and voltmeter to the spot. The positive power supply lead goes to the 7-cm circle, placing it at a potential of +15 V. Repeat Procedures 5 and 6 for the region between the electrodes, that is, between the circle and the spot. Again, it is convenient to move the voltmeter probe along a horizontal line through the centimeter marks on the grid as in Procedure 5, but symmetry should tell you that the equipotentials will be circles in the present case. You may be guided by this idea in looking for the various points at a particular voltage, and you should be able to use a drawing compass to draw the best circle through your set of points for any particular potential. Be particularly careful in determining the 3-V points, which are quite critical. You should thus obtain four circles as your equipotentials, one for each of the four voltages, 12, 9, 6, and 3 V.

8. Replace the resistive paper with the concentric circle configuration of Fig. 24.3(b) with the paper having the two circular spots of Fig. 24.3(c). Connect the power supply and voltmeter as before, and find the equipotentials for 12, 9, 6, and 3 V by repeating Procedures 5 and 6 for this configuration.

DATA

Parallel Electrode Configuration (a)

Equipotential pair	Distance between equipotentials	Deviation
15–12 V		
12–9 V		
9–6 V		
6–3 V		
3–0 V		
Averages		

Value of electric field between
the electrodes _____

Concentric Circle Configuration (b)

Equipotential	15-V	12-V	9-V	6-V	3-V	0-V
Radius, cm	7					1

Equipotential pair	Distance between equipotentials	Midpoint between equipotentials	Reciprocal of midpoint radius	Field at midpoint
15–12 V				
12–9 V				
9–6 V				
6–3 V				
3–0 V				

CALCULATIONS

1. On your graph of equipotentials for the parallel-electrode configuration, sketch several field lines. To draw a field line, start at some point along one electrode edge, draw a line perpendicular to that edge (which is itself an equipotential), and continue the line so that it crosses each successive equipotential at right angles, finally meeting the other electrode edge at right angles. Do your field lines in the central region between the electrodes (that is, not near the ends) go straight across from one electrode to the other?

2. Along a field line in the central region noted in Calculation 1, measure the distances separating adjacent equipotentials and record them in the boxes in your data table.

3. Your parallel electrodes drawn as two bands on paper represent an edge view of a pair of parallel plates in three dimensions. According to theory, the electric field in the central region between such parallel, oppositely charged plates should be uniform so that equipotentials differing by the same voltage should be equally spaced. See how close you are to realizing this prediction by averaging the five separations recorded in Calculation 2, computing a deviation for each, and averaging the deviations to obtain an a.d. Record your results in the spaces provided in the data sheet.

4. Adjacent equipotentials in the present case all differ by 3 V; hence, the value of the electric field between the plates is this voltage divided by the average spacing found in Calculation 3. Find this field and record it, attaching the error obtained from the a.d. of the spacing.

5. The concentric circle configuration of Fig. 24.3(b) represents a pair of concentric cylinders in three dimensions. According to theory, the equipotentials between such cylinders are cylindrical surfaces (hence circles in our two-dimensional experiment) having radii that increase logarithmically with voltage. Read the radii of the equipotential circles on your concentric-circle graph sheet and record these readings in the pertinent data table. Then plot them against voltage on the semi-logarithmic graph sheet provided. Note that you have a zero-volt equipotential at 1 cm and a 15-V equipotential at 7 cm. Draw the best straight line through your points. Refer to the section on plotting graphs in the Introduction for details on this type of plot.

6. On your concentric-circle graph sheet, measure and record the distances between each pair of adjacent equipotentials. Then find and record the radial distance to the midpoint between each equipotential pair and calculate the reciprocal of each of these distances.

7. Although the field between concentric cylinders is not uniform, a good approximation over small voltage intervals may still be obtained by dividing the potential difference between adjacent equipotentials by their radial separation and calling that the field at a radius midway between them. Calculate these field values for the radii halfway between each of your five equipotential pairs and enter them in your data table.

8. According to theory, the field between oppositely charged coaxial cylinders goes inversely with the radial distance from the axis; hence a plot of E against $1/r$ should be a straight line. Again, refer to the instructions on the use of graphs in the Introduction and plot the values obtained for the electric field in Calculation 7 against the reciprocals of the radial locations assigned to them from Calculation 6. Draw the best straight line through your points.

9. On your graph sheet with the two circular spots, draw several field lines, constructing each one perpendicular to successive equipotentials as described in Calculation 1. Do your field lines and equipotentials seem to have any particular shape for this electrode configuration?

QUESTIONS _____

1. (a) Why was an electronic voltmeter (EVM) used in this experiment as opposed to an ordinary moving coil (d'Arsonval) voltmeter? (b) What would have been the effect of substituting the latter type?

2. What is the significance of the slope of the line plotted in Calculation 5? the line in Calculation 8?

3. The areal resistivity of the carbon-impregnated paper you used in this experiment is about 10,000 Ω. Explain what this statement means, showing in particular why the units of areal resistivity are the same as the units of resistance between the terminals of a conductor.

4. Show why the field between two large but closely spaced parallel plates carrying equal but opposite charges should be uniform in the region not too near the edges. Use any method you can understand and explain.

5. Derive a relation between the potential V and the magnitude of the field $\mathbf{E}$ at a radial distance r from the axis of a very long uniformly charged rod of radius a $(r > a)$.

6. The two circular spots of the configuration of Fig. 24.3(c) represent two long, parallel cylinders in three dimensions. Look up and tell in your own words what the field lines and equipotential surfaces of such a pair of cylinders carrying equal and opposite charges should look like. How does your answer to this question compare with your results in Calculation 9? If your two answers do not correspond, can you explain why?

The Force Between Electric Currents 25

Although magnets appear to have poles that attract or repel other poles, in reality no magnetic poles are found to exist in the sense that electric charges do. What, then, is the source of the magnetic field if there are no poles or "magnetic charges" to fill this role? Magnetic fields result from the *motion* of electric charges, that is, from electric currents. Permanent magnets produce magnetic fields because the electrons going around in their orbits in the atoms of the material constitute tiny loops of electric current, each of which is a magnetic field source. In materials like iron, nickel, and cobalt, out of which permanent magnets are made, it is possible to line up the atoms so that their fields are collinear and add up to the large fields found around such magnets. This process is called magnetization, and the atoms will remain lined up in permanent magnets provided the temperature is not high enough for the resulting random thermal motion to upset the arrangement (the "Curie temperature").

The most elementary source of a magnetic field is a long, straight wire carrying an electric current. The magnetic lines of force due to such a current are simply concentric circles centered on the wire, as shown in Fig. 25.1. In addition, because there are no magnetic-pole sources of the magnetic field, magnetic forces are not properly described as the action of a magnetic field on a pole. Instead, a force is exerted by a magnetic field on a moving electric charge, that is, on an electric current. Thus, in describing magnetic fields and forces, we consider only moving electric charges and do not think in terms of poles at all. In particular, when an electric

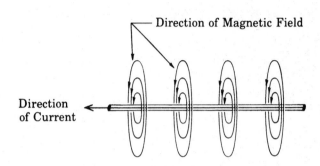

Figure 25.1 *Magnetic field direction*

charge moves in a magnetic field, it feels a force in a direction perpendicular both to the direction of the field lines (magnetic lines of force) and to the direction of motion. Clearly, the charge will feel no force if it moves parallel to the field lines, and the force it feels will increase from zero to a maximum as the direction of motion is changed from parallel to perpendicular to these lines. In Fig. 25.1, if a second wire is laid parallel to the one shown, a current in it will be perpendicular to the field, and a force pushing it toward or away from the first wire (depending on the current direction) will result. This is the most fundamental arrangement possible for the study of magnetic forces. The force law between parallel current-carrying wires may be regarded as the fundamental law of magnetostatics, just as Coulomb's law is regarded as the fundamental law of electrostatics. The purpose of the present experiment is to study this law.

THEORY

The magnetic field of a long, straight wire carrying current I is directed as shown in Fig. 25.1 and has the value

$$B = \frac{\mu_0 I}{2\pi r} = 2 \times 10^{-7} \frac{I}{r} \tag{25.1}$$

where B is the magnetic field strength in teslas, I is the current in amperes, r is the perpendicular distance in meters from the wire to the point where B is being evaluated, and μ_0 is a constant called the *permeability of free space* and has the value $4\pi \times 10^{-7}$ Wb/A · m. Notice that for the electromagnetic work in this experiment, MKS (SI) units must be strictly adhered to. Therefore, r in Equation 25.1 must be expressed in meters even though it will often be so small that the meter will not be the most convenient unit to use.

If a long, straight wire carrying current I' is in a magnetic field of strength B, the force on a length L of the wire is given by

$$F = BI'L \sin \theta \tag{25.2}$$

where F is the force in newtons, B is the field strength in teslas, I' is the current in amperes, L is the length in meters of the wire on which force F is exerted, and θ is the angle between the directions of the current and the field. The force is perpendicular to these two directions, but its sense remains to be determined. This is most easily done with the aid of Fig. 25.2, which shows the directions of a current I' and a magnetic field B by means of arrows. If your right hand is then placed so that the fingers curl in the direction needed to rotate the current arrow into the field arrow, the extended thumb will point in the direction of F. This procedure, sometimes called "the right-hand rule," is also useful in determining the field direction around the wire of Fig. 25.1. In this case, the wire is grasped with the right hand with the thumb extended in the direction of current flow. The fingers will then curl naturally around the wire in the direction of the magnetic field lines.

According to Equation 25.1, if a second wire carrying current I' is laid parallel to the one shown in Fig. 25.1 at a distance d away from it, the value of B at this wire due to the current I in the first wire will be $2 \times 10^{-7} I/d$. Moreover, Fig. 25.1 shows that the direction of B at the position of the second wire will be perpendicular to the wire and hence to I'. Equation 25.2 then says that the force on a length L of the second wire will be $BI'L$ and will be directed either toward or away from the first wire. Consideration of Fig. 25.2 shows that the two wires will attract each other if the currents I and I' are in the same direction and will repel if these currents are oppositely directed.

Combining Equations 25.1 and 25.2 for the case of parallel wires yields

$$F = \frac{\mu_0 I' I L}{2\pi d} = 2 \times 10^{-7} \frac{I' I L}{d} \qquad (25.3)$$

for the force of attraction or repulsion between the wires. Note that it makes no difference which wire is considered the source of the field and which the one on which F acts. Equation 25.3 is obtained in either case. In an actual experiment, one of the two parallel wires will be movable so that the force on it can be measured. The other, which is fixed in position, may then be conveniently taken as the field source.

In the present experiment, a sensitive balance is used to measure the force between two parallel wires.

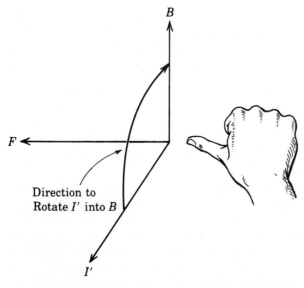

Figure 25.2 *The direction of the magnetic force*

High sensitivity is needed because the factor 2×10^{-7} in Equation 25.3 results in very small forces unless inconveniently large currents and/or small values of the separation distance d are used. The apparatus, which is shown in Fig. 25.3, consists of a heavy base on which one straight wire is mounted by means of insulating posts. The second or movable wire forms one side of a rectangular frame, which itself forms one arm of a beam balance. The other arm consists of a threaded rod on which counterweights may be screwed back and forth to balance the beam assembly. The beam rotates on knife edges under the side of the rectangle opposite the wire, and a mirror is mounted on this side so that small rotations of the beam assembly may be easily observed. This is done by means of an optical lever as shown in Fig. 25.3. A laser shines a light ray on the mirror, which reflects the ray to a wall several feet away. This reflected ray is the "lever," the displacement of the light spot on the wall being twice the angle through which the mirror rotates multiplied by the ray's length (mirror-to-wall distance). As we shall see, an actual measurement of the mirror's rotation will be unnecessary, but the mirror-to-wall distance, although it need not be known, should be sufficient to insure detection of even small rotations of the beam balance. In addition, the beam assembly is provided with a damping system consisting of a conducting blade moving in the field of a permanent magnet and a beam lift that raises the assembly off the knife edges to prevent damage during transportation and to position it correctly on the knife-edge supports. Finally, the movable wire carries a small pan in which calibrated weights may be placed to measure the force exerted on this wire by the magnetic field of the fixed wire.

The earth's magnetic field is about 5×10^{-5} T (0.5 G) and therefore cannot be neglected in comparison with the fields that can be produced at the movable wire by reasonable currents in the fixed one. Although the apparatus can and should be oriented with the wires parallel to the earth's field (roughly north-south) to minimize its effect, we cannot count on eliminating it entirely, and steps must be taken to compensate for it. Suppose that the component of the earth's field perpendicular to the plane of the two wires is B_0. Then the total field to be used in Equation 25.2 is $B + B_0$, where B is the field due to current I in the fixed wire, and Equation 25.3 becomes

$$F = \left(\frac{\mu_0 I}{2\pi d} + B_0 \right) I' L \qquad (25.4)$$

The apparatus is now connected with the two wires in series so that $I = I'$. Equation 25.4 then becomes

$$F = \left(\frac{\mu_0 I^2}{2\pi d} + B_0 I \right) L \qquad (25.5)$$

If the current I is then reversed, the force is given by

$$F = \left(\frac{\mu_0 I^2}{2\pi d} - B_0 I \right) L \qquad (25.6)$$

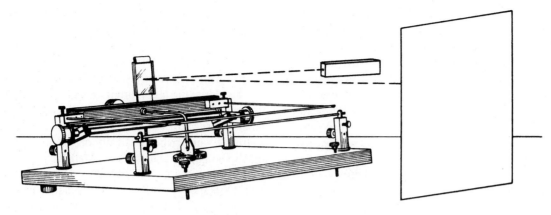

Figure 25.3   *Current balance*

in which the force due to the current is unchanged because I has been reversed in both wires (the sign of I^2 doesn't change), but the force due to the earth's field has been reversed by the reversal of the current in the mov-

able wire. Clearly, the average of the two forces measured with the current in each of the two directions will be the force due to the current alone, or that given by Equation 25.3 with $I = I'$.

APPARATUS

1. Current balance with calibrated weights
2. Laser with support stand and necessary clamps
3. Level
4. Micrometer caliper
5. Meter stick
6. $\frac{1}{8}$-in. and $\frac{1}{4}$-in. spacer blocks
7. Power supply capable of delivering 15 A at 10 V dc
8. Compass

9. Variable line transformer, about 150-W rating ("Variac" or "Powerstat"), if not included in above power supply
10. Ammeter, 0–15 A dc
11. Resistor, 0.5 Ω, 100 W
12. Two double-pole, double-throw knife switches
13. Roll of white calculator paper
14. Masking tape

PROCEDURE

1. Set up the current balance on the laboratory bench with the wires extending in a north-south direction as indicated by the compass. Carefully lift the beam assembly out of the apparatus and lay it aside. Using the level provided, level the base by means of the two thumbscrews at its front.

2. While the beam assembly is out of the apparatus, measure the length L of the movable wire with the meter stick. This measurement is made between the centers of the supporting side wires in the beam assembly. Record your result in meters, estimating the measurement to a tenth of a millimeter (0.001 m). Also measure the diameters of both the movable and fixed wires with the micrometer caliper. Make two measurements in different places on each wire and record your results *in meters*.

3. Note the two tapered screws on the lifting mechanism and the matching tapered holes in the bottom of the beam assembly near the knife edges. Rotate the lifting shaft so that the tapered screw points project upward and carefully place the beam assembly on these screws so that

the points enter the tapered holes. Then gently lower the assembly onto the knife-edge bearing surfaces. Check that the movable wire and the fixed wire below it are in the same vertical plane. If they are not, loosen the set screws holding the wire sides of the rectangle in the knife edges and push the sides in or out as required. Be sure to tighten the set screws after making this adjustment.

Now lift the beam assembly off its supports with the beam lift, noting that an important function of the lifting mechanism is to position the wire rectangle correctly. Again lower the assembly onto its supports and recheck the position of the movable wire. Note that, when viewed from the front, the wires should be straight and parallel. Bring them together by loading some weights onto the pan on the movable wire and check that no light is visible between the wires when a piece of white paper is held behind them. Adjustment for parallelism may be made by loosening the set screws retaining the lower (fixed) wire in its mounting posts and setting it parallel to the upper (movable) wire. However, even after this adjustment has been made, the wires will seldom be so straight as to touch at every point along their full length, but perfect straightness is not essential for good results. Consult your instructor if the wires appear to be seriously bent.

4. Use the beam lift to position the balance assembly correctly and adjust the horizontal counterweights until the assembly is balanced and pivots freely on its knife edges. *Note:* The counterweights that hang vertically downward have already been set and should not be touched. Check that the damping blade rides in the middle of the hole cut for it in the base and does not rub on any of the damping magnet poles. If the blade comes too close to one side of the hole, lift the beam assembly off its supports and *gently* press the blade right or left as required. Then replace the assembly, positioning it properly with the lifting mechanism. If either damping magnet is too close to the blade, loosen that magnet's retaining screw and adjust its position as necessary.

5. Wire the apparatus as shown in Fig. 25.4. Notice that switch SW_1 reverses the current through both wires of the current balance, whereas SW_2 reverses the current in one wire with respect to the other in order to change from repulsive to attractive force measurements. The neutral position of either switch opens the circuit, but reducing the current to zero will usually be done with the power supply control. Note that if this control is not built into your supply, you will need to plug the supply into a separate variable transformer. Have your instructor check your setup, and be sure the power supply switch is off and the control or variable transformer is turned all the way down before plugging it into the power line.

6. Measure the thickness of the $\frac{1}{8}$-in. spacer block with the micrometer caliper. Note that the $\frac{1}{8}$ in. is a nominal value only. You need much greater precision than this; hence you should make two independent measurements with the micrometer and record the results *in meters* to four significant figures.

7. Place the $\frac{1}{8}$-in. spacer between the wires and put some weights in the pan on the movable wire to bring this wire down on the spacer hard enough to retain it in position. Lift the balance assembly momentarily with the beam lift to assure that it is correctly positioned and check that the spacer block remains properly positioned to separate the wires by just its thickness.

8. Mount the laser, turn it on, and position it so that its emitted ray strikes the mirror on the apparatus and is reflected back to a wall or screen at least 2 m away. Tape a strip of white paper vertically on the wall so that the light spot falls at about its center. Mark the position of this spot on the paper. CAUTION: The low-power laser used in this experiment cannot hurt you even if its light falls directly on your skin, *but do not allow anyone, including yourself, to look directly into the emitted light ray.*

9. Remove the weights from the pan on the movable wire, remove the spacer from between the wires, and adjust the horizontal counterweights until the beam assembly balances with the light spot at the marked position on the paper strip. This is a critical adjustment and should be made with great care. Lift the balance assembly and return it to its supports to assure correct positioning several times and check that the spot returns to the marked position. Note that this procedure assures that the wires are spaced by exactly the thickness of the spacer block.

10. Set switch SW_2 for oppositely directed current in the two wires. Turn on the power supply and close SW_1 in either direction, the direction you choose to be called "forward" hereafter. Increase the power supply output to a current of about 5 A, noting that the resulting repulsive force between the wires causes the light spot to move upward along the paper strip. Now carefully add weights to the pan to bring the spot back down approximately to the marked position. Then make a fine adjustment of the current so that the spot comes to the marked position exactly. As the beam assembly is now in the same position it occupied with no weight and no current, the repulsive force between the wires must be just equal to the added weight. Record the mass in the pan and the final current as read on the ammeter.

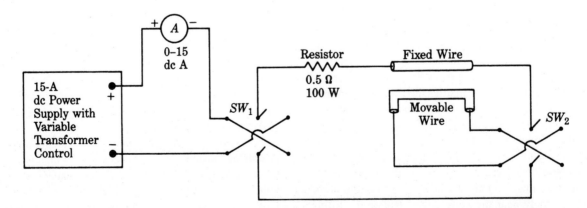

Figure 25.4 *Circuit for the current balance*

11. Reverse the current through the apparatus by throwing switch SW_1 to the opposite or "reverse" position. Readjust the current to return the spot to the exact position of the mark on the paper strip. Record this new current value.

12. Repeat Procedures 10 and 11 with additional weight and a current of about 10 A.

13. Repeat Procedures 10 and 11 with still more weight and a current of close to 15 A.

14. Reduce the current through the apparatus to zero by turning the power supply control all the way down and also opening switch SW_1 (neutral position). Adjust the horizontal counterweights to return the light spot to the marked position *with the weights of Procedure 13 still in the pan*. Throw SW_2 to give currents in the same direction in the two wires so that the force between them will be attractive.

15. Lift the beam assembly momentarily with the beam lift and return it to its supports to insure that it is positioned correctly and balances with the spot on the mark as noted in Procedure 14. Then carefully remove the weights from the weight pan without disturbing the apparatus. Throw switch SW_1 to the forward position and slowly increase the current through the wires while watching the spot move back down the paper strip toward your mark. Your aim is to find a current at which

the beam assembly just balances with the spot at the marked position. This will be a very delicate procedure, as the spot will tend to overshoot, the movable wire being drawn into contact with the fixed one. If this happens, reduce the current to zero, check the positioning of the balance assembly, and try again. Note that when you have the assembly balanced with the spot on your mark, the attractive force between the wires is equal to the weight you removed from the pan. Record the current read on the ammeter and the mass removed.

16. Reduce the current through the wires to zero and throw SW_1 to the reverse position. Again carefully increase the current to bring the light spot to the marked position as in Procedure 15 and record the current so obtained.

17. Repeat Procedures 14–16 with the weights used in Procedure 12 in the pan for the initial balancing.

18. Repeat Procedures 14–16 with the weights used in Procedure 10 in the pan for the initial balancing.

19. Repeat Procedures 6–18 using the $\frac{1}{4}$-in. spacer block. The strip of paper you already have taped on the wall may continue to serve with a new mark on it corresponding to the new wire spacing, but label this mark to avoid confusion with the one you were using with the $\frac{1}{8}$-in. spacer.

DATA

Length L of movable wire	_____	Micrometer zero reading	_____
Radius r_1 of movable wire	_____	Thickness of $\frac{1}{8}$-in. spacer	_____
Micrometer readings:		Micrometer readings:	
Trial 1	_____	Trial 1	_____
Trial 2	_____	Trial 2	_____
Average	_____	Average	_____
Radius r_2 of fixed wire	_____	Thickness of $\frac{1}{4}$-in. spacer	_____
Micrometer readings:		Micrometer readings:	
Trial 1	_____	Trial 1	_____
Trial 2	_____	Trial 2	_____
Average	_____	Average	_____

Wire separation d with $\frac{1}{8}$-in. spacer   _____

Relative direction	Current			Mass on pan	Magnetic force
	Forward	Reverse	Squared		
Opposite (repulsive force)					
Same (attractive force)					

Wire separation d with $\frac{1}{4}$-in. spacer   _____

Relative direction	Current			Mass on pan	Magnetic force
	Forward	Reverse	Squared		
Opposite (repulsive force)					
Same (attractive force)					

Value of μ_0 with $\frac{1}{8}$-in. spacer   _____   Average value of μ_0 _____

Value of μ_0 with $\frac{1}{4}$-in. spacer   _____   Percent error   _____

CALCULATIONS _____

1. Compute the average of the two diameter readings for each wire. Subtract the zero reading of the micrometer to obtain the diameter in each case and divide your results by 2 to obtain the radii r_1 and r_2 of the two wires.

2. Calculate the actual thickness of the $\frac{1}{8}$-in. spacer by averaging your two measurements and subtracting out the micrometer zero reading. Then find the center-to-center wire separation d for this case by adding r_1, r_2, and the thickness just determined.

3. Compute the square of the current readings obtained in Procedures 10–13 and 15–18.

4. From your data on the mass in the weight pan obtained in Procedures 10–13, find the repulsive magnetic force between the wires for each current. This calculation requires multiplication of the mass in the pan *in kilograms* by the acceleration of gravity in meters per second per second.

5. From your data on the mass removed from the weight pan in Procedures 15–18, find the attractive magnetic force between the wires for each current. This calculation requires multiplying the mass removed in each case by the acceleration of gravity. Express the masses in kilograms and the acceleration of gravity in meters per second per second to obtain the forces in newtons.

6. Plot a curve using the magnetic forces obtained in Calculations 4 and 5 as abscissas and the corresponding values of the square of the current as ordinates. Use an entire sheet of graph paper for this curve. Place the origin near the center of the sheet and consider attractive forces and the corresponding values of I^2 to be negative. Distinguish between points representing current through the apparatus in the forward direction and those for current in the reverse direction by using different colored pencils or marking the points in some distinctive way (for example, circles around points for forward current and squares around points for reverse current). Draw the best straight line through your points, remembering that the origin (zero force, zero current) should be a point on this line.

7. Calculate a value of μ_0 from the slope of the graph obtained in Calculation 6.

8. Repeat Calculations 2–7 for the wire separation obtained with the $\frac{1}{4}$-in. spacer and the associated data called for in Procedure 19.

9. Compute an average of your two values of μ_0 and compare this result with the accepted value by finding the percent error.

QUESTIONS _____

1. What conclusions do you draw from your two graphs?

2. Explain why reversing switch SW_2 changes the sign of I^2 but reversing switch SW_1 does not.

3. The Theory section points out that the effect of the earth's magnetic field can be eliminated by averaging the two force values obtained with opposite flow directions of a particular current. Explain why this was not done in the present experiment and what was done instead. Discuss how the relative positions of the points representing forward and reversed current on your graphs relate to this matter.

4. Estimate the smallest angle of rotation of the beam assembly that you could detect by observing the consequent motion of the light spot on the wall. Note that this will require you to estimate the mirror-to-wall distance (you didn't measure it but you must know approximately what it was), and remember that when a mirror turns through a given angle, the reflected ray is turned through twice that angle. In your answer, show why this is so.

5. What is the purpose of the damping blade and magnet arrangement? How does it work?

6. Suppose the current balance was oriented so that a horizontal component of the earth's magnetic field equal to 0.17 G (1.7×10^{-5} T) remained perpendicular to the movable wire. What mass would have to be added to the weight pan to balance out the effect of this field when a current of 10 A flows in the wire?

7. Let the fixed and movable wires be spaced by a distance d between their centers and let y be the coordinate of any point between the wires measured from the center of the fixed wire ($y = 0$). Derive expressions for the total magnetic field between the wires ($r_2 < y < d - r_1$) due to a current I in each wire when (a) the current flows in the same direction in the two wires and (b) the current flows in opposite directions.

8. A length of No. 14 bare copper wire (diameter $= 0.163$ cm) is laid flat on a table and connections are made to its ends with flexible leads. A second wire is mounted 1 cm above and parallel to the first one and is connected in series with it so as to produce an attractive force when a current is passed through the two wires. Find the current necessary to just raise the first wire off the table. Do you think the No. 14 wire could handle this current without overheating?

Ohm's Law 26

The most important principle in direct-current electrical circuit theory is Ohm's law, named after its discoverer, Georg Simon Ohm (1787–1854). This principle states that the current flowing through a conductor (that is, the rate at which electrical charge passes through it) is proportional to the voltage, or potential difference, between that conductor's ends. In the light of our present-day knowledge of electrical conduction, this does not seem too surprising, as a water analogy is easily visualized. Thus, if we have a pipe of a certain diameter, we would expect that a bigger pressure difference between the ends (the analog of voltage between the ends of a conductor) would push more water through per second, that is, give rise to a bigger water current. This linear relation between voltage and current is obeyed for most ordinary metals, although when we study electronics we will find that there are devices for which the voltage-current relationship is anything but linear. In this experiment, however, we will limit ourselves to materials that obey Ohm's law, we will observe how the linear relation between voltage and current holds in these cases, and we will study what happens when conductors obeying Ohm's law are connected in series and in parallel.

THEORY

Ohm's law states that the voltage across a conductor is proportional to the current through it, the proportionality constant being the *resistance* of the conductor. Although Ohm himself established his law on purely empirical grounds, we can give a theoretical derivation that makes it plausible and helps us understand why most ordinary metals obey it. Consider the bar of conducting material shown in Fig. 26.1. Such a material is a conductor because it contains charged particles, called *charge carriers,* that are free to move under the influence of an electric field. Electrons are the most common charge carriers. It is characteristic of ordinary metals such as copper, silver, and aluminum for the valence or outer atomic electrons to be loosely bound to their respective atoms and thus free to move through the structure of the material with relatively little resistance. This, of course, is a necessary characteristic for a conductor and is what makes metals good conductors. In contrast, insulators, in which all electrons are tightly bound to their respective atoms and cannot move about, do not conduct.

Let us then assume that the conducting bar of Fig. 26.1 contains a charge-carrier density of n such particles per unit volume. In consequence, the slice of thickness Δx appearing in this figure will contain $nA\Delta x$ charge carriers, where A is the cross-sectional area of the bar and hence $A\Delta x$ is the volume of the slice. If we now assume that the charge on each carrier is the electronic charge e (1.6×10^{-19} C), the total amount of movable charge in the slice of thickness Δx is

$$\Delta q = neA\Delta x \tag{26.1}$$

Now if a battery or power supply maintains a voltage V between the ends of the bar, there will be an electric field E along the bar equal to V/L, where L is the bar's length, with the result that each charge carrier will be urged in this direction with a force $eE = eV/L$. The charge carriers will then be accelerated, but although they are free to move in the bar, they do not do so without encountering some resistance. We now make the crucial assumption about the nature of ordinary conductors that leads to Ohm's law: When charge carriers move through the material, they feel an opposing force similar to the force of viscous friction encountered by an object moving through a fluid: for example, a ball bearing sinking in molasses. We further assume that this retarding force is proportional to velocity, so that the net force felt by each charge carrier is $eV/L - Kv$, where v is the so-called drift velocity of the charge carrier and K is the constant of proportionality between this velocity and the retarding force just mentioned. Clearly, a *terminal velocity* will soon be reached as the charge carriers accelerate; at this velocity the net force on each carrier will

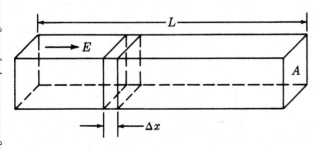

Figure 26.1 *A conducting bar*

be zero. The drift velocity v will thereafter be constant at the value

$$v = \frac{eV}{LK} \qquad (26.2)$$

required to give a zero net force ($eV/L - Kv = 0$).

Now consider a time interval $\Delta t = \Delta x/v$ in which charge carriers at the back of the slice in Fig. 26.1 just make it to the front. Clearly, in this time *all* the charge carriers that were initially in the slice will have made it out through the front face (to be replaced, of course, by other charge carriers coming in from behind). Thus, the charge Δq given by Equation 26.1 will have passed through the front face of the slice in time Δt. The charge passing this point per unit time is then

$$I \equiv \frac{\Delta q}{\Delta t} = neA \frac{\Delta x}{\Delta t} = neAv \qquad (26.3)$$

where I is the current in amperes, which is by definition the amount of charge passing a given point on a conductor per unit time. Substituting Equation 26.2 for v yields

$$I = \left(\frac{ne^2}{K}\right)\left(\frac{A}{L}\right)V \qquad (26.4)$$

This is, in effect, a statement of Ohm's law, for it says that the current through our conducting bar is proportional to the voltage between its ends. The proportionality constant ne^2A/KL is called the *conductance G* of the bar, so that Equation 26.4 may be written

$$I = GV \qquad (26.5)$$

It is more usually written with the voltage as the dependent variable:

$$V = \frac{1}{G} I = RI \qquad (26.6)$$

where $R = 1/G$ is the resistance of the bar and is stated in ohms, the SI unit of resistance. Equation 26.6, which is the mathematical statement of Ohm's law, gives consistent numerical results when I is in amperes, R is in ohms, and V is in volts.

A portion of a direct-current (dc) circuit that contains no voltage sources but only an arrangement of various resistors with wires coming out at two points for connection to the rest of the circuit is called a *two-terminal passive network*. It is two-terminal because we connect to it at only two points, and it is passive because there are no batteries or other sources of voltage or current in it. The voltage between the ends (terminals) of such a network is proportional to the current through it, so that the entire network may be replaced by a single resistor whose value is that of the proportionality constant. Two very usual resistor combinations of this sort are the *series* connection shown for the two resistors in Fig. 26.2 and the *parallel* connection shown in Fig. 26.3.

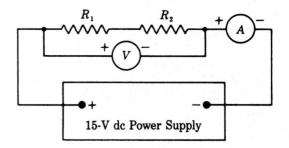

Figure 26.2   *Resistances in series*

When resistors are connected in series, the current is the same in each because the current in any one must all go on to the next one. The total voltage across the network, however, is the sum of the voltage drops across each resistor. Thus,

$$I = I_1 = I_2$$

and

$$V = V_1 + V_2$$

where I and V are the current through and the voltage across the network as a whole, I_1 and V_1 refer to resistor R_1, and I_2 and V_2 to resistor R_2. But by Ohm's law, $V_1 = I_1R_1$ and $V_2 = I_2R_2$, so that

$$V = I_1R_1 + I_2R_2 = IR_1 + IR_2 = I(R_1 + R_2)$$

Here we see that the voltage across the whole network is proportional to the current through it with proportionality constant

$$R = R_1 + R_2 \qquad (26.7)$$

so that the series combination may be replaced by a single resistor R whose resistance is the sum of those of the individual resistors R_1 and R_2

The situation is simply reversed for the parallel connection. In this case, it is the total current that is the sum of the currents through the individual resistors and the voltage across the network as a whole that is equal to that across each resistor:

$$V = V_1 = V_2$$

and

$$I = I_1 + I_2$$

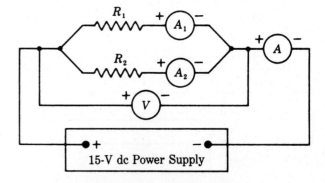

Figure 26.3   *Resistances in parallel*

This time we apply Ohm's law to the two resistors in the form $I_1 = G_1V_1$ and $I_2 = G_2V_2$ so that we get

$$I = G_1V_1 + G_2V_2 = G_1V + G_2V = (G_1 + G_2)V$$

This result shows that I is proportional to V with proportionality constant

$$G = G_1 + G_2 \qquad (26.8)$$

which looks exactly like Equation 26.7 except that it is a relation among conductances. We conclude that the parallel combination may be replaced by a single resistor whose *conductance* is the sum of the *conductances* of the two individual resistors. However, although this is quite

correct, it is unusual to talk about conductances. Consequently, Equation 26.8 is commonly written in the form

$$\frac{1}{R} = \frac{1}{R_1} + \frac{1}{R_2} \qquad (26.9)$$

which states that the parallel combination of resistors may be replaced by a single resistor such that the reciprocal of its resistance will be equal to the sum of the reciprocals of the individual resistances R_1 and R_2. In each case, R is the value of the *equivalent* resistor that can replace the complete network without altering the relation between the network current I and its voltage drop V.

APPARATUS _____

1. Tubular rheostat (about 100 Ω)
2. Tubular rheostat (about 200 Ω)
3. 15-V regulated dc power supply
4. Volt-ohm-milliammeter with 0–500 mA scale
5. dc voltmeter (0–15 V)

PROCEDURE _____

CAUTION: This experiment introduces you to a very useful and versatile instrument, the volt-ohm-milliammeter, often called a V-O-M or a multimeter. It consists of a high-quality meter mounted in a case with appropriate circuitry and switching so that by turning a switch and/or plugging the connecting leads into different jacks, voltages, currents, and resistances may be measured on various scales. Of particular interest is the use of this device as an ohmmeter, which requires impressing a known voltage across the resistor under test and measuring the current passing through it. A battery inside the V-O-M provides the voltage, and the circuit is arranged and an appropriate scale on the meter face calibrated to allow a direct resistance readout in ohms. Note, however, that *this meter is extremely sensitive, so that an erroneous connection in your circuit or switching it to the wrong scale can cause it to be instantly and permanently ruined as soon as power is applied.* Do not plug the power supply into the line or turn it on until your circuit has been approved by your instructor. Any changes in the circuit are to be made only after the power supply has been turned off. Have your circuit approved by your instructor every time you make any changes. Remember, you are using expensive equipment; good experimenters always treat their apparatus with great care.

1. Connect the two rheostats in series as shown in Fig. 26.2. Use the 100-Ω rheostat as R_1 and set it at about half resistance. Use the 200-Ω rheostat as R_2 and set it at maximum. Switch the V-O-M to its 500-mA scale and connect it at A in Fig. 26.2. The voltmeter is connected across the two rheostats in series as shown. Have your circuit approved by your instructor.

2. Plug in the power supply, turn it on, and adjust its voltage control so that the voltmeter reads exactly 15 V.

Then, without touching anything else, reconnect the voltmeter across R_1 alone. Record the current through the circuit and the voltage across R_1 in the first pair of boxes in the first table of your data sheet.

3. Increase the current through the circuit in four approximately equal steps by reducing the resistance of R_2, the final setting of this rheostat being zero. Record the current through and the voltage across R_1 for each setting of R_2 in the subsequent boxes in your table. Do not alter the setting of R_1 during this procedure.

4. Turn off the power supply. Reset R_2 to a resistance near its midrange value and reconnect the voltmeter across the two ends of this rheostat without disturbing the rest of the circuit. Turn on the power supply and record the readings of the voltmeter and the V-O-M as V_2 and I_2, respectively, in the appropriate boxes of the second table in your data sheet. Note that these are *measured* values.

5. With the power off, reconnect the voltmeter across the two rheostats in series as shown in Fig. 26.2. Apply power to your circuit and record both meter readings. These are the measured values of V and I for your series circuit.

6. Turn off the power supply and disconnect the rheostat used as R_1 and the volt-ohm-milliammeter from the circuit. Have your instructor explain the use of the V-O-M as an ohmmeter. Then switch this instrument to the proper scale, zero it, and use it to measure R_1. Record your result in the space provided on the data sheet.

7. Wire the circuit of Fig. 26.3, which shows the two rheostats in parallel. The three meter symbols marked A, A_1, and A_2 show different possible connections for the V-O-M. Switch this instrument back to its 500-mA scale and connect it at position A, the other two positions representing simply direct connections for this step. Do not change the setting of either rheostat when you reconnect them in this circuit.

8. Before applying power, have the circuit approved by your instructor. Record the total current through the circuit and the voltage drop across the parallel resistance combination as your measured I and V for the parallel case.

9. Switch off the power supply and reconnect the V-O-M at point A_1 without disturbing the rest of the circuit. Apply power and record the current through R_1 as your measured I_1 and the voltmeter reading as your measured V_2.

10. Repeat Procedure 9 for R_2, connecting the V-O-M at A_2 instead of A_1 and recording your results as I_2 and V_1. Turn off and unplug the power supply when you have finished your experimental work.

DATA

Current, mA					
Voltage, V					

Resistance value from graph _____   Resistance value from ohmmeter _____

		Current, mA			Voltage, V			Resistance, Ω		
		I	I_1	I_2	V	V_1	V_2	R	R_1	R_2
Series connection	Measured									
	Calculated									
Parallel connection	Measured									
	Calculated									

CALCULATIONS

1. Plot a curve using the values of current obtained in Procedures 2 and 3 as abscissas and the corresponding values of voltage as ordinates. Remember that the origin is a sixth point on this curve.

2. From the slope of your graph, find the resistance at which R_1 was set in Procedure 1. Remember that your currents are in milliamperes.

3. Compare the value of R_1 just obtained from your graph with that measured with the ohmmeter in Procedure 6. Decide which value you think is best and enter it as the measured value of R_1 for the series connection in the second table of your data sheet. Explain why you decided as you did.

4. Compute the value of R_2 and the value of the total resistance R in the series combination by the application of Ohm's law, that is, from the measured values of the current and the voltage drops in the data of Procedures 4 and 5. Record these results as your calculated values of R and R_2 for the series connection. Then calculate R_1 from your rule for series resistors (Equation 26.7). Enter this result in the appropriate box in your table and compare it with the measured value recorded in Calculation 3.

5. Find the values of I_1 and V_1 for your series circuit by referring to the discussion of such circuits in the Theory section. Record these values as your calculated I_1 and V_1 (series connection).

6. Calculate the values of the two resistances R_1 and R_2 in the parallel connection and also of the total resistance R by the application of Ohm's law, that is, from the measured values of the currents and voltage drops found in Procedures 8–10. Although these results are calculated from your voltage and current measurements, record them as your measured values for the parallel connection.

7. As you did not change the rheostat settings, the values of R_1 and R_2 should be the same for the parallel connection as they were for the series. Copy the values found in Calculation 4 as your calculated values for these resistances in parallel, and use Equation 26.9 to obtain a calculated value of R. Compare your calculated and measured values of the three resistances for your parallel case.

8. Using your knowledge of parallel circuits, determine calculated values of I and V from the values measured for I_1, I_2, V_1, and V_2 in your parallel setup. Compare these calculated values with the measured values.

QUESTIONS

1. What is an ohm; that is, what combination of fundamental SI units does it represent?

2. Explain how your graph demonstrates Ohm's law.

3. Show how the comparisons between your measured and calculated values check the formulas for the current, voltage, and resistance in series and parallel circuits.

4. In Figs. 26.2 and 26.3, voltmeter V is shown connected so as not to include ammeter A in the network. What error in finding the network resistance from the readings of V and A results from this connection? What error would be made if V were connected to the other side of A? How can these errors be minimized?

5. The resistance included between two points in a circuit is 10 Ω. How much resistance must be placed in parallel to make the total resistance 4 Ω?

6. A coil whose resistance is 60 Ω and one whose resistance is 40 Ω are connected in parallel to a 120-V line. (a) What is the resistance of the combination? (b) What is the current in each coil? (c) What would be the current if they were connected in series? (d) What then would be the voltage drop across each coil?

7. A lamp is connected to a 120-V line. When the switch is first turned on and the filament is cold, the current through the lamp is 8 A, but this drops quickly to 0.5 A when the filament is hot. By what factor is the lamp's resistance increased due to the change in temperature?

8. Prove that the resistance of n identical resistors in series is n times the resistance of one resistor and that the resistance of n identical resistors in parallel is $1/n$ times the resistance of one resistor.

9. Express the formula for series resistors in terms of conductances instead of resistances. Does it look familiar? Find out what the unit of conductance is commonly called and write it below.

10. You are given three identical resistors, each of 100 Ω resistance. Show how by using them singly or in series, parallel, or series-parallel combinations you can obtain resistances of $33\frac{1}{3}$, 50, $66\frac{2}{3}$, 100, 150, 200, or 300 Ω.

11. Show that in a series combination of resistors the total resistance can never be less than that of the largest resistor whereas in a parallel combination the total resistance can never be more than that of the smallest.

12. Referring to the circuit shown below, complete the associated table using Ohm's law and your knowledge of series and parallel circuits. Note that the numbers 1, 2, 3, and 4 in the table refer to the resistance R of, the current I through, and the voltage V across the four resistors R_1, R_2, R_3, and R_4, respectively.

	R, ohms	I, amperes	V, volts
1		0.500	
2			4
3	18		
4			10

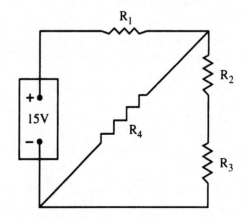

The Wheatstone Bridge and Specific Resistance

27

The most accurate method of measuring resistances of widely different values is by means of the Wheatstone bridge. The purpose of this experiment is to learn to use the slide-wire form of the Wheatstone bridge and to measure the resistance of several metallic conductors with it. Experiments show that if the temperature of a substance is kept constant, then its resistance depends only on its dimensions. Thus the resistance of a wire varies directly with its length and inversely with its cross-sectional area. Again a water analogy is useful. You'd expect it to be easier to push water through a fat pipe (one with a large cross section) than a thin one, and you would also expect more resistance to the passage of water through a long pipe than a short one. To eliminate the dependence of resistance on the dimensions of a conductor, let us standardize on a piece of the material in question having unit length and unit cross-sectional area. The resistance between the ends of this sample (which could be a cube, in which case the resistance between opposite faces is meant) is called the *specific resistance* or the *resistivity* of the material of which the sample is made and is seen to depend only on the nature of that material. We can thus speak of the resistivity of copper or steel without reference to any dimensions. A further purpose of this experiment will thus be to use our Wheatstone bridge to determine the specific resistance of the metals making up some sample wires.

THEORY

A Wheatstone bridge is a circuit consisting of four resistors arranged as shown in Fig 27.1. It is used for finding the value of an unknown resistance by comparing it with a known one. Three known resistances are connected with the unknown resistance, a galvanometer, a dry cell, and a switch, as shown in Fig. 27.1.

For a condition of balance, no current flows through the galvanometer. Hence the current through R_1 is the same as the current through R_2, and the current through R_3 is the same as that through R_4. Moreover, because there is no current through G, there must be no voltage across it; hence, the potential drop across R_1 must be equal to that across R_3. This requires that

$$i_1 R_1 = i_2 R_3$$

Similarly, the potential drop across R_2 must equal that across R_4, so that

$$i_1 R_2 = i_2 R_4$$

Dividing the first equation by the second yields

$$\frac{R_1}{R_2} = \frac{R_3}{R_4} \qquad (27.1)$$

Therefore, if three of the resistances are known, the fourth may be calculated from Equation 27.1.

In the slide-wire form of the bridge, as shown in Fig. 27.2, the resistances R_3 and R_4 are replaced by the uniform wire AB with a sliding contact key at C. Since the wire is uniform, the resistances of the two portions are proportional to their lengths; hence, the ratio R_3/R_4 is

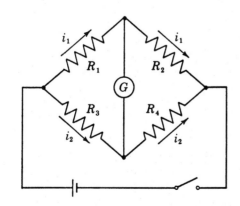

Figure 27.1 *The Wheatstone bridge circuit balanced*

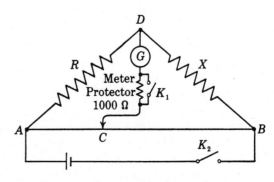

Figure 27.2 *The slide-wire Wheatstone bridge*

217

equal to the ratio AC/CB. If R_1 is represented by a known resistance R, and R_2 by the unknown resistance X, Equation 27.1 becomes

$$\frac{R}{X} = \frac{AC}{CB} \qquad (27.2)$$

From this, the value of X may be calculated.

As already noted, the resistance of a conductor depends on the material of which it is made, its length, its cross-sectional area, and the temperature. For a further discussion of this, refer to the Theory section of Experiment 26 on Ohm's law and in particular to Equations 26.4, 26.5, and 26.6. These show that the resistance R of a conductor is given by

$$R = \left(\frac{K}{ne^2}\right)\left(\frac{L}{A}\right) = \rho \frac{L}{A} \qquad (27.3)$$

where L is the length and A the cross-sectional area of the conducting bar, rod, or wire involved. The proportionality constant $\rho = K/ne^2$ is called the *resistivity* or *specific resistance* of the conducting material and depends only on the temperature and the nature of the material itself. The units of resistivity are clearly ohm-meters, and this quantity may be thought of as the resistance between opposite faces of a 1-m cube of the material in question. Resistivity may be found by determining the resistance of a measured length of a wire made of the given material. The cross-sectional area must also be determined from a measurement of the wire's diameter, after which ρ may be calculated from Equation 27.3. Because K and n in this equation, and hence ρ, are somewhat temperature-dependent, the resistivities that we will obtain in this experiment for several common conducting materials will be those corresponding to the laboratory temperature.

APPARATUS

1. Slide-wire form of Wheatstone bridge
2. Dry cell (1.5 V, No. 6 size)
3. Galvanometer
4. Standard decade resistance box (1000 Ω)
5. Single-pole, single-throw switch
6. Key
7. 1000-Ω, 1-W carbon resistor
8. Board with eleven lengths of wire of unknown resistance. The board is about 1 m long and the wires are mounted between binding posts. The first nine lengths are a single piece of copper wire, the tenth is a piece of manganin wire, and the eleventh is a piece of nichrome.
9. Micrometer caliper
10. Meter stick with caliper jaws

PROCEDURE

1. Hook up the slide-wire form of the bridge as shown in Fig. 27.2. Let X be the resistance of the nine lengths of copper wire in series, and let R be the standard resistance box. Use the key at K_1 and the single-pole, single-throw switch at K_2. Before closing any switches, have the circuit approved by the instructor.

2. Measure the resistance of the copper wire with your bridge. To do this, first set the decade box R to 5 Ω and move sliding contact C to the middle of the slide wire. Note that contact C is a knife edge mounted on a slider and equipped with a knob so that it can be pushed down to contact the slide wire at a precise point. Never move the slider while the knife edge is pressed down on the wire.

Now turn your apparatus on by closing switch K_2. Notice that key K_1 is normally open and should remain so for the moment so that the 1000-Ω protective resistor is in series with the galvanometer. Press contact key C gently down on the slide wire and note whether the galvanometer shows a deflection. If it does, release contact C, change the value of R, and repeat. Continue this procedure until the least possible galvanometer deflection is

observed. Then move the sliding contact to a position at which no deflection occurs when key C is tapped to bring the knife edge into contact with the slide wire.

3. For the most sensitive adjustment of the bridge, the 1000-Ω protective resistor is short-circuited with key K_1. When the galvanometer shows no deflection on tapping the contact key C, close K_1 and hold it down. Adjust the sliding contact very carefully, moving it no more than a fraction of a millimeter at a time, until the galvanometer again shows no deflection when key C is tapped. Record the setting of the standard resistance R and the distances AC and CB from the knife-edge contact to the ends of the slide wire. Open switch K_2 when you have finished your measurements.

4. Repeat Procedures 2 and 3 with the tenth length of wire (the manganin wire) connected at X in place of the copper wire.

5. Repeat Procedures 2 and 3 with the eleventh length of wire (the nichrome wire) substituted at X.

6. Measure the length of each wire by means of the meter stick and caliper jaws, adjusting the jaws so that they measure the distance between the binding posts where the wire is mounted. Remember that the total length of copper wire is the sum of the nine measured lengths.

7. Measure the diameter of each type of wire at four different points with the micrometer caliper. For the copper wire, the four points should be spread out along the total length. Record all four readings of the diameter of each wire, estimating to one tenth of the smallest scale division. The readings should be to 0.0001 cm.

8. In a similar manner, take and record four independent readings of the zero setting of the micrometer to 0.0001 cm. Be sure to note the algebraic sign of the zero correction.

DATA

Wire	R, Ω	AC, cm	CB, cm	X, Ω	Resistivity, $\Omega \cdot m$	Resistivity from Table VIII
Copper						
Manganin						
Nichrome						

Wire	Micrometer readings					Diameter, m	Area, m²	Length, m
	1	2	3	4	Average			
Copper								
Manganin								
Nichrome								
Zero reading								

CALCULATIONS

1. Compute the average of the readings for the diameter of each wire, as well as for the zero reading of the micrometer. Calculate and record the diameter of each wire, properly corrected for the zero reading and converted to meters.

2. Compute the cross section of each wire in square meters.

3. Calculate the values of the resistances measured using Equation 27.2.

4. Compute the specific resistance for each metal used.

5. Look up the values of the specific resistances in Table VIII at the end of the book and record them. Why might they differ somewhat from the ones you have just measured?

QUESTIONS _____

1. Why is the bridge method not adapted to measuring the hot resistance of an electric lamp?

2. Name three sources of error in using a slide-wire bridge.

3. Why is it more accurate to set the sliding contact near the center of the bridge instead of near one end when balance is obtained?

4. How is the operation of the Wheatstone bridge affected by changes in the cell voltage? Would there be an advantage in using a higher voltage? What limits the voltage that can be used?

5. Assuming that there is an error of 1 mm in locating the position of the contact on the slide wire, calculate the percent error in determining the value of X when the contact was (a) at 50 cm, (b) at 5 cm. Suppose that $R = 10.0 \ \Omega$ in both cases, and $AB = 1.0$ m.

6. Draw a diagram of the bridge as actually used in one of your measurements.

7. How does the resistance of a wire vary with the resistivity?

8. What would be the effect of a change in temperature on the resistivity?

9. Calculate the resistance of 40 mi of No. 22 B&S gauge copper wire. The diameter of the wire is 0.0644 cm. The value of the specific resistance is given in Table VIII.

10. What is the resistance of a copper bus bar 10 m long and 0.3×0.3 cm in cross section?

11. Assume that there was an error of 0.001 cm in measuring the diameter of the copper wire used in the experiment. Calculate the error that this would introduce into your value of the specific resistance.

12. The temperature coefficient of resistance for copper is 0.00393 per degree Celsius. Assuming room temperature to be 22°C, compute the specific resistance of copper at 0°C from your measurements.

Meter Sensitivity: Voltmeter and Ammeter 28

It is extremely important to understand the operation of electrical measuring instruments because they are constantly used. A galvanometer is an instrument for measuring very small electric currents. It forms the basis of all non-digital electrical measuring instruments; hence an understanding of its working principle and the type of measurements that can be made with it is essential for quantitative electrical work. In the present experiment the resistance and sensitivity of a galvanometer will be measured, and the instrument will be converted into a voltmeter that gives full-scale deflection for 3 V. In addition, it will be converted into an ammeter giving full-scale deflection for 500 mA (0.5 A).

THEORY

A galvanometer is an instrument for measuring small currents. It consists essentially of a coil of fine wire mounted so that it can rotate about a diameter in the field of a permanent magnet. When a current flows through the coil, magnetic forces are produced that exert a torque on it. It therefore rotates until equilibrium is established between the torque due to the current and that exerted by a restoring spring. The angle of rotation and hence the deflection of the attached indicating needle is directly proportional to the current flowing through the coil, provided the magnetic field is arranged so that the torque it produces depends only on coil current and does not vary as the coil rotates through its full-scale range. The magnet's poles are shaped so as to bring this about.

The sensitivity of a galvanometer is usually expressed as the current that must be passed through the instrument to produce a deflection of one scale division. The so-called full-scale current sensitivity, which is the current necessary for full-scale deflection, is also often used. If the full-scale sensitivity is known, the sensitivity in amperes per division can be found by dividing the full-scale sensitivity by the number of divisions. This result is called the galvanometer constant K. Clearly,

$$i = Kd$$

where i is the current passing through the galvanometer in amperes and d is the galvanometer deflection in scale divisions. Notice that the more sensitive a galvanometer is, the less current is required to cause a given deflection and the smaller the galvanometer constant K.

Another important characteristic of a galvanometer is the resistance r of its coil. Both K and r will be measured in the present experiment.

A galvanometer may be converted into an ammeter by placing a suitable shunt in parallel with the coil to permit the flow of a larger current through the complete ammeter than the galvanometer coil will carry alone.

This also makes the meter read in amperes with a predetermined full-scale value. Let S be the resistance of such a shunt (see Fig 28.2) and r the resistance of the galvanometer. Now let I_g be the current through the galvanometer and I the total current through the complete ammeter. Then, since r and S are in parallel, the current through the galvanometer is given by

$$I_g = I\frac{S}{r + S} \tag{28.1}$$

If the ammeter is to display a deflection of n divisions when I amperes pass through it, then I_g must equal Kn, and we get

$$I\frac{S}{r + S} = Kn \tag{28.2}$$

Then if I is the full-scale current reading desired and n the total number of divisions on the meter scale, Equation 28.2 can be used to calculate the necessary shunt resistance S. In most cases, where a sensitive galvanometer (small K) is being used to measure a large current I, S will be much smaller than r and can thus be neglected in the denominator of this equation. S is then given to a very good approximation by

$$S \approx \frac{rKn}{I} \tag{28.3}$$

A galvanometer may also be converted into a voltmeter by placing a large resistor (called a multiplier) in series with the coil. If R is the multiplier resistance, then

$$I_g = \frac{V}{R + r} \tag{28.4}$$

where V is the voltage impressed across the complete voltmeter (galvanometer and multiplier) terminals. Note that the voltmeter draws current I_g from the source

225

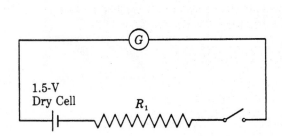 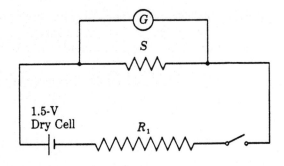

Figures 28.1 and 28.2   *Circuits for measuring the resistance of a galvanometer*

whose voltage is being measured and that because the galvanometer coil and the multiplier resistor are in series, I_g flows through both. Now if V is the desired full-scale voltage reading, the corresponding I_g must equal Kn, and Equation 28.4 becomes

$$Kn = \frac{V}{R + r} \qquad (28.5)$$

from which the necessary multiplier resistance can be calculated. In most cases R will be much greater than r, and so Equation 28.5 may be approximated by

$$R \approx \frac{V}{Kn} \qquad (28.6)$$

This simple formula is just Ohm's law, saying that the resistance R is given by the voltage V across it divided by the corresponding current Kn through it, the small additional series resistance r having been neglected.

   Much of the above theory requires a knowledge of the galvanometer constant K and coil resistance r, and in this experiment a special method will be used to measure these parameters. The problem is to measure them without passing an excessive current through the galvanometer coil. This is done by connecting a known voltage V in series with a large resistance R_1 and the coil as shown in Fig. 28.1. The value of R_1 is adjusted so that an exact

full-scale deflection is obtained. Then by Ohm's law

$$I_g = Kn = \frac{V}{R_1 + r} \approx \frac{V}{R_1} \qquad (28.7)$$

so that the galvanometer constant is given by

$$K \approx \frac{V}{nR_1} \qquad (28.8)$$

A shunt resistance S is now connected in parallel with r (see Fig. 28.2) and set so that the new deflection is exactly one-half scale. Under these conditions

$$I_g = \frac{1}{2}Kn = \frac{SV}{R_1 (r + S)} \qquad (28.9)$$

   The insertion of the shunt does not alter the fact that $Kn \approx V/R_1$, and substitution of this relation from Equation 28.7 in Equation 28.9 leads to the conclusion that $S = r$. This result is not surprising if we note that the combination of R_1 and the voltage source constitutes what is called a "current source," a source of a fixed current V/R_1. The current is fixed only to the approximation that resistances such as r or the parallel combination of r and S inserted in series with R_1 are so much smaller than R_1 that they make a negligible change in the total circuit resistance. This approximation is usually very good, however.

APPARATUS

1. Portable galvanometer
2. dc voltmeter (0–3 V)
3. Volt-ohm-milliammeter with 0–500-mA scale
4. Dry cell (1.5-V, No. 6 size)
5. Decade resistance box (10,000 Ω)

6. Decade resistance box (1000 Ω)
7. Single-pole, single-throw switch
8. 15-V regulated dc power supply
9. 200-Ω, 50-W rheostat
10. Roll of No. 28 copper wire

PROCEDURE AND CALCULATIONS

   1. Measure the voltage of the dry cell with the voltmeter and record it.

   2. Connect the dry cell in series with the 10,000-Ω resistance box, the galvanometer, and the switch as shown in Fig. 28.1. The resistance box, which serves as R_1, should be set at 3000 Ω.

3. Before closing the switch, have the circuit approved by the instructor. Now close the switch and note the galvanometer deflection. Carefully adjust R_1 until the galvanometer reads exactly full scale. Do not allow the galvanometer deflection to exceed the full-scale limit. Record the value obtained for R_1.

4. Use Equation 28.8 to find the galvanometer constant K.

5. Open the switch. Connect the 1000-Ω resistance box in parallel with the galvanometer as shown at S in Fig. 28.2, making it equal to 300 Ω and leaving R_1 connected and set as before. Close the switch and adjust S until the galvanometer reads exactly half scale. Record this value of S as your galvanometer resistance r.

6. Calculate the series resistance R (the multiplier) that will convert the galvanometer into a voltmeter giving full-scale deflection for 3 V.

7. Connect the galvanometer in series with the 10,000-Ω decade resistance box and set the box to your calculated value of R. This combination is now a voltmeter. Use it to measure the voltage of the dry cell, making sure that your series multiplier resistance R is included in the circuit. Record this voltage and compare it with that measured in Procedure 1.

8. Calculate the shunt resistance S that will convert the galvanometer into an ammeter having a full-scale reading of 0.5 A (500 mA). If you have not already learned how to use the volt-ohm-milliammeter (V-O-M) as an ohmmeter in Experiment 26, have your instructor explain its use to you now. Tell whether you think it could be used to measure the shunt resistance just calculated.

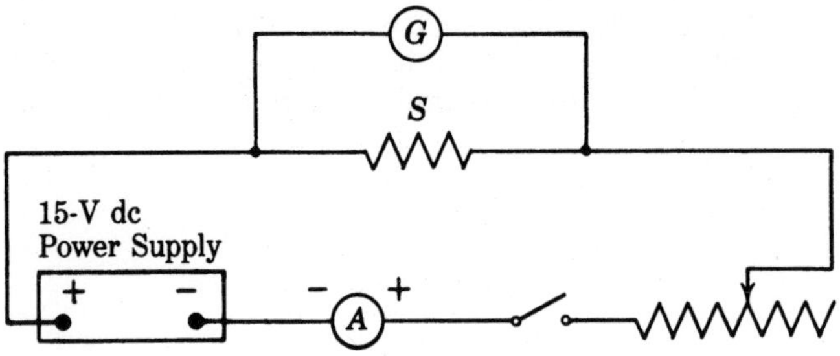

Figure 28.3 *Circuit for using a galvanometer as an ammeter*

9. Look up the resistance of No. 28 copper wire in Table XIV at the end of the book. Cut slightly over 15 m of this wire from the roll supplied and clean the insulation from the ends. With the ohmmeter, measure the resistance of exactly 15 m by connecting the V-O-M probes to the bare wire ends at points exactly 15 m apart. Compare your result with the value you would expect for 15 m of No. 28 wire from the data of Table XIV.

10. When you have established a best value for the resistance of the 15-m length of your No. 28 wire, compute the length needed to give the shunt resistance S found in Procedure 8. Cut a length 2 cm greater than this from the supply roll and clean the insulation from 1 cm at each end for connection to the galvanometer terminals. Connect your shunt in parallel with the galvanometer by squeezing the bared 1-cm lengths under the terminal screws. Make sure the full 1 cm is used at each terminal so that the effective length of the shunt is that just computed. Also check that the connections are tight so that no extraneous resistance is introduced. This combination is now an ammeter.

11. Connect your ammeter combination in series with the V-O-M set to its 500-mA range, the rheostat set with all of its resistance in the circuit, the switch, and the power supply, as shown in Fig. 28.3. When making connections to your meter, be careful not to disturb the shunt connections made in Procedure 10. Before closing the switch, have your circuit approved by the instructor.

12. Turn on the power supply, set it for an output of about 15 V, and close the switch. Adjust the rheostat until the V-O-M reads 0.1 A, and record the reading of your ammeter. Now adjust the rheostat to make the V-O-M read 0.2, 0.3, 0.4, and 0.5 A in turn, recording your ammeter's reading each time. Should your ammeter read off scale (due to your having made S a little too large) when the V-O-M indicates 0.5 A, try to estimate the reading, or, if necessary, note that the galvanometer needle is against the stop at the high end.

DATA

Galvanometer Characteristics

Dry cell voltage V _____

Value of R_1 for full-scale deflection _____

Number of scale divisions n _____

Galvanometer constant K _____

Galvanometer resistance r _____

Voltmeter

Resistance R for full-scale deflection
   of 3 V _____

Reading of this voltmeter for the
   voltage of the dry cell _____

Ammeter

Shunt resistance S for full-scale
   deflection of 0.5 A _____

Length of No. 28 copper wire
   required _____

Reading of V-O-M	0.1 A	0.2 A	0.3 A	0.4 A	0.5 A
Reading of galvanometer used as ammeter					

QUESTIONS _____

1. Derive Equation 28.1.

2. Does your data justify neglecting r in comparison with R_1 in Equation 28.7?

3. (a) Derive Equation 28.9. (b) Show that when $Kn = V/R_1$ is substituted in this equation, the conclusion $S = r$ follows.

4. The sensitivity of a voltmeter is often expressed in "ohms per volt," which means the resistance of the complete meter (galvanometer and multiplier in series) divided by the full-scale voltage reading. Derive a relation between sensitivity in ohms per volt and the galvanometer constant K, and find the ohms-per-volt sensitivity of the meter used in this experiment.

5. Calculate the series resistance R that will convert your galvanometer into a voltmeter reading 150 V for full-scale deflection.

6. Calculate the shunt resistance S that will convert your galvanometer into an ammeter reading 0.3 A for full-scale deflection.

7. How should an ammeter be connected in a circuit? A voltmeter?

8. Why could you not have measured the resistance r of your galvanometer with the V-O-M set to the appropriate ohms range?

9. A certain 3-V voltmeter requires a current of 10 mA to produce a full-scale deflection. How may it be converted into a voltmeter with a range of 150 V?

10. Find the sensitivity of the voltmeter of Question 9 in ohms per volt.

11. A 1.5-V dry cell furnishes current to two resistances of 2 Ω and 16,000 Ω in series. A galvanometer whose resistance is 200 Ω is connected across the 2-Ω resistance. The deflection produced is 10 scale divisions. Calculate the current through the galvanometer and the galvanometer constant.

12. Design and draw the schematic wiring diagram of a switching circuit that includes the galvanometer, the multiplier resistor R determined in Procedure 6, the shunt resistor S calculated in Procedure 8, and an appropriate switch that allows you to switch your meter from being a 3-V voltmeter to being a 500-mA ammeter. In other words, give the design for a circuit that will make your galvanometer into a simple multimeter having scales of 0–3 V and 0–500 mA.

Joule's Law 29

Whenever an electric current flows through a conductor, a certain amount of electrical energy is transformed into heat energy by the conductor's resistance, just as mechanical energy may be transformed into heat energy by friction. The object of this experiment is to measure the mechanical equivalent of heat by the electrical method. The electrical energy used may be calculated by measuring the current, the voltage, and the time during which this current flows. The number of calories liberated in the process is measured by means of a calorimeter. The importance of this experiment lies in the fact that heat may be supplied to a body either mechanically (see Experiment 22) or electrically, and the same relation between joules and calories is found in either case.

THEORY

Energy exists in many forms and is measured in many different units. The purpose of this experiment is to find the ratio between the mechanical unit and the heat unit of energy by using an electric current as a source of heat and equating the energy supplied by the current, expressed in joules, to the same energy expressed in calories.

Whenever a current flows through a conductor, a certain amount of heat is developed. The work done by the electric current is given by

$$W = VIt \qquad (29.1)$$

where W is the work expressed in joules, V is the potential difference between the terminals of the conductor in volts, I is the current in amperes, and t is the time during which the current flows in seconds. But, by Ohm's law,

$$V = IR \qquad (29.2)$$

where R is the resistance of the conductor. Equation 29.1 may therefore be written as

$$W = I^2Rt \qquad (29.3)$$

This relation is known as Joule's law of heating. Expressed in words, the law states that when an electric current flows through a pure resistance, the quantity of heat produced in the conductor is proportional to the time of current flow, to the resistance, and to the square of the current. Note that W is the electrical energy dissipated in the conductor expressed in joules. This energy appears as heat, which can be measured in calories by calorimetric methods. Note also that the products VI in Equation 29.1 and I^2R in Equation 29.3 give the *power* in *watts* supplied to the conductor. The watt, named after James Watt (1736–1819), is a joule of energy delivered per second. In other words, power is defined as energy supplied per unit time; hence the multiplication by t in Equations 29.1 and 29.3 to give the total energy W delivered.

From the principle of conservation of energy, whenever work is done in developing heat, the quantity of heat produced is always proportional to the work done. Thus,

$$W = JH \qquad (29.4)$$

where W is the work expressed in joules; H is the heat energy produced, in calories; and J is the mechanical equivalent of heat, which is the number of joules required to produce one calorie as discussed in Experiment 22.

If the heat energy is used to raise the temperature of the water contained in a calorimeter, then

$$H = (m_wc_w + m_1c_1 + C_2)(T_2 - T_1) \qquad (29.5)$$

where H is the heat delivered to the water, m_w is the mass and c_w the specific heat of the water, m_1 is the mass and c_1 the specific heat of the calorimeter, and C_2 is the heat capacity of the immersion heater. The temperatures T_1 and T_2 are the initial and final temperatures, respectively. Note that in this experiment it will be convenient to measure all masses in grams, not only because your balance will be calibrated in these units but also because you can then set c_w equal to 1. The temperatures must of course be in degrees Celsius, the specific heats will be in calories per gram-degree Celsius, and the heat capacity C_2 in calories per degree Celsius.

Since the immersion heater is immersed in the water in the calorimeter, all the heat produced in the heater resistance R must pass into the water. Thus Equation 29.4 holds with W given by Equation 29.1 (or Equation 29.3, but we will be measuring I and V in this experiment) and H given by Equation 29.5. Hence,

$$J = \frac{W}{H} = \frac{VIt}{(m_wc_w + m_1c_1 + C_2)(T_2 - T_1)} \qquad (29.6)$$

As all quantities in the expression on the right in this equation can be measured, J can be found just as it was in Experiment 22. In fact, a comparison of Equations 29.6 and 22.3 shows that the basic difference between this experiment and Experiment 22 is that in the latter the heat was provided by the dissipation of mechanical

233

energy by friction, whereas in the present case it is provided by the dissipation of electrical energy in a resistance. The analogy between friction in mechanics and resistance in electrical work should thus be obvious. Indeed, a very close parallel can be drawn.

APPARATUS

1. Immersion heater for 12-V operation
2. Calorimeter
3. 0–50°C thermometer
4. Stopwatch or stop clock
5. dc voltmeter (0–15 volts)
6. dc ammeter (0–5 A)
7. Triple-beam balance
8. Power supply capable of delivering at least 5 A at 10 V dc

PROCEDURE

1. Weigh the empty calorimeter and record its mass m_1 in grams. Then fill it about two thirds full of cold water (as cold as it will run from the tap) and determine the mass of the calorimeter plus the water $(m_1 + m_w)$. Replace the calorimeter in its containing jacket, put on the calorimeter cover, and immerse the heater. Make sure that the heater is fully immersed, but take care that neither it nor its connecting leads touch the sides or bottom of the calorimeter cup (see Fig. 29.1).

2. Wire the circuit as shown in Fig. 29.1. Check that the output control on the power supply is turned all the way down and that its power switch is off. Have your setup approved by your instructor before plugging the power supply in.

3. Insert the thermometer into the water through the hole in the calorimeter cover and stir the water with it, being careful not to knock the heater around. Then turn on the power supply, set the heater voltage to exactly

10 V, quickly note and record the ammeter reading, and return the power switch to off.

4. Again stir the water gently with the thermometer and read the temperature to the nearest tenth of a degree. Record this reading as your initial temperature T_1.

5. Turn on the power supply and simultaneously start the stopwatch. Keep the voltage constant at 10 V with the output control and stir the water gently from time to time with the thermometer. When the stopwatch reads 1 min, note and record the ammeter reading.

6. Note and record the ammeter reading at 2 min, 3 min, and 4 min, keeping the voltage constant at 10 V with the power supply control and stirring the water occasionally with the thermometer.

7. When the stopwatch reads *exactly* 5 min, shut off the power. Continue stirring the water until the maximum temperature is reached. Record this temperature to the nearest tenth of a degree as your final temperature T_2.

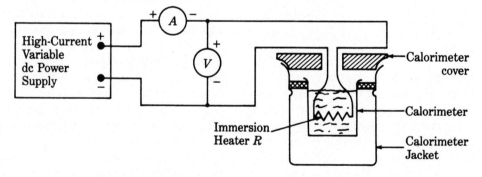

Figure 29.1 *Arrangement for electrical measurement of the mechanical equivalent of heat*

DATA

Mass of the empty calorimeter	_____	Heat capacity of the immersion heater	_____
Mass of the calorimeter with water	_____		
Mass of water	_____	Initial water temperature T_1	_____

Time	Initial	1 min	2 min	3 min	4 min		Average
Current							

Final water temperature T_2 _____   Temperature rise $T_2 - T_1$ _____

Mechanical equivalent of heat from
   Equation 29.6   _____

Mechanical equivalent of heat from
   Table I at end of book   _____

Percent error   _____

CALCULATIONS _____

1. Subtract the mass of the empty calorimeter from the mass of the calorimeter with water to obtain the mass m_w of the water alone. Also find the average value of I from your five ammeter readings.

2. Compute your value of the mechanical equivalent of heat from Equation 29.6. The value of C_2, the heat capacity of the immersion heater, will be given to you by the instructor. Remember that $V = 10$ V and the time t is 5 min but must be converted to seconds. Your instructor will tell you the material of which the calorimeter is made so that you can look up its specific heat in Table VII at the end of the book.

3. Compare your value of the mechanical equivalent of heat with the accepted one by calculating the percent error.

QUESTIONS _____

1. Discuss the principal sources of error in this experiment and justify the difference between your value of the mechanical equivalent of heat and the correct value.

2. Using your data, calculate how long it would take to heat the water from 20°C to 100°C by setting the power supply to send a current of 2 A through the same immersion heater.

3. In Question 2, how long would it take if a current of 4 A were used? Remember that the resistance R of the immersion heater is fixed at a value you should be able to calculate from your data.

4. State the factors on which the heating effect of an electric current depends and how it varies with each.

5. Give three examples of desirable heating effects of an electric current and three examples of undesirable ones.

6. How long after it is turned on will a 100-W electric heater take to bring a quart of water to a boil from room temperature (20°C)? Assume normal sea-level atmospheric pressure. Watch your units.

7. What must be the resistance of a coil of wire in order that a current of 2 A flowing through the coil may produce 1000 cal of heat per minute?

8. What is the purpose of starting with the temperature of the water lower than room temperature and ending, as you probably did, at about the same amount above room temperature?

9. Discuss the possibility of using the water itself as the heating element by just sticking the two wires (which must be bare to make contact with the water) down in the calorimeter without having the immersion heater shown in Fig. 29.1 connected between them. What would be the advantages and disadvantages of doing this?

10. Explain how a microwave oven heats food and tell why no metallic objects should be put in such an oven. Do you think microwave ovens are dangerous?

One of the most important instruments used in the electrical laboratory is the potentiometer, an instrument for measuring voltages, or potential differences. The potentiometer possesses a great advantage over the voltmeter because under operating conditions it draws no current from the source whose voltage is being measured and thus does not disturb the value of that voltage. The potentiometer is capable of very high precision and is easy to handle. One of its most important uses is the calibration of electrical measuring instruments such as voltmeters and ammeters. The purpose of this experiment is to study the slide-wire form of the potentiometer and to measure several voltages with it. Note that modern electronic voltmeters also draw little or no current from the source being measured, but many of them have the equivalent of a potentiometer circuit built in.

THEORY

The so-called open-circuit voltage of a dry cell, generator, or other voltage source is the potential difference that the source exhibits between its terminals under "no-load" conditions, that is, when no current is being drawn from it. Practical voltage sources are never "ideal" in the sense that they cannot maintain their output voltage constant at the open-circuit value regardless of the current drawn from them. Their terminal voltage always falls somewhat below this value whenever a load (some circuit drawing current) is connected. Fig. 30.1 illustrates what happens. A less-than-ideal voltage source is represented by the series combination of an ideal voltage source giving the open-circuit voltage V_{oc} and a resistor R_t called the *internal resistance* of the source. Ohm's law tells us that the current i_0 around the circuit will be $V_{oc}/(R_t + R_L)$ and that therefore the actual voltage delivered by the source to the load R_L will be

$$V_o = i_o R_L = V_{oc} \frac{R_L}{R_t + R_L} = V_{oc} \frac{1}{1 + R_t/R_L} \quad (30.1)$$

Clearly, V_o is less than V_{oc} by a fraction that approaches unity as the internal resistance R_t of the source is made small compared to the load resistance R_L.

Although in many cases the condition $R_t \ll R_L$ is realized, the open-circuit voltage of a voltage source cannot be measured accurately with an ordinary voltmeter because the voltmeter itself draws some current. This problem may be minimized by using a voltmeter of high sensitivity or even an electronic voltmeter (EVM). In the latter, the actual meter movement is driven by a separate power source, and only the very high input resistance (of the order of 10 million ohms in ordinary EVMs) of the associated electronic amplifier is connected across the voltage to be measured.* All such devices require calibration, however. The potentiometer measures the voltage of a given source by a null method, so that no current is drawn from the source when a balance is reached and the reading is being taken. Standard cells are available that produce an accurately known open-circuit voltage. The potentiometer allows precise measurements of other voltages to be made by comparing them with the known standard voltage.

Basically, a potentiometer is a voltage divider with a circuit such as the ones shown schematically in Fig. 30.2. Battery voltage V_{in} causes current $i = V_{in}/(R_a + R_b)$ to flow through the two series resistors in Fig. 30.2(a) so that the output voltage V_o is given by

$$V_o = i R_b = V_{in} \frac{R_b}{R_a + R_b} \quad (30.2)$$

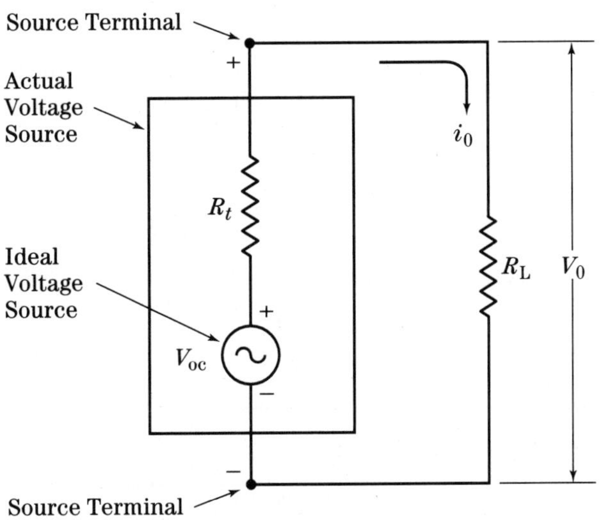

Figure 30.1 *Schematic diagram of a practical voltage source and load*

---

*See also the discussion of EVMs in the Theory section of the next experiment.

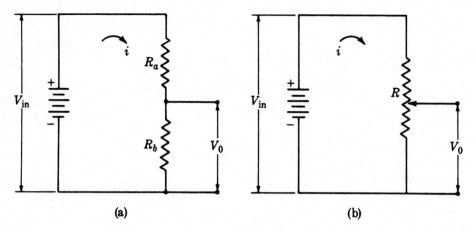

Figure 30.2 *Voltage dividers (a) fixed and (b) variable*

and is thus the fraction $R_b/(R_a + R_b)$ of V_{in}. We can say that the circuit has divided the voltage V_{in} by the factor $(R_a + R_b)/R_b$. In Fig. 30.2(b) this factor has been made continuously variable by making R_a and R_b into a single resistor R with a sliding contact. Thus R is always equal to $R_a + R_b$, and the fraction by which V_{in} is multiplied to give V_o is just R_b/R and is continuously variable from 0 to 1. Thus any voltage from 0 to V_{in} may be selected.

In the slide-wire form of the potentiometer, resistor R is a uniform bare wire provided with a sliding contact and through which a steady current is sent from an appropriate power supply. The voltage source to be measured is connected through a galvanometer and a protective resistance between one end of the wire and this contact in such a way that its polarity opposes that of the wire's potential difference. For balance, the contact is moved along the wire until no current flows through the galvanometer. The voltage V_o across this portion of the wire is now equal to that of the source being measured. This process is repeated with a standard cell. Because the wire is uniform, the length of the wire spanned is proportional to R_b and hence to the potential drop V_o. Thus, the open-circuit voltage (sometimes called the *electromotive force* or emf) of any voltage source can be compared with that of a standard cell.

The circuit used in this experiment is shown in Fig. 30.3. E_s is the standard cell, E_x is a cell whose emf is to be measured, K_2 is a double-throw switch that can connect either cell in the circuit, G is the galvanometer, R_1 is the protective resistance, and R_2 is a rheostat used to adjust the current in the slide wire. Because the electromotive force of the standard cell is equal to the potential drop in the length of the wire spanned for a condition of balance and the same is true for the unknown cell, the emf's are proportional to the lengths of wire spanned. Thus,

$$\frac{E_x}{E_s} = \frac{L_x}{L_s}$$

The unknown emf is then given by

$$E_x = E_s \frac{L_x}{L_s} \tag{30.3}$$

where E_x is the unknown emf, E_s is the emf of the standard cell, L_x is the length of the wire (AC) used for the unknown cell, and L_s is the length of the wire used for the standard cell for a condition of balance.

The scale measuring length along slide wire AB can be made to read voltage directly by proper adjustment of

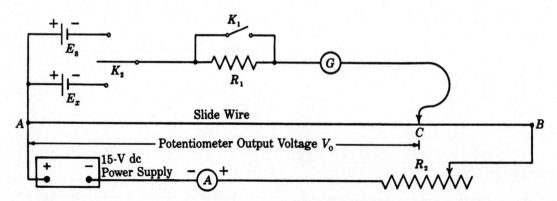

Figure 30.3 *The potentiometer circuit*

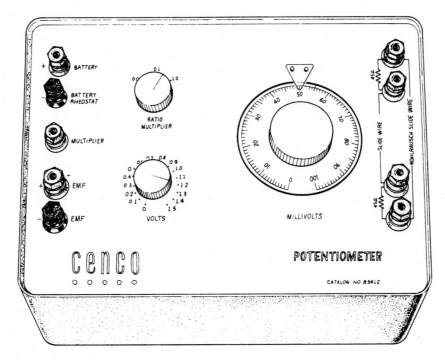

Figure 30.4   *Typical student potentiometer (Cenco)*

the rheostat R_2. Thus, if the total length AB of the slide wire is 1 m and R_2 is set so that the emf across this length is 1 V, output voltage V_o will be given by

$$V_o = V_{in}\frac{R_b}{R} = V_{in}\frac{kAC}{kAB} = \frac{1kAC}{k(1)} = AC \quad (30.4)$$

where k is the proportionality constant between the length and resistance of the uniform slide wire, and we have inserted $V_{in} = 1$ V and $AB = 1$ m. An unknown voltage can then be read directly from the measured position of contact C at balance. To establish exactly 1 V across AB, the standard cell is selected with switch K_2, contact C is set at a scale reading equal to the known

standard cell voltage, and balance is obtained by adjustment of R_2.

If the potential across AB is set at 1 V, the potentiometer is limited to measuring voltages equal to or less than 1 V. Since in this experiment voltages up to 1.5 or 1.6 V are to be measured, the voltage across AB will be set at 2 V and unknown voltages read from the scale by taking the scale reading at contact C and doubling it.

Commercial potentiometers are available with slide wires calibrated directly in volts and with various auxiliary resistors designed to give the unit several voltage ranges included in a single instrument assembly. A simple example of such an instrument is shown in Fig. 30.4.

APPARATUS

1. Slide-wire potentiometer
2. Galvanometer
3. 4700-Ω, 1-W carbon resistor
4. 100-Ω, 50-W rheostat
5. 15-V regulated dc power supply
6. Standard cell

7. Good dry cell, or dry cell that tests 1.5 V
8. Old dry cell, or dry cell that tests about 1.0 V
9. dc ammeter (0–1 A)
10. dc voltmeter (0–3 V)
11. Single-pole, single-throw switch
12. Single-pole, double-throw switch

PROCEDURE

1. Connect the apparatus as shown in Fig. 30.3 using the good dry cell for E_x. R_1 is the 4700-Ω resistor and R_2 the rheostat. Keep all switches open while connections are being made. Have your circuit approved by the instructor.

2. *Leave switch K_1 open.* Set R_2 for maximum resistance and then turn on the power supply. Set its output to

about 15 V and adjust the current as read on the ammeter to about 0.6 A, keeping it constant at this value for the next four steps by means of the control rheostat. Read the exact value chosen.

3. Close switch K_2 in the direction to put the standard cell E_s in the circuit and find the setting of the contact point C for a condition of balance. For this initial adjustment, the galvanometer is connected in series with a high protective resistance. Contact C is pressed and the galvanometer deflection observed. This contact is then moved along the wire until there is no galvanometer deflection when the key is pressed. CAUTION: Do not slide contact C along the wire while pressing it into contact. This will cause the knife edge to scrape the wire and thereby damage it.

4. Final adjustments are made with the protective resistance R_1 short-circuited by switch K_1. When the galvanometer shows no deflection on pressing C, close K_1. Again locate C so that the galvanometer shows no deflection. Only a *very small* displacement of the contact point will be needed to do this. Closing K_1 increases the sensitivity of the galvanometer many times and allows point C to be located more accurately. Record the final setting of the contact point and the known value of the standard cell voltage.

5. *Open switch K_1.* Move K_2 to put the good dry cell E_x in the circuit instead of the standard cell. Obtain a balance as described in Procedures 3 and 4 to measure the emf of the good dry cell. Record the final setting of contact point C.

6. Again *open switch K_1.* Replace the good dry cell with the old dry cell and repeat Procedure 5 to measure the old dry cell's emf. Record the final contact point setting.

7. *Open switch K_1* and return switch K_2 to its original position putting the standard cell in the circuit. Set contact point C to a scale reading exactly equal to one half the known emf of the standard cell and adjust the rheostat for zero galvanometer deflection. Close K_1 for final adjustments of the rheostat but *be very careful* not to drive the galvanometer off scale when the protective resistance is short-circuited. Record the ammeter reading and make sure it does not change for the remainder of the experiment.

8. *Open switch K_1.* Move K_2 to put the old dry cell back in the circuit. Obtain a balance as described in Procedures 3 and 4. Read contact point C's position on the scale and record it.

9. Again *open switch K_1.* Replace the old dry cell with the good dry cell and repeat Procedure 8. Record the contact point setting so obtained.

10. Measure the terminal voltage of the good dry cell and of the old dry cell with the voltmeter and record the values. Do not measure the voltage of the standard cell with the voltmeter because the current taken by the meter will cause the cell to change so that it will no longer be standard.

DATA

Slide-wire current, Procedure 2 _____   Slide-wire current, Procedure 7 _____

Cell used	emf from Equation 30.3		emf direct from scale		Voltmeter reading
	AC, cm	emf, V	AC, cm	emf, V	
Standard cell					
Old dry cell					
Good dry cell					

CALCULATIONS

1. Compute the emf of the good dry cell from the data of Procedures 4 and 5.

2. Compute the emf of the old dry cell from the data of Procedure 6.

3. Obtain the emf's of the good and the old dry cells again by multiplying the distances *AC* measured in Procedures 8 and 9 by 2.

QUESTIONS _____

1. Compare the emf of the good dry cell with that of the old dry cell. Compare their voltages as measured with the voltmeter. Explain these results.

2. If you use a 1-m wire of uniform cross section having a resistance of 5 Ω with a 6-V storage battery whose internal resistance is 0.2 Ω, what resistance must be placed in series with the wire in order that the potential drop per millimeter will be exactly 1 mV?

3. The power supply voltage was set only approximately and was not measured or used in the calculations. Why is the value of this voltage not needed?

4. A cell whose emf is 1.52 V is connected to a voltmeter that records 1.48 V. If the resistance of the voltmeter is 200 Ω, what is the internal resistance of the cell?

5. Find the resistance of your slide wire from data taken in this experiment.

6. Eight small flashlight cells in series will give 12 V, just as the storage battery in your car does. Could you start your car with the flashlight cells? Explain.

7. Suggest modifications to the circuit of Fig. 30.3 that would increase the voltage range of the potentiometer.

8. Find, by any method you can think of, what value R_L should have in Fig. 30.1 so that the power delivered to it by the actual (non-ideal) voltage source will be maximized.

9. In Fig. 30.2 (b), *R* represents a potentiometer, and in Fig. 30.3, R_2 is called a rheostat. From looking at these two figures, determine and state the difference between a potentiometer and a rheostat.

The *RC* Circuit 31

When a resistor is connected in parallel with a charged capacitor, a current flows through the resistor, draining charge out of the capacitor. The capacitor is thus discharged over a period of time. The relation between the voltage remaining across the capacitor and the time is a very useful one in electronics and electrical engineering,

and the purpose of this experiment is to study it. In doing so, we will find that this relation depends on the resistance R of the resistor and the capacitance C of the capacitor only through their product RC and that we thus can determine C if R is known.

THEORY

Fig. 31.1 shows a capacitor C and a resistor R connected in parallel. Let the capacitor be initially charged to a voltage V_0 by some external voltage source that does not appear in this figure and has been removed at time $t = 0$. The capacitor then discharges through resistor R, the voltage V across the capacitor at any particular time causing a current $i = V/R$ to flow at that instant according to Ohm's law. Current i is the rate in amperes at which charge is leaving the capacitor (recall that an ampere is a coulomb per second), so that voltage V decreases steadily. Note, however, that the smaller V gets, the smaller i is and hence the slower the rate of discharge. Situations like this in which the rate at which something changes is proportional to its value at every instant are common in physics and are always described by the exponential function, which is characterized by the property of having its rate of change proportional to its value. The exponential function is simply the base of natural logarithms e raised to a power proportional to the time. For example, in the absence of inhibiting effects, the rate of growth of the population of any given species is proportional to the number already present. The population as a function of time is then described by an exponential in which the exponent is t times a constant expressing the probability that any particular specimen will reproduce. In the present case, the voltage across the capacitor *de-*

creases as time goes on, which means that its rate of change is negative. Consequently, the exponent in the exponential must be negative. It can be shown that the voltage across the capacitor is given as a function of time by

$$V = V_0 e^{-t/RC} \qquad (31.1)$$

where V is the voltage in volts at time t, V_0 is the initial voltage (at $t = 0$), R is the resistance of the resistor in ohms, C is the capacitance of the capacitor in farads, and t is the time in seconds measured from the instant at which V was equal to V_0. It is interesting to note that R and C occur only as the product RC. Moreover, the exponent as a whole cannot have a dimension, and so RC must have the units of time. Indeed, multiplication of ohms by farads gives seconds. The product RC is therefore called the circuit's *time constant*. Note that when $t = RC$, $V = V_0/e = 0.368V_0$ or somewhat less than half of the original voltage. Equation 31.1 shows that theoretically V never gets to zero (the discharge rate getting slower and slower the smaller V becomes), but in a practical case V becomes negligibly small after a period equal to a few time constants has elapsed.

The validity of Equation 31.1 will be investigated in this experiment by plotting the discharge of a capacitor by a parallel resistor as a function of time. Since practical capacitors have capacitances measured in microfarads, the discharging resistor must have a value of several megohms (millions of ohms) if a time constant of a few seconds is to be realized. Although electronic equipment often uses RC circuits with time constants in the microsecond range, we are not now prepared to measure such short intervals and must therefore have time constants sufficiently large to time the voltage decrease with a stopwatch. To measure V, however, a voltmeter must be connected across the capacitor, and an ordinary voltmeter will itself have a resistance less than the several megohms desired. To overcome this difficulty, an electronic voltmeter or EVM (sometimes called a VTVM for vacuum-tube voltmeter) will be used. An electronic voltmeter contains an amplifier that drives

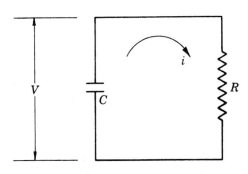

Figure 31.1 *The RC circuit*

the actual meter with power obtained either from batteries or from the power line. Ideally this means that the meter actuating current does not have to come from the circuit under test so that the EVM input terminals can have an infinite resistance between them. Practical EVMs usually have an input resistance of about 10 MΩ, but this is already so large that most circuits are not affected by it. Thus, the EVM, like the potentiometer, has the advantage of drawing negligible current from the circuit under test, but it is much more convenient for general voltage-measuring purposes.

In this experiment, an EVM will be used both to measure the voltage across a capacitor and to serve as the parallel resistor. A plot of *V* against *t* will be made on semilogarithmic paper and should turn out to be a straight line if Equation 31.1 is correct. The time constant *RC* can be obtained from the slope of this line, as discussed in the section of the Introduction on plotting exponentials. The values of *R* and *C* cannot, however, be found individually. Another plot will therefore be made using a different value of *R* obtained by connecting a large known resistance in parallel with the EVM, and a new value of *RC* will be calculated from its slope. Since the two values of *R* are related by the formula for resistances in parallel, both the value of *C* and the EVM input resistance can be found from the data. The nominal value of *C* is marked on the capacitor and serves as a check on your results.

APPARATUS

1. Power supply capable of delivering 300 V dc
2. Electronic voltmeter with 0–300 V or higher scale
3. 4-μF capacitor
4. 10-MΩ precision resistor (1% tolerance)
5. Single-pole, single-throw switch
6. Stopwatch or stop clock

PROCEDURE

1. Wire the power supply, capacitor, electronic voltmeter, and switch in the circuit shown in Fig. 31.2. Omit the 10-MΩ resistor R_s in this step. Be sure the EVM is set to an appropriate range (0–300 V or higher). Leave switch K_1 open and do not plug in or turn on the power supply until your circuit has been approved by the instructor. CAUTION: The output of the power supply is high enough to give a painful shock. Be careful around the apparatus and do not make changes in your circuit

with the power on. Notice that opening switch K_1 is not enough. The power supply itself must be turned off with its own switch or unplugged from the power line before any wiring changes are attempted.

2. After the instructor has checked your circuit, turn on the power supply, close switch K_1, and adjust the power supply output to 300 V as read on the EVM.

3. Open K_1 and at the same instant start the stopwatch. Stop the stopwatch at the instant the EVM passes through 250 V. Record the time thus measured.

4. Close switch K_1 and readjust the power supply if necessary to bring the voltage across the capacitor to 300 V. Then again open K_1 and start the stopwatch simultaneously. Again stop the stopwatch when the EVM passes through 250 V and record the indicated time. Repeat this procedure a third time to give you three independent measurements of the time required for the capacitor voltage to fall from 300 to 250 V.

5. Repeat Procedures 3 and 4, stopping the stopwatch at 200, 150, 100, and 50 V.

6. Connect the 10-MΩ resistor at R_s in Fig. 31.2. Repeat Procedures 2–5 with this setup.

7. Be sure to turn the power supply off and unplug it when you have finished the experiment.

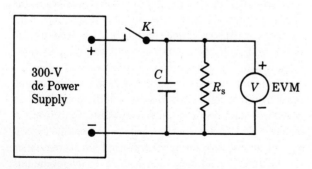

Figure 31.2 *Circuit for measuring time constants*

DATA _____

Voltage across the capacitor	Ratio V/V_0	Time, EVM alone				Time, EVM in parallel with R_s			
		Trial 1	Trial 2	Trial 3	Average	Trial 1	Trial 2	Trial 3	Average
300	1 .00	0.00	0.00	0.00	0.00	0.00	0.00	0.00	0.00
250									
200									
150									
100									
50									

Value of R_mC from graph _____   Calculated value of capacitor C _____

Value of R_1C from graph _____   Nominal value of C _____

Value of meter resistance R_m _____

CALCULATIONS _____

1. Divide each of your values of the capacitor voltage V by the initial voltage V_0 to obtain the ratio V/V_0 in each case. Note that this will allow you to plot Equation 31.1 in the form

$$\frac{V}{V_0} = e^{-t/RC}$$

2. Calculate the average of each set of three time measurements made in Procedures 3–5 and enter your results in the appropriate boxes in the data sheet.

3. On semilogarithmic paper, plot the average times obtained in Calculation 2 against the corresponding values of the voltage ratio V/V_0. Plot this ratio on the logarithmic scale and the time on the linear one, choosing your scale so as to spread your graph over as much of the sheet as possible. Read the instructions for plotting exponential functions in the Introduction. Draw the best straight line through your plotted points.

4. From the slope of your graph, calculate R_mC where R_m is the resistance of the EVM.

5. Calculate the average of each set of three time measurements made in Procedure 6 and enter your results in the appropriate boxes in the data sheet.

6. On the second sheet of semilogarithmic graph paper, plot the average times obtained in Calculation 5 against the corresponding values of V/V_0. Use the logarithmic scale for the voltage ratio and the linear scale for time, spreading your graph out as much as you can, as before. Draw the best straight line through this set of plotted points.

7. From the slope of the graph drawn in Calculation 6, compute R_1C, where R_1 is the resistance of R_m and R_s in parallel.

8. You now have values for R_mC and R_1C, and you know that $(1/R_1) = (1/R_m) + (1/R_s)$ where the value of R_s is given. On this basis, calculate the capacitance C and the meter resistance R_m.

9. Compare your calculated value of C with the nominal value marked on the capacitor. *Note:* Commercial capacitors have a tolerance of $\pm 10\%$ or more from the nominal value. You should therefore not expect remarkably good agreement between this and your measured capacitance. The instructor may, however, give you a more precise value of C with which to compare your results.

QUESTIONS _____

1. Explain how your graphs check the validity of Equation 31.1.

2. Why was it better to plot the ratio V/V_0 rather than simply the voltage V against t in Calculations 3 and 6?

3. Show that the product RC has the dimensions of seconds when R is in ohms and C is in farads.

4. Let $R_mC = A$ and $R_1C = B$, where A and B are the measured values of these time constants obtained from the slopes of your graphs in Calculations 4 and 7. Recalling that R_1 is the resistance of R_m and R_s in parallel, derive expressions for R_m and C in terms of the known quantities A, B, and R_s.

5. When the capacitor in this experiment is initially charged to voltage V_0, the energy stored is $\frac{1}{2} CV_0^2$. What happens to this energy as the capacitor discharges? Can you show quantitatively that your answer is correct?

6. A charged 2-μF capacitor is connected in parallel with a 500,000-Ω resistor. How long after the connection is made will the capacitor voltage fall to (a) 50% of its initial value; (b) 30% of its initial value; (c) 10% of its initial value; (d) 5% of its initial value?

7. (a) Show by any method you know that Equation 31.1 may be approximated by

$$\frac{V}{V_0} \approx 1 - \frac{t}{RC}$$

if $t \ll RC$. (b) The above equation is linear, and a plot of V/V_0 against t on ordinary graph paper would therefore be a straight line as long as this equation was valid. To what fraction of V_0 can V fall before the error made in using this approximation reaches 10%?

1.0

0.9

0.8

0.7

0.6

0.5

0.4

0.3

0.2

0.1

Electromagnetic Induction: The Transformer

32

The production of induced currents is one of the most important branches of the study of electricity. First discovered by Michael Faraday (1791–1867) and then exploited by Nicola Tesla (1856–1943), it is the basis of operation of the transformer, whose ability to shift voltage from one value to another with high efficiency and without moving parts sparked the changeover of the entire electrical industry from direct current (dc) to alternating current (ac). This simple means of stepping voltage up and down has, on the one hand, made long-distance transmission of electric power possible, and on the other, allowed any household appliance, whatever it may need for its circuits, to be plugged into the common 120-volt power line. It is no surprise that so many appliances are marked "ac only," for the likelihood is that the first thing the power cord goes to is a transformer that changes the line voltage to the value the appliance actually requires.

The present experiment deals with some aspects of electromagnetic induction. Much of the work is qualitative, so that a complete record of the observations must be kept. The object of the experiment is to study Faraday's law of electromagnetic induction, Lenz's law, and the general features of a magnetic circuit. In particular, simple examples of a transformer will be studied. In each case, the voltage induced in a coil of wire by a varying magnetic flux will be investigated, and we shall see why a transformer operates on ac only.

THEORY

A voltage is induced in a loop of wire whenever there is a *change* in the magnetic flux passing through the loop. Magnetic flux is the product of the magnetic field intensity B and the area bounded by the loop through which this field passes. If the direction of B is not perpendicular to the plane of the loop, then the perpendicular component must be taken. Mathematically the flux is given by

$$\Phi = B_\perp A \qquad (32.1)$$

where Φ is the flux in webers, $B_\perp$ is the perpendicular component of the magnetic field in teslas, and A is the area bounded by the loop. Faraday's law of electromagnetic induction then states that the voltage induced in a loop by a changing magnetic flux is given by

$$V_\ell = -\frac{\Delta\Phi}{\Delta t} \qquad (32.2)$$

where V_ℓ is the average value of the induced voltage in volts; $\Delta\Phi$ is the change in flux in webers; and Δt is the time interval in seconds during which the change in flux takes place. If instead of a one-turn loop we have a coil of N turns, voltage V_ℓ is induced in each turn, and since the coil is in effect N turns in series, the total voltage at the coil terminals is

$$V = NV_\ell = -N\frac{\Delta\Phi}{\Delta t} \qquad (32.3)$$

From the definition of flux, we note that a change in Φ may be produced in any of three different ways: the area of the loop may be changed, thus changing the factor A in Equation 32.1; the strength of the magnetic field B may be altered; or the coil may be rotated so that the field's perpendicular component changes. All of these methods of changing Φ are used in practical applications of Faraday's law.

The minus sign in Equation 32.2 gives the polarity of the coil voltage relative to the sign of the change in flux. It is best understood in terms of Lenz's law, which states that the polarity of V will be such as to oppose the change in Φ that produced it. Thus, suppose the terminals of a certain coil were connected to each other so that an induced voltage would cause a current to flow around the coil. Now suppose that a magnetic field directed through the coil from an external source were increased. Lenz's law requires that the current induced in the coil flow in such a direction as to result in a magnetic field opposing the applied one, that is, bucking the increase in B that induced it. Similarly, should the externally applied field decrease, the induced current will flow around the coil in the other direction so as to produce a field in the same direction as the applied field and thus oppose the decrease. Note that the relation between the magnetic field produced by a current flowing around a coil and the direction of this current is given by the right-hand rule. When the fingers of the right hand curl around the coil in the direction of the current flow, the extended thumb points in the direction of the resulting magnetic field.

An important application of electromagnetic induction is the transformer. A transformer consists of two interwound coils called the primary and secondary. These coils are often wound on a core of magnetic material, and this core is extended to form one or more complete loops of the material passing through and around the coils. Such a loop forms a magnetic circuit that concentrates the magnetic field due to current in either coil in the loop of material passing through both coils.

If a current is now sent through the primary coil, a magnetic field passing through the plane of both coils and hence a magnetic flux through these coils will be produced. Any change in the current will result in a change in this flux. The primary will therefore have induced between its terminals a voltage

$$V_1 = -N_1 \frac{\Delta \Phi}{\Delta t} \qquad (32.4)$$

and a voltage

$$V_2 = -N_2 \frac{\Delta \Phi}{\Delta t} \qquad (32.5)$$

will be induced between the terminals of the secondary, where V_1 and V_2 are in volts, N_1 is the number of turns in the primary, and N_2 the number of turns in the secondary. Note that even though the primary is wound of heavy copper wire so that it has negligible resistance, a voltage nevertheless develops across it when a current is sent through it, *but only when the current is changing*. Hence, with direct current, once the current is on, $\Delta \Phi$ and therefore both V_1 and V_2 will be zero. This explains why transformers work only on alternating current, which is by definition a current that is continually changing. The voltage V_1 appearing across the primary as a result of the changing primary current is sometimes called a *back emf* because of the opposing polarity given it by Lenz's law as expressed by the minus sign in Equation 32.4.

The most important feature of the transformer is revealed when we divide Equation 32.4 by Equation 32.5 to obtain

$$\frac{V_1}{V_2} = \frac{N_1}{N_2} \qquad (32.6)$$

This result says that the ratio of the primary to the secondary voltage is equal to the turns ratio. Thus, in an ac circuit a transformer may be used to change the voltage to any desired value by simply winding the coils with the appropriate numbers of turns. This simple arrangement is used throughout the electrical industry, from household appliances to huge transmission-line substations. Much of the electrical equipment in common use today would not work without it, and the fact that the transformer works on ac only is a principal reason for the widespread use of ac rather than dc power.

To demonstrate the operation of the transformer, we shall use two different setups. The first is simply a pair of coils, one of which can be slid inside the other to form a primary, with the outer coil being the secondary. We shall see the effect of using a core of magnetic material as noted above, and in particular the effect of providing a closed magnetic circuit by extending the core to form a complete loop of iron. The second setup consists of a demountable model transformer—a close approximation to actual, commercially available transformers—powered by ac and connected to meters and a load resistance R_L as shown in Fig. 32.2. This setup will not only check the validity of Equation 32.6 but will also illustrate a further characteristic of transformers, that of *impedance matching*.

Impedance is a term we have not yet encountered but that will be treated in detail in Experiment 34, where we will see that in ac circuits it plays a role similar to resistance in the dc case. For our present purposes, we need only note that when alternating current is applied to a coil, the coil's impedance takes care of the fact that a voltage appears across it due not only to its (small) resistance but also to electromagnetic induction producing the back emf already mentioned. The impedance is thus the ratio of the (ac) voltage across the coil to the (ac) current through it. Similarly, with the load resistance R_L in place across our transformer's secondary, the voltage V_2 induced in the secondary will cause current I_2 to flow around the circuit thus formed. Note carefully, however, that a transformer is not an energy source, nor does it (at least in the ideal case, which commercial transformers approach quite closely) dissipate any energy. Hence the power fed into the primary, which, in the ideal case, we can write as $I_1 V_1$, where I_1 and V_1 are the primary current and voltage, respectively, must (again in the ideal case) be equal to $I_2 V_2$, the product of the secondary current and voltage. In other words, to maintain the power constant, if the voltage is stepped up (or down) by some factor, the current must be stepped down (or up) by the same factor. According to Equation 32.6, however, $V_2 = (N_2/N_1)V_1$, N_2/N_1 being the stepping factor. We therefore conclude that, since $I_1 V_1 = I_2 V_2 = I_2(N_2/N_1)V_1$, then

$$I_2 = \frac{N_1}{N_2} I_1 \qquad (32.7)$$

and, using Z to represent the impedance as already defined, we have $Z_1 = V_1/I_1$ and

$$Z_2 \equiv \frac{V_2}{I_2} = \frac{\frac{N_2}{N_1}V_1}{\frac{N_1}{N_2}I_1} = \left(\frac{N_2}{N_1}\right)^2 \frac{V_1}{I_1} = \left(\frac{N_2}{N_1}\right)^2 Z_1 \quad (32.8)$$

Equation 32.8 states that the *impedance ratio* Z_2/Z_1 *is equal to the square of the turns ratio*. This means that if we have a source of ac power that wants to work into a load Z_1 but in fact we want to drive a load Z_2, we can use a transformer whose turns ratio is chosen so that its square is equal to the ratio of the two impedances involved. An example of this is a typical high-fidelity audio amplifier whose output terminals are marked "8 ohms," so that you know you must use a loudspeaker that has an 8-ohm voice coil. The chances are that the audio output circuit in this amplifier would like to work into an impedance considerably higher than 8 ohms and that an *output transformer* is used to *match* the impedance that is proper for that circuit to the 8 ohms of the loudspeaker. Note that the audio signal voltage coming from the amplifier is a perfectly good example of ac, although the frequency is not limited to 60 hertz as is that of the power line. Indeed, 60 hertz is a perfectly good audio frequency and, if made audible by a loudspeaker connected (through a suitable step-down transformer) to the power line, would be heard as a low tone often referred to as a hum.

Equation 32.8 also implies that if we open-circuit the secondary by disconnecting R_L, Z_2 becomes infinite (because we've made I_2 zero), and so Z_1 should also be infinite, requiring I_1 to be zero. In other words, open-circuiting the secondary circuit should result in the primary circuit being open also, since there is no current flowing in the primary coil. This is obviously impossible in a practical case, for although we could conceive of this coil developing a back emf that was just equal and opposite to the applied voltage, thus preventing current flow, there has to be some flow in order for the back emf to be developed. Engineers postulate a so-called *ideal transformer* in which this current is negligibly small, and practical transformers approach the ideal rather closely as long as reasonable loads remain connected. The model transformer used in this experiment is not particularly ideal, but it may at least serve to approximate the performance we would expect from an ideal one.

APPARATUS

1. Primary coil of about 100 turns of copper wire
2. Secondary coil of about 1000 turns of copper wire
3. U-shaped piece of iron
4. Straight piece of iron
5. Portable galvanometer
6. ac voltmeter (0–15 V)
7. ac ammeter (0–0.5 A)
8. dc ammeter (0–1 A)
9. Multimeter with ac voltage ranges
10. Transformer delivering 6.3 V at 3 A
11. Model transformer with 200-, 400-, and 800-turn coils
12. Power supply capable of delivering at least 5 A at 10 V dc
13. Tubular rheostat (about 100 Ω)
14. Decade resistance box (1000 Ω)
15. Double-pole, double-throw reversing switch
16. Permanent bar magnet

PROCEDURE

1. Connect the secondary coil to the portable galvanometer. Thrust the north pole of the bar magnet into the coil. Record the magnitude and direction of the deflection of the pointer for a fast speed of motion and for a slow speed. Observe whether the deflection is to the left or to the right; a deflection of the pointer to the right means that the current enters the galvanometer at its + terminal. Determine the direction of the current induced in the coil by noting the direction of the current through the galvanometer and the direction in which the coil is wound.

2. Repeat Procedure 1, withdrawing the north pole from the coil after having inserted it as in that procedure. Observe the direction of the current induced in the coil when the north pole is being withdrawn.

3. Repeat Procedures 1 and 2 using the south pole of the magnet.

4. Connect the primary coil in series with the rheostat, the ammeter, the double-pole, double-throw reversing switch, and the power supply, as shown in Fig. 32.1. Have the circuit approved by your instructor. Mount the two coils in a complete square of iron made up of the U-piece and the straight piece. Using a current of 0.2 A through the primary coil, observe and record the magnitude and direction of the deflection when the primary circuit is opened and when it is closed. Repeat the observations using primary currents of 0.4, 0.6, 0.8, and 1.0 A.

5. Using a current of 1 A, observe the effect of reversing the current in the primary coil. Record the

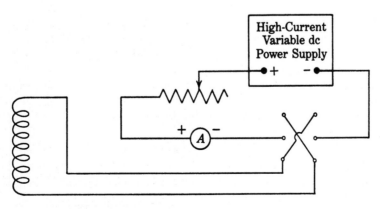

Figure 32.1 *Primary circuit used in studying electromagnetic induction*

direction of the induced current when the primary circuit is opened and when it is closed for each direction of the primary current.

6. Keep the current through the primary coil at some constant value, say 0.5 A. Observe the galvanometer deflection upon opening and closing the primary circuit, using the complete square of iron.

7. Repeat Procedure 6 using the straight piece only.

8. Repeat Procedure 6 using no iron at all.

9. Repeat Procedure 8 placing the two coils twice as far apart as in 8.

10. Test your two coils as a transformer by connecting the primary coil to the 6.3-V ac output of the commercial transformer and the secondary coil to the multimeter. CAUTION: Multimeters, or volt-ohm-milliammeters as they are often called, are easily destroyed by being connected in circuits while switched to an improper range. If you have not done Experiment 26, read the discussion of multimeters in the Procedure section of that experiment. Always check the position of the range switch on any multimeter you are using before connecting the meter leads. In the present case, be sure your multimeter is switched to its highest ac voltage range before connecting it in the circuit, and leave the line cord from the 6.3-V transformer unplugged until your setup has been checked by the instructor.

11. Begin by mounting the two coils in the complete square of iron made up of the U-piece and the straight piece. Plug in the 6.3-V transformer and adjust the multimeter range switch to the lowest ac voltage range for which the pointer remains on scale. Record the reading. Now use the voltmeter to measure the primary voltage. This should be close to 6.3 V, but the commercial transformer rating is nominal and obviously depends on the line voltage that happens to be present in the laboratory at the time you do the experiment. The primary voltage must therefore be checked so that you'll have the actual value for your data-sheet entry.

12. Repeat Procedure 11 using the straight piece only.

13. Repeat Procedure 11 using no iron at all.

14. Study the demountable model transformer, noting that it consists of a rectangle of iron that forms a complete magnetic circuit over the sides of which various primary and secondary coils can be placed. One side of the rectangle is removable to allow the coils to be slipped onto the two opposing sides but is retained tightly in place by a rod with a threaded end as shown in Fig. 32.2. Care should be taken when assembling the transformer to be sure that the removable side is aligned exactly on the ends of the two sides it rests on, that the mating surfaces are clean, and that the retaining rod is tightened down securely by means of its knurled knob. Begin by using the two 400-turn coils (giving a turns ratio of 1:1), and wire them in the circuit of Fig. 32.2. Use the ac ammeter at A, the ac voltmeter at V_1, the multimeter switched to its 10-volt ac range at V_2, and the decade resistance box at R_L.

15. Set the decade box to 200 ohms and have your instructor check your setup before plugging the commercial transformer into the power line. Record the ammeter and the two voltmeter readings (I_1, V_1, and V_2).

16. Unplug the commercial transformer from the power line, disassemble the model transformer, and replace the 400-turn secondary coil with the 800-turn one, giving you a turns ratio of 1:2. Carefully reassemble the transformer and reconnect the power line after checking your connections. Vary R_L by readjusting the decade box until the ammeter reading is the same as it was in Procedure 15. CAUTION: Be careful not to short-circuit the secondary coil by letting all the decade box dials go to zero at once. Change their settings carefully, making sure that at least one of them is away from its zero position. With your 800-turn secondary, R_L will be *larger* than the 200-ohm setting used in Procedure 15. Record the value thus found for R_L and the accompanying value of V_2. The data table also provides spaces for V_1 and for the ammeter reading I_1, but note that these should not have changed from the values previously recorded.

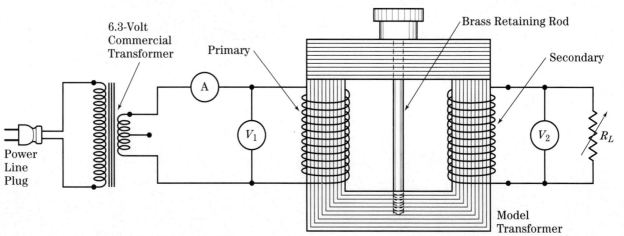

Figure 32.2 *Setup for testing the model transformer*

17. Repeat Procedure 16 with the 200-turn coil substituted as the secondary (turns ratio of 2:1). You will now be *reducing* R_L to make the primary current I_1 the same as that previously recorded, so be careful not to re- duce the decade box setting too far. Again record the value of R_L thus found along with the three meter readings. Don't forget to unplug the commercial transformer when you have finished the experiment.

DATA

Procedures 1–3

	Direction of current	Galvanometer deflection	
		Slow motion	Fast motion
North pole into the coil			
North pole out			
South pole into the coil			
South pole out			

Procedures 4–9

Primary current	Direction of secondary currents		Galvanometer deflection		Magnetic circuit
	Opening	Closing	Opening	Closing	
0.2 A					Complete square of iron
0.4 A					Complete square of iron
0.6 A					Complete square of iron
0.8 A					Complete square of iron
1.0 A					Complete square of iron
1.0 A					Complete square of iron
0.5 A					Complete square of iron
0.5 A					Straight piece only
0.5 A					No iron
0.5 A					No iron, double distance between coils

Procedures 10–13

Core material	ac voltage		Ratio primary/secondary
	Primary coil	Secondary coil	
Complete square of iron			
Straight piece only			
No iron			

Number of turns on primary coil _____   Ratio primary turns/secondary turns _____

Number of turns on secondary coil _____

Procedures 14–17

No. of secondary turns N_2	Ratio N_2/N_1	Ratio $(N_2/N_1)^2$	Primary voltage V_1	Primary current I_1	Load R_L	Secondary voltage V_2	Ratio (V_2/V_1)
200							
400							
800							

CALCULATIONS

1. Plot the data of Procedure 4 using primary currents as abscissas and galvanometer deflections as ordinates. Note that the *magnitude* of these deflections for a given primary current should be the same when the switch is opened as when it is closed.

2. Calculate the ratios of the primary and corresponding secondary voltages measured in Procedures 11–13. Compare these voltage ratios with the turns ratio of your two coils, which you should calculate if the number of turns on each coil is known.

3. Calculate the turns ratios and the squares of these ratios for each secondary used in the model transformer in Procedures 14–17. Remember that you had a 400-turn primary in each case.

4. Repeat Calculation 2 for the data of Procedures 14–17 and compare these voltage ratios with the respective turns ratios found in Calculation 3.

5. Plot the squares of the turns ratios found in Calculation 3 against the respective values of the load resistance R_L. Although this gives you only three points, they should, according to theory, lie on a straight line. Draw the best one you can through your plotted points.

QUESTIONS _____

1. How does the experiment verify Lenz's law?

2. Show how the laws of electromagnetic induction were verified.

3. Discuss the essentials of a good magnetic circuit from the observations of Procedures 6–9 and 11–13.

4. Name three practical applications of electromagnetic induction.

5. By using Lenz's law, how can you predict the direction of flow of current in a secondary coil when the current in the primary is turned on? Assume a direction for the primary current and consider the two coils to be coaxial.

6. (a) Can an induced voltage be present in a circuit without an induced current flowing? (b) Can an induced current flow without an induced voltage?

7. A circular coil of 200 turns of fine wire has a diameter of 4 cm. It is held in a magnetic field of 0.5 T with its axis aligned with the field direction (its plane perpendicular to this direction). If it is then pulled out of the field in 0.1 s, what is the induced voltage?

8. Explain what happens when a transformer is plugged into a dc line by mistake.

9. You probably found the voltage ratios obtained for the model transformer in Calculation 4 to be a bit smaller than the corresponding turns ratios. Explain why you might expect this.

10. The coil cores and cross pieces (the complete assembly is often referred to as the transformer's core) making up the magnetic circuit in the model transformer are not solid iron bars but rather stacks of appropriately shaped iron plates called *laminations*. Explain why the core is made this way.

11. The long rod holding the removable side of the model transformer's core in place must be made of a nonmagnetic material such as brass. Why?

12. (a) How does the plot you made in Calculation 5 check the theoretical conclusion that a transformer's secondary-to-primary impedance ratio is equal to the square of its turns ratio? (b) Should the origin be a point on this plot?

13. If you took the primary impedance of your model transformer to be V_1/I_1 and its secondary impedance to be R_L (which is necessarily equal to V_2/I_2), would you find this ratio to be $(N_2/N_1)^2$? If not, why not?

14. As noted in the Theory section, if the secondary of an ideal transformer is open-circuited—that is, not connected to anything—the primary presents an open circuit to any voltage source connected to it even though it is wound with very low resistance copper wire. It does this by producing a back emf that just opposes the applied voltage, making the net voltage and hence the current around the primary circuit zero. As was mentioned, this condition cannot be realized in a practical transformer, but it can be approached by proper design. What do you think you could do in designing an actual transformer to make it approach the ideal as closely as possible?

The Cathode Ray Oscilloscope 33

The cathode ray oscilloscope is one of the most important instruments used in electrical and electronic research, development, and testing, and is therefore encountered everywhere in laboratories and shops dealing with this kind of work. Basically, it is a device for automatically and rapidly plotting one voltage, the so-called *horizontal voltage,* as the abscissa against another, the *vertical voltage,* as the ordinate. If the vertical voltage has a functional dependence on the horizontal voltage, that function is displayed electronically as a graph on the instrument screen. In addition, most oscilloscopes contain circuits that produce a voltage that increases linearly with time. When applied to the horizontal input, this voltage converts the screen's *x* axis to a time axis in the resulting graph, and the time dependence of any applied vertical voltage is plotted. It is the purpose of this experiment to become familiar with a typical cathode ray oscilloscope by using it to observe some characteristics of varying voltages.

THEORY

As noted above, the cathode ray oscilloscope is essentially a device for electronically plotting a graph of one variable against another, the two having a functional relationship and being represented by two voltages that can be applied to the oscilloscope's horizontal and vertical inputs. The complete instrument is built around a cathode ray tube—a long, evacuated glass bulb with an electron-emitting cathode in the small end followed by a series of electrodes used for accelerating the electrons and focusing them into a beam directed at the fluorescent screen that makes up the big end (see Fig. 33.1). This screen gives off light when struck by electrons, so that a bright spot appears at the point where the beam strikes. This spot may be moved around by deflecting the beam, and two sets of deflection plates mounted at right angles to each other are placed in the tube to do this. When a voltage is impressed across the pair of plates in one of these sets, the electrons in the beam are attracted to the positive plate and repelled by the negative one, and are accordingly deflected in their flight to the screen. A vertical displacement of the spot of light proportional to the voltage across the vertical deflection plates can thus be obtained, and similarly a voltage on the horizontal plates produces a proportional horizontal deflection. Thus, if a varying voltage v_x is connected to the horizontal plates and a voltage v_y, which depends on v_x according to some function, is put on the vertical plates, the spot will trace out a graph of that function on the screen.

The oscilloscope instrument contains the power supplies necessary for the operation of the cathode ray tube and vertical and horizontal amplifiers whose outputs are connected to the vertical and horizontal deflection plates, respectively, and that serve to amplify small signal voltages up to the reasonably high levels required for appreciable deflection of the beam. Gain controls, some calibrated by the manufacturer and some requiring calibration by the user, are provided so that the input voltage required for a given deflection may be set. The calibration is conveniently given in centimeters per volt, and many oscilloscopes have a grid of l-cm squares, often called a *graticule*, superposed on the screen, so that the complete display looks like an actual graph. When a voltage that is a function of time is to be studied, the horizontal axis is made a time axis by applying a voltage proportional to *t* to the horizontal input. Such a voltage is graphed in Fig. 33.2. It rises linearly with time until it reaches a value corresponding to full beam deflection and then drops quickly back to zero and starts over. For obvious reasons, it is often called a "sawtooth" voltage, and special circuits are provided in the oscilloscope to generate it. When applied to the horizontal deflection plates, it causes the spot to move across the screen at a constant rate and then snaps it back and "sweeps" it across again. During a sweep, the horizontal or *x* deflection is proportional to time, so that this axis may be calibrated in seconds. If the voltage to be studied is applied

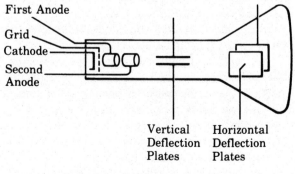

Figure 33.1 *The cathode ray tube*

267

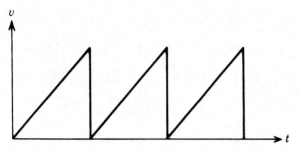

Figure 33.2 *A sawtooth voltage*

to the vertical plates while the spot is moving linearly across the screen, a plot of this voltage as a function of time will be traced out. Furthermore, if this voltage is periodic and the sawtooth is made to have the same period and to start at the same instant as a cycle does, a plot of one complete cycle will be traced over and over again. If the frequency is high enough so that one trace does not have a chance to fade from the screen before the next one starts (period short compared to the screen's *persistence time*), the plot of one cycle will appear to remain stationary and without flickering. To keep the sweep in step with the voltage under study, the oscilloscope includes *synchronizing* circuits, which deliver pulses to the sweep oscillator telling it when to start each sweep. The complexity of these circuits and the elaborateness of the associated controls vary greatly among different oscilloscopes; the details of the instruments provided in your laboratory will be explained by the instructor. In general, however, all oscilloscopes have the following basic controls:

Intensity (or Brightness): Controls the voltage on the cathode ray tube grid and hence the intensity of the electron beam and the brightness of the resulting spot or pattern on the screen.

Focus: Controls the voltage on the first anode and hence the potential difference between it and the second anode, which is at a fixed accelerating voltage. This potential difference has the effect of bringing the electron beam to a focus and may thus be adjusted to put the focal point at the screen.

Vertical Centering: Controls the dc component of the voltage between the vertical deflecting plates and therefore allows centering of the spot on the screen in the vertical direction. A complete pattern may also be centered to fill the screen evenly in this direction.

Horizontal Centering: Has the same function as vertical centering but in the horizontal direction.

Vertical Gain: Controls the gain of the vertical amplifier and hence the oscilloscope's vertical deflection sensitivity, that is, the number of centimeters of vertical deflection that will be produced by a given input voltage. On some instruments, this control is calibrated directly in centimeters per volt, but on the less expensive models you must make your own calibration.

Horizontal Gain: Same as vertical gain, but applies to the horizontal amplifier. Note that when a horizontal

time axis is desired, the output of the sweep generator is switched to the horizontal amplifier input, and the horizontal gain then controls the amplitude of the sawtooth applied to the horizontal deflection plates. In this case the gain control sets the width of the pattern.

Sweep Frequency: Controls the sawtooth repetition rate so that it may be set equal to the repetition rate of a periodic voltage applied to the vertical channel for study. The sawtooth repetition rate may also be set equal to a submultiple of that of the voltage under study, in which case several cycles will appear on the screen. In high-quality oscilloscopes, this control is called *sweep speed* and is calibrated directly in seconds for the spot to travel 1 cm horizontally. Such an instrument provides a horizontal axis calibrated in time from which the period and hence the frequency of the vertical signal voltage may be read directly. Less expensive instruments do not have this feature and thus do not provide for measurement of the period of the observed alternating current.

Synchronization Controls: These vary in complexity with different oscilloscopes, but in general serve to control the introduction of pulses into the sweep oscillator that tell it when to start each sawtooth. Simple instruments merely have a gain control that sets the level of these pulses. It is advanced from zero until synchronization is obtained but should not be advanced too far, as "oversynchronization" can produce distortion in the pattern by interfering with the linear nature of the sawtooth sweep. There is also a switch to select the source of the pulses. Since the voltage under study and the sweep ordinarily are to have the same period, a very usual source is the voltage under study itself, and the *internal* position of the selector switch makes the necessary connection inside the oscilloscope. For studying voltages derived from the ac power line, a *60-Hz* or *line* position allows pulses taken from the power line to be introduced into the sweep generator. Finally, an *external* position is provided for cases when synchronization is to be effected by pulses from a separate source. Input terminals are provided on the oscilloscope panel for bringing in such pulses. For further details on the instruments provided in your laboratory, consult your instructor.

Alternating Current We have already seen in Experiment 14 that simple harmonic motion is sinusoidal. Indeed, it is an interesting fact (and the subject of a topic in advanced mathematics called Fourier analysis) that when we speak of something oscillating with a certain frequency (as opposed to the more general term "repetition rate") like the pendulum or the mass on a spring of Experiment 14 we automatically imply that the motion is sinusoidal. The electrical case is similar to the mechanical one, and in fact an electric generator rotated at constant speed produces a sinusoidal voltage whose frequency is directly related to the angular velocity of the generator shaft; hence the recurrence of the symbol ω in what follows. Thus the term "alternating current" (or "ac") means not just any current that flows first in one direction and then in the other but more specifically a sinusoidally varying current such as would be obtained by applying a sinusoidal voltage to a resistor. Such a voltage (often called an ac voltage even though it's not a current) may be written as a function of time as

$$v = V_m \sin(\omega t + \phi_v) \qquad (33.1)$$

where v is the value of the voltage at time t, V_m is the maximum value or *amplitude* of this voltage, ω is the so-called *angular frequency* in radians per second, and ϕ_v is the *phase angle* in radians. Notice that the value of v at the instant $t = 0$ is $V_m \sin \phi_v$, so that the purpose of the phase angle is to let Equation 33.1 give the correct value of v at whatever instant the experimenter chooses to call $t = 0$. Fig. 33.3 shows v plotted as a function of t for the particular case of $v = 0$ when $t = 0$, so that $\phi_v = 0$ and $v = V_m \sin \omega t$. Notice that the voltage is positive until $t = \pi/\omega$, when it goes through zero and becomes negative until returning to zero at $t = 2\pi/\omega$. Thereafter, v repeats the performance it has just gone through in the interval 0 to $2\pi/\omega$, and this complete oscillation, which is repeated over and over, is called a *cycle*. The time $2\pi/\omega$ seconds required for one cycle is called the *period* of the ac voltage, and its reciprocal $\omega/2\pi$, which is the number of cycles completed in one second, is known as the *frequency f*. Equation 33.1 is therefore often written

$$v = V_m \sin(2\pi ft + \phi_v) \qquad (33.2)$$

where f is in cycles per second (cps) or, according to currently accepted terminology, hertz (Hz).

If the voltage v is impressed across a resistance R, Ohm's law holds at every instant so that a current

$$i = \frac{v}{R} = \frac{V_m}{R} \sin 2\pi ft = I_m \sin 2\pi ft \qquad (33.3)$$

flows through the resistance. Here we have arbitrarily chosen v to be zero at $t = 0$ so that the phase angle is zero and have written I_m for V_m/R, as this is clearly the amplitude of the resulting sinusoidal current. Notice that the power dissipated in the resistance is $iv = (V_m^2/R) \sin^2 2\pi ft$ and is a function of time. We therefore cannot simply multiply it by a time interval to get the energy delivered in that interval because, although power is the rate at which energy is delivered, this power is not constant. Hence an average value must be calculated for this purpose, and the average of $\sin^2 2\pi ft$ over one cycle (and thus over any integral number of cycles) is $\frac{1}{2}$. Consequently,

$$P_{av} = \frac{1}{2} \frac{V_m^2}{R} \qquad (33.4)$$

This average power can be multiplied by a time to give an energy delivered just as the power V^2/R can in the dc case. For this reason a quantity V_{rms} called the root mean square voltage (because it is the square root of the average of the square of v) is defined as $V_m/\sqrt{2}$. The root mean square (rms) voltage can then be used to calculate average power and energy delivered in the same way that a dc voltage can, Equation 33.4 taking the form $P_{av} = V_{rms}^2/R$. This concept is so useful that it is the rms value of ac voltages and currents that is usually quoted. Thus, when the ordinary household ac power line is said to be 120 V, it is the rms value, not the amplitude, that is meant.

Lissajous Figures A very important application of the oscilloscope is the study of the pattern resulting from the application of sinusoidal voltages of the same frequency to both the vertical and horizontal inputs. Such a pattern is called a *lissajous figure*. In general, a lissajous figure is an ellipse and is a stationary pattern on the oscilloscope screen if the two frequencies are indeed equal.

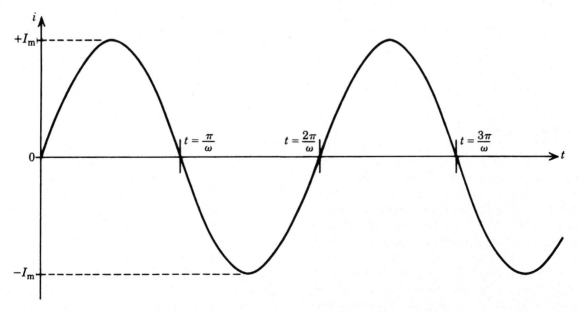

Figure 33.3 *A sinusoidal current as a function of time*

Thus, if one signal comes from a calibrated variable-frequency source and the other is unknown, tuning the calibrated source until a stationary ellipse is obtained allows the frequency of the unknown to be read from the calibrated source's dial. The nature of the ellipse can also reveal the phase difference and the ratio of the amplitudes of the two applied sinusoidal signals. Thus, if these signals are represented by

$$v_h = A \sin 2\pi ft \qquad (33.5)$$

and

$$v_v = B \sin (2\pi ft + \phi) \qquad (33.6)$$

where v_h is the voltage applied to the horizontal input, A is its amplitude, v_v is the vertical input voltage, B is its amplitude, and ϕ is the phase angle between v_h and v_v, then the lissajous figure will be an ellipse whose major and minor axes make an angle θ with the horizontal and vertical axes given by

$$\tan 2\theta = \frac{2 \cos \phi}{(A/B) - (B/A)} \qquad (33.7)$$

Two interesting special cases deserve particular attention. The first is the very simple one obtained when $A = B$ and $\phi = 0$. Then $v_h = v_v$ and the pattern on the oscilloscope screen is a straight line through the origin at an angle $\theta = 45°$. The second is that of $A = B$ and $\phi = 90°$. In this case, $v_v = A \cos 2\pi ft$, and the relation between the vertical and horizontal voltages is therefore

$$v_h{}^2 + v_v{}^2 = A^2 \qquad (33.8)$$

This is the equation of a circle of radius A. In the more general case of $A = B$ and an arbitrary phase angle, an ellipse is obtained from which the phase angle can be determined as follows (see Fig. 33.4): If the horizontal deflection is given by Equation 33.5, it will be zero at

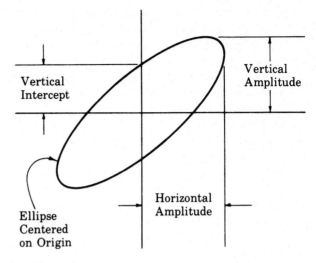

Figure 33.4 *Phase measurement with a lissajous figure*

$t = 0, 1/f, 2/f$, etc., and at these times the vertical deflection, which is given by Equation 33.6, is $B \sin \phi$. The vertical deflection at zero horizontal deflection is the vertical intercept marked in Fig. 33.4 and can thus be read from the oscilloscope pattern. If A is made equal to B, the vertical intercept is $A \sin \phi$ and thus gives $\sin \phi$ when divided by the vertical amplitude. We conclude that the phase angle can be obtained from the relation

$$\sin \phi = \frac{\text{vertical intercept}}{\text{vertical amplitude}} \qquad (33.9)$$

These characteristics of lissajous figures will be useful in our study of the performance of resistors, inductors, and capacitors in ac circuits (Experiment 34). Our aim in the present experiment is to become familiar with the oscilloscope and to see how it is used to observe the ac signals we shall encounter in our future work.

APPARATUS

1. Cathode ray oscilloscope
2. Transformer delivering 6.3 V at 3 A
3. Signal generator
4. 1.5-V dry cell
5. Multimeter with rms ac voltage ranges
6. Assorted hookup wires and connectors

PROCEDURE

1. Connect the output (secondary) of the transformer to the vertical input of the oscilloscope and set the horizontal circuits for a time sweep appropriate for viewing a 60-Hz voltage. Before plugging in the transformer, have the instructor check your arrangement and explain the operation of your particular oscilloscope.

2. Plug in the transformer and adjust the oscilloscope to produce a stationary pattern of one cycle of a sine curve. Learn to use the sweep frequency and synchronizing controls on your instrument by readjusting them to get stationary patterns of two and three cycles.

Note that since you are viewing a 60-Hz signal, you may use either "internal" or "line" synchronization.

3. Calibrate the vertical direction on the oscilloscope screen in centimeters per volt at the vertical input terminals. The method appropriate for your oscilloscope will be described by the instructor. Then measure the transformer secondary voltage (which is also the vertical input voltage) with the multimeter. CAUTION: Multimeters, or volt-ohm-milliammeters as they are often called, are easily destroyed by being connected in circuits while switched to an improper range. If you have not done Experiment 26, read the discussion of multimeters in the Procedure section of that experiment. Always check the position of the range switch on any multimeter you are using before connecting the meter leads. In the present case, be sure your multimeter is switched to the range appropriate for measuring the 6-V ac output of the transformer before connecting it. Record the meter reading. Using the vertical calibration just obtained, measure and record the voltage amplitude by converting the amplitude of the sinusoidal pattern displayed on the oscilloscope to volts.

4. If your oscilloscope has a calibrated time base (horizontal sweep giving you a known time per centimeter along the screen's *x* axis), use it to measure the period of one cycle of your transformer's secondary voltage. To do this with optimum precision, set the sweep speed to display two cycles of the ac voltage, adjust the horizontal centering control to bring one peak of the pattern exactly on a vertical graticule line, and measure the distance to the next peak. This will be in centimeters, which can then be converted to seconds using the chosen time base calibration (seconds per centimeter selected by the sweep speed switch on oscilloscopes so equipped).

5. If your oscilloscope has an "ac-dc" switch, notice that it does not set the instrument for operation from an ac or dc power line but controls the response of the vertical amplifier. In the dc position, any voltage, including a constant one, that is connected to the vertical input terminals will cause a vertical deflection of the trace on the screen. With this switch in the ac position, however, only ac signals will produce a vertical deflection, the oscilloscope's response cutting off when the frequency gets down to 3 or 4 Hz. Throw this switch back and forth while observing the two-cycle pattern displayed in Procedure 4 and note below whether you see any change.

6. Disconnect the transformer secondary from the vertical input, connect it to the synchronizing signal input, and set your oscilloscope for external synchronization. This will give you a horizontal line on the screen. Now connect the 1.5-V dry cell to the vertical input terminals. Throw the "ac-dc" switch back and forth as in Procedure 5 and record below what you see.

7. Use your observations in Procedure 6 and your vertical calibration to measure the voltage of the dry cell. Switch the multimeter to the proper scale and measure the dry-cell voltage with it. Record the two readings and compare them.

8. Disconnect the dry cell and the transformer secondary, unplug the transformer, and plug in the signal generator instead. Connect the latter's output to the vertical input of the oscilloscope. Set the signal generator frequency to about 400 Hz and obtain a two-cycle pattern of the generator output on the screen. Internal synchronization should be used in this case. Vary the generator output and also the generator frequency and describe your observations in the space below.

9. Switch off the oscilloscope's time base so that the instrument is set for an external horizontal input. CAUTION: Leave the vertical deflection due to the signal generator as it was in Procedure 8, so that a vertical line is displayed. Oscilloscopes should not be left with the beam on but no horizontal or vertical deflection, as the continuous electron bombardment at one point on the screen will soon burn away the fluorescent material there. Connect the transformer secondary to the *horizontal* input terminals, have the instructor check your setup, and then plug the transformer into the ac line. Set the signal generator frequency to about 30 Hz and adjust the horizontal and vertical gain controls to obtain a lissajous figure filling most of the screen. Vary the signal generator frequency carefully in the neighborhood of 30 Hz until you obtain a stationary pattern having two loops, one above the other. Record in the Data section the exact reading of the generator dial when the pattern is stationary.

10. Shift the signal generator frequency to about 60 Hz and again move the dial slowly until a stationary lissajous figure is obtained. Notice that this may be a circle, an ellipse, or a straight line depending on the phase difference between the transformer secondary voltage and the generator output and on their relative amplitudes. Try to get a circle by making the generator's output amplitude equal to that of the transformer secondary voltage and shifting its frequency very carefully to bring about a 90° phase difference. Then try to obtain a straight line at 45° to the horizontal by bringing the phase difference to zero degrees. Finally, with the pattern stationary, observe and record the reading on the generator's frequency dial.

11. Increase the generator frequency to about 120 Hz, again adjust it carefully to obtain a stationary lis-sajous figure, and record the dial setting. Note that this time the pattern should have two loops, one beside the other.

12. Repeat Procedure 11 with the signal generator set for around 180 Hz. A three-loop stationary pattern should be obtained.

13. When you have finished the experiment, switch off the oscilloscope and signal generator and make sure the transformer is unplugged.

DATA

rms transformer secondary voltage
(from multimeter) _____

Calculated amplitude of transformer
secondary voltage _____

Distance between peaks of line
voltage pattern _____

Frequency of line voltage _____

Dry-cell voltage from oscilloscope _____

Amplitude from oscilloscope _____

Percent error _____

Period of line voltage _____

Percent error _____

Dry-cell voltage from multimeter _____

Frequency, Hz	30	60	120	180
Generator dial reading				

CALCULATIONS

1. Compute the amplitude of the transformer secondary voltage from the rms value measured with the multimeter in Procedure 3. Compare this result with the amplitude as determined from the oscilloscope by calculating the percent error.

2. Calculate the period of the ac wave observed in Procedure 4. Then calculate its frequency and compare it with the "true" value of 60 Hz by finding the percent error. Note that the 60-Hz line frequency is not a nominal value but very precise (about 1 part in 10,000), as the accuracy of certain kinds of electric clocks depends on it.

3. Calculate the voltage of the dry cell from the displacement of the horizontal line observed in Procedure 7. Note the difference between the voltage so obtained and that measured with the multimeter.

4. Plot a calibration curve for your signal generator in the range 20 to 200 Hz using the data of Procedures 9–12. Draw the best smooth curve through your four plotted points, extrapolating it to 20 Hz at one end and 200 Hz at the other.

QUESTIONS

1. Calculate the amplitude of the 120-V ac line voltage.

2. What is the time standard (analogous to the balance wheel in a windup watch or the pendulum in a pendulum clock) in an electric clock that does *not* use the ac power line for this purpose? *Hint:* What is meant by a "quartz" watch or clock?

3. In high-quality oscilloscopes having the sweep frequency control calibrated directly in seconds per centimeter, the horizontal gain control is automatically switched out of the circuit when the sweep system is being used so that the pattern width is permanently preset rather than being adjustable. Why is this so?

4. Show that if there is no phase difference between the two sinusoidal voltages applied to an oscilloscope in a lissajous figure observation, the angle θ between the pattern axis and the horizontal axis is given by

$$\tan \theta = \frac{B}{A}$$

where A is the horizontal and B the vertical amplitude. What does the pattern look like under these circumstances? Does Equation 33.7 agree with this result when $\phi = 0$?

5. A certain cathode ray tube (CRT) has deflection plates of length $l = 2$ cm and spacing $d = 0.5$ cm located a distance $L = 30$ cm (measured from the center of the plates) from the fluorescent screen. Electrons are accelerated by a potential difference $V_{acc} = 1000$ V and focused in a beam directed down the axis of the tube between the deflection plates and onto the center of the screen (see Fig. 33.1). (a) Derive a formula for the deflection sensitivity of the CRT, that is, the distance D through which the spot on the screen is deflected for each volt applied to the deflection plates. (b) Put the given values of l, d, L, and V_{acc} into your formula to obtain a numerical result for the deflection sensitivity. (c) On the basis of your answer to Part (b), explain why vertical and horizontal amplifiers are needed.

34

Because of the usefulness of the transformer, which does not work on direct current, alternating current has dominated the electrical industry for many years. Strictly speaking, the term "alternating current" should mean any current that varies with time so that it flows sometimes in one direction and sometimes in the other. However, as was seen in Experiment 33, the term usually refers specifically to just one form of such variation—namely, a sinusoidal dependence on time. In this experiment, only currents and voltages that are sinusoidal functions of time will be dealt with, and the performance of circuit elements under these conditions will be studied. It should be emphasized that our conclusions about ac circuits apply *only* when sinusoidal voltages and currents are being used.

THEORY

An ac current that is a sinusoidal function of time may be written

$$i = I_m \sin(\omega t + \phi_i) \tag{34.1}$$

where i is the instantaneous value of the current in amperes, I_m is the maximum value or *amplitude* of this current, ω is the so-called *angular frequency* in radians per second, t is the time in seconds, and ϕ_i is the *phase angle* in radians. The purpose of the phase angle is to set the value of i at $t = 0$. Thus, at this initial instant, $i = I_m \sin \phi_i$, which may have any value from 0 to $\pm I_m$ depending on the value chosen for ϕ_i. Fig. 34.1 shows i plotted as a function of t for the particular case of $i = 0$ at $t = 0$, so that $\phi_i = 0$ and $i = I_m \sin \omega t$. Notice that the current returns to zero after flowing in the positive direction when $t = \pi/\omega$ and again after its reverse (negative) flow when $t = 2\pi/\omega$. After $t = 2\pi/\omega$, the current repeats the perform-

ance it has just gone through in the interval 0 to $2\pi/\omega$, and this complete oscillation, which is repeated over and over, is called a *cycle*. The time $2\pi/\omega$ seconds required for one cycle is called the *period* of the alternating current, and its reciprocal $\omega/2\pi$ is the number of cycles per second and is known simply as the frequency f. Equation 34.1 is therefore often written

$$i = I_m \sin(2\pi f t + \phi_i) \tag{34.2}$$

where f is in cycles per second (cps) or, according to currently accepted terminology, hertz.

The question now arises as to what happens when a sinusoidal current passes through a circuit element such as a resistor. For alternating current, two other circuit elements, the inductor and the capacitor, are of interest. Notice that for direct current the ideal inductor, a coil of resistanceless wire, is simply a short circuit, whereas an

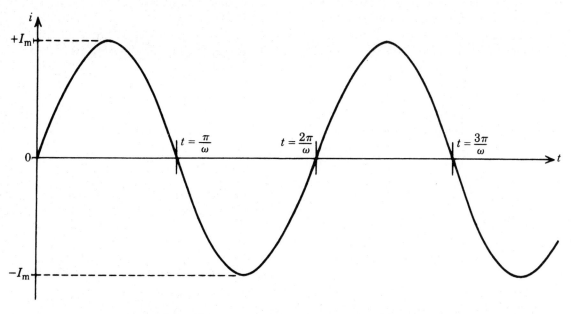

Figure 34.1 *A sinusoidal current as a function of time*

ideal capacitor, two conducting plates separated by a perfect insulator, is an open circuit. With alternating current, however, the current through the inductor changes; hence the magnetic field changes and an induced voltage appears across the coil terminals. In the case of the capacitor, charge flows onto its plates with one polarity when the current is in one direction and the voltage across the capacitor builds up, but then the current reverses, discharging the capacitor and recharging it in the opposite direction. We will see that in each case there results a sinusoidal voltage across the circuit element that is proportional to the current (just as in the case of the resistance) so that a sort of Ohm's law holds for inductors and capacitors as well when the voltages and currents involved are sinusoidal. The proportionality constant, however, is called the reactance instead of the resistance in the case of inductors and capacitors. The reason for this will become apparent.

Resistance in ac Circuits

If the current i flows through a resistance R, Ohm's law holds at every instant so that the voltage across the resistance is

$$v_R = iR = I_m R \sin 2\pi ft = V_{mR} \sin 2\pi ft \quad (34.3)$$

where we have taken zero phase angle for the current and have written $V_{mR} = I_m R$ since V_{mR} is clearly the amplitude of the sinusoidal voltage that develops across the resistance. Notice that the power dissipated in the resistance is $iv_R = I_m^2 R \sin^2 2\pi ft$ and is a function of time but is never negative. This is because the current varies but energy is transformed into heat in the resistor no matter which way the current goes through it. Power is the rate at which energy is delivered, but in this case we cannot simply multiply $I_m^2 R \sin^2 2\pi ft$ by a time interval to get the total energy because this power is not constant. An average value of the time-varying power must be calculated for this purpose, and the average of $\sin^2 2\pi ft$ over one cycle (and hence over any integral number of cycles) is $\frac{1}{2}$. Thus,

$$P_{av} = \frac{1}{2} I_m^2 R \quad (34.4)$$

This average power can be multiplied by a time to give an energy delivered just as the power I^2R can in the dc case. For this reason, a quantity I_{rms} called the root mean square current (because it is the square root of the average of the square of i) is defined as $I_m/\sqrt{2}$. The root mean square (rms) current can then be used to calculate average power and energy delivered in the same way that direct current can. This concept is so useful that it is the rms value of ac currents and voltages that is usually quoted. Thus, when the ordinary household ac power line is said to be 120 V, it is the rms value, not the amplitude, that is meant. All this has already been considered in Experiment 33 where we noted that P_{av} could be written as V_{rms}^2/R. Similarly, and again paralleling the dc case, Equation 34.4 may be written as $P_{av} = I_{rms}^2 R$.

Inductance in ac Circuits

When an inductor (a coil of wire) carries a current, a magnetic field is produced. If the current and hence the field varies, there is a rate of change in the magnetic flux through the coil and an induced voltage appears across its terminals as seen in Experiment 32 even though the resistance of the coil may be zero. If the current through an inductor is given by Equation 34.2 with $\phi_i = 0$, a flux proportional to this current will result, so that we can write

$$\Phi = lI_m \sin 2\pi ft \quad (34.5)$$

where l is the proportionality constant and has the dimensions of webers per ampere. Now Faraday's induction law states that the induced voltage around a single loop is the negative of the rate of change of magnetic flux through that loop (see Equations 32.2 and 32.3). The flux given by Equation 34.5 is certainly changing with time, and the rate of change of the sine of an angle is known to be the cosine of that angle multiplied by the angle's own rate of change. Thus, the voltage induced around one turn of the coil is $-2\pi flI_m \cos 2\pi ft$ and across the coil terminals is $-2\pi fNlI_m \cos 2\pi ft$, where N is the number of turns. A voltage equal and opposite to this must therefore be applied to the coil terminals to keep the current flowing, and this will be

$$v_L = 2\pi fLI_m \cos 2\pi ft = V_{mL} \sin (2\pi ft + 90°) \quad (34.6)$$

where $L = Nl$ is called the *inductance* of the coil and is measured in henrys, in honor of the American physicist Joseph Henry (1797–1878), who was one of the early workers in electromagnetism. The inductance is a quantity that characterizes the coil and depends on the permeability of the core material (if any), the number of turns, and geometrical factors. Equation 34.6 suggests that L is the proportionality constant between the rate of change of current through an inductor and the resulting voltage induced across the inductor's terminals, and it is often so defined.

The important thing about the result stated by Equation 34.6 is that the voltage across the inductor is sinusoidal just as the current through it is. The amplitude V_{mL} of the voltage is equal to $2\pi fLI_m$ and is thus proportional to the amplitude I_m of the current. The proportionality constant $X_L = 2\pi fL$ is called the *inductive reactance* of the coil and appears to play the role of resistance in relating I_m and V_{mL}. Indeed, it has the dimensions of ohms. There are several differences, however, between this case and that of the resistor. Here, while the current goes as $\sin 2\pi ft$, the voltage goes as $\cos 2\pi ft$, that is, as the current with a phase shift of 90° or $\pi/2$ rad. Thus, while the current through and the voltage across a resistor are *in phase*, the current and voltage for an inductor are 90° *out of phase with the voltage leading the current*. Plots of i and v_L are shown in Fig. 34.2. Notice that when i is at its maxima and minima so that for an instant it and the resulting magnetic flux are not changing, v_L is zero, whereas v_L has its maxima and minima at the instants

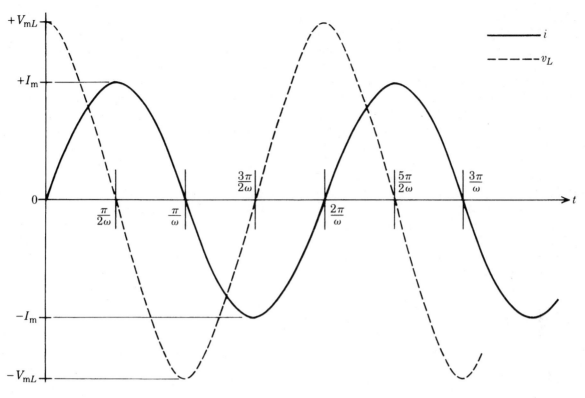

Figure 34.2 *Voltage and current in an inductor*

when i is zero, the points where the sine curve changes most rapidly. Note also that, unlike R, X_L is proportional to the frequency and becomes zero when $f = 0$. This is to be expected since a coil of resistanceless wire is a short circuit for direct current. Finally, the power delivered to the inductor cannot be obtained from the expression $i^2 X_L$ because v_L is not equal to iX_L. Instead, the power delivered must be calculated from the original definition of electric power—namely, the current times the voltage. But if i is multiplied by v_L we get $2\pi f L I_m^2 \sin 2\pi ft \cos 2\pi ft$, which is sometimes negative and sometimes positive. Moreover, the expression $\sin 2\pi ft \cos 2\pi ft$ averages to zero, showing that on the average no energy is delivered to the inductor. This is hardly surprising, since energy delivered to the inductor has to go somewhere, and the only place it can go is into the magnetic field. This is in fact what happens when the factor $\sin 2\pi ft \cos 2\pi ft$ is positive. When it is negative, the magnetic field is collapsing and the stored energy is being returned to the source. Thus, unlike the resistor, the inductor does not dissipate but rather temporarily stores energy.

Capacitance in ac Circuits When an electric current flows into a capacitor, the plates become charged and the terminal voltage rises according to the relation

$$v_C = \frac{Q}{C} \qquad (34.7)$$

where Q is the charge on the capacitor in coulombs, C is the capacity of the capacitor in farads, and v_C is the

capacitor voltage. If the charging current is dc, the capacitor voltage will simply build up to the source voltage, after which current flow will cease, for the capacitor is in fact an open circuit. However, if alternating current is used, the capacitor will charge up until the current reverses, after which it will discharge and then charge again with the opposite polarity. Thus current flows alternately into and out of the capacitor through the connecting wires, so that alternating current appears to pass through it, although no charges can actually travel through the dielectric from one plate to the other.

The situation is illustrated in Fig. 34.3. Looking at the first half-cycle of the current, we note that i is positive throughout that half-period, and so the capacitor will be charging. It must reach its maximum voltage at the end of the half-cycle because thereafter i reverses and the capacitor discharges. This process continues during the second half-cycle, the capacitor attaining maximum charge in the reverse direction and hence maximum negative voltage at the end of a full cycle of the ac current. A detailed study shows that the capacitor voltage, like the voltage across the inductor, is sinusoidal but is shifted by $-90°$ ($-\pi/2$ rad) with respect to the current. In other words, the phase shift is again 90°, but the current leads the voltage. Thus v_C will be proportional to $\sin (2\pi ft - 90°)$ or $-\cos 2\pi ft$.

The amplitude V_{mC} of the capacitor voltage must surely be proportional to I_m, for this represents the maximum charging rate. Moreover, Equation 34.7 shows that V_{mC} should be inversely proportional to C, since the

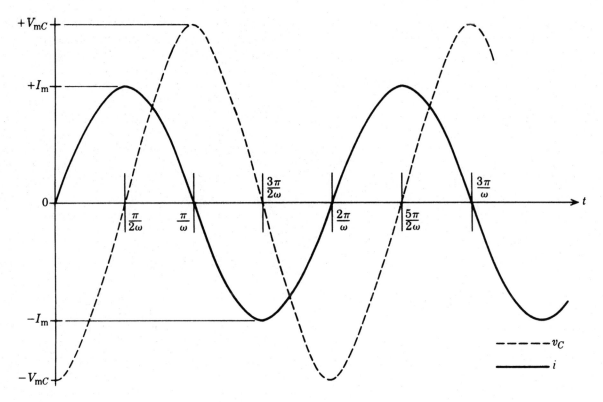

Figure 34.3 *Voltage and current in a capacitor*

greater the capacity the lower the voltage developed by a given amount of charge. We also expect the voltage amplitude to be proportional to the period of the ac current, for half of this period is just the time allowed for charging the capacitor in any one direction. In terms of the frequency, the reciprocal of the period, we should find $V_{mC} \sim 1/f$. Detailed calculation yields

$$v_C = -\frac{I_m}{2\pi fC} \cos 2\pi ft = V_{mC} \sin(2\pi ft - 90°) \quad (34.8)$$

The quantity $1/(2\pi fC)$ that multiplies I_m to give V_{mC} is called the *capacitative reactance* of the capacitor. Like the inductive reactance, it is the ratio of the voltage amplitude to the current amplitude and therefore has the dimensions of ohms. It is also dependent on the frequency but goes to infinity rather than zero as the frequency decreases, indicating that the capacitor is an open circuit rather than a short circuit for direct current. Finally, as in the case of the inductor, the 90° phase difference between the voltage and the current results in energy being alternately delivered to and returned from the capacitor. In the capacitor, it is the electric field between the plates rather than the magnetic field of the coil that stores the energy; but as in the inductor, no energy is dissipated.

The Series *RLC* Circuit Suppose a resistor, an inductor, and a capacitor are connected in series and a current $i = I_m \sin 2\pi ft$ is sent through them, as shown in Fig.

34.4. The total voltage v across the combination must then be the sum of the voltages across the three series elements, and from Equations 34.3, 34.6, and 34.8 we have

$$v = v_R + v_L + v_C = I_m[R \sin 2\pi ft + (X_L - X_C) \cos 2\pi ft] \quad (34.9)$$

This expression is not in a convenient form because it does not display either the amplitude or the phase of v, and it does not really show that v is sinusoidal. It can, however, be rewritten in the form

$$v = I_m Z \sin(2\pi ft + \phi_v) \quad (34.10)$$

That Equations 34.9 and 34.10 are equivalent is easily shown by expanding the latter with the usual formula for the sine of the sum of two angles. Doing this gives

$$v = I_m (Z \cos \phi_v \sin 2\pi ft + Z \sin \phi_v \cos 2\pi ft) \quad (34.11)$$

Equations 34.11 and 34.9 are identical if

$$R = Z \cos \phi_v \quad (34.12)$$

and

$$(X_L - X_C) \equiv X = Z \sin \phi_v \quad (34.13)$$

where X is simply defined as the reactance of the circuit and ϕ_v is the phase angle of the voltage with respect to the current.

The quantity Z is called the *impedance* of the circuit. It multiplies the current amplitude I_m to give the amplitude $I_m Z$ of the voltage across the complete circuit. Like the resistance and reactance, it is measured in ohms

$i = I_\mathrm{m} \sin 2\pi ft$

R

$v_R = RI_\mathrm{m} \sin 2\pi ft$

v

L

$v_L = X_L I_\mathrm{m} \cos 2\pi ft$

C

$v_C = -X_C I_\mathrm{m} \cos 2\pi ft$

Figure 34.4 *The series* RLC *circuit*

and is given by

$$Z = \sqrt{R^2 + X^2} \qquad (34.14)$$

This solution for Z is obtained by squaring and adding Equations 34.12 and 34.13. These equations can also be solved for ϕ_v by dividing the second by the first. This gives

$$\tan \phi_v = \frac{X}{R} \qquad (34.15)$$

These last two results suggest that Z can be represented by a vector drawn on a set of coordinate axes designated X and R as shown in Fig. 34.5. X is positive on the left because $X_L > X_C$, and we say that the net reactance is inductive. On the right is a case of net capacitative reactance (X negative). A most interesting possibility is that of $X = 0$ by reason of X_L and X_C having the same magnitude. In this case, which is called *series resonance*, the circuit acts as if only R were present. The reason is that $v_L = X_L I_\mathrm{m} \cos 2\pi ft$ and $v_C = -X_C I_\mathrm{m} \cos 2\pi ft$, so that if $X_L = X_C$ these two voltages will be equal in magnitude but opposite in polarity at every instant and will thus cancel each other out. Note, however, that neither v_C nor v_L is zero. This illustrates the most important point that in a series ac circuit the amplitudes of the voltages across the individual elements may not be added, for these voltages will in general not have their maximum values at the same time due to their having different phase angles. It is therefore the instantaneous values that must be added, as was done in obtaining Equation 34.9. In particular, in the series *RLC* circuit, the voltage across the inductor is maximum and positive when the voltage across the capacitor is at its maximum negative value. Making the two reactances equal makes these two maxima equal, so that the total voltage across the inductor and capacitor in series is zero. Since as the frequency is increased X_L gets larger while X_C gets smaller, there will surely be a particular frequency f_0 for which they are equal. This is easily found to be

$$f_0 = \frac{1}{2\pi \sqrt{LC}} \qquad (34.16)$$

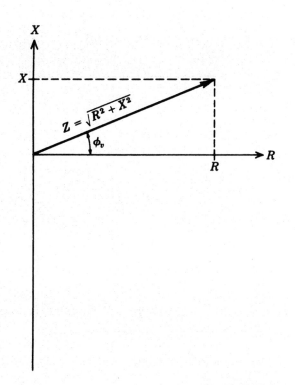
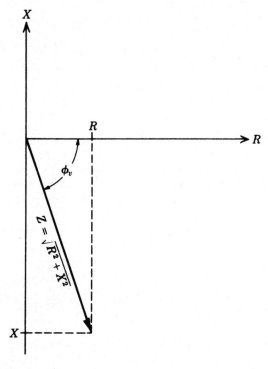

Figure 34.5 *Impedance as a vector*

APPARATUS

1. Cathode ray oscilloscope
2. Transformer delivering 6.3 V at 3 A
3. Iron-core inductor (about 1 H)
4. Capacitor decade with a range of up to at least 8 μF
5. Carbon resistor, 1-W, 5% tolerance, 2200 Ω
6. Decade resistance box (0–10,000 Ω)
7. Multimeter with rms ac voltage ranges
8. Assorted hookup wires and connectors

PROCEDURE

1. Wire the transformer, oscilloscope, carbon resistor, and decade box in the circuit diagrammed in Fig. 34.6. If you have not done Experiment 33, read the Theory section of that experiment to learn about the oscilloscope, and have your instructor explain the operation of the ones in your laboratory to you. The carbon resistor should be used in the position marked "Element Under Test." Notice that this element and the decade box are in series across the 6.3-V source (the oscilloscope inputs have very high impedance and thus draw negligible current) so that the same current flows in both and that the voltage across the element under test is applied to the oscilloscope's vertical input and the voltage across the decade resistance box to the horizontal input. These two voltages obviously have the same frequency, and so a stationary lissajous figure will be displayed on the oscilloscope screen. Note, however, that the "high" terminal of the horizontal input is connected to the "low" rather than the "high" terminal of the resistance box. This means that the polarity of the voltage applied to the horizontal input is reversed relative to that applied to the vertical input, so that the pattern will be reversed left-to-right as compared with that shown in Fig. 33.4 of the last experiment. This will have no effect on your measurements, however.

2. Set the oscilloscope to display the lissajous figure; that is, switch the horizontal channel to the input

terminals rather than the sweep generator. CAUTION: Be sure you have a pattern rather than just a spot on the screen. Oscilloscopes should not be left with the beam on but no horizontal or vertical deflection, as the continuous electron bombardment at one point on the screen will soon burn away the fluorescent material there. Set the horizontal and vertical gain controls so that the horizontal and vertical sensitivities are equal and are such as to give a pattern filling most of the screen. To do this, set the resistance box to zero and temporarily shift the "high" horizontal input lead to the "high" vertical input terminal so that both oscilloscope inputs receive the same voltage—namely, that across the resistor under test. Then adjust the gain controls so that a straight line extending almost all across the screen and at exactly 45° to the horizontal is obtained. The centering controls should be adjusted until this line passes through the origin, that is, the center of the screen. The angle of 45° is assured if the horizontal extent of the line (its horizontal component) is exactly equal to its vertical extent (the vertical component). Do not touch the centering or gain controls after these adjustments have been made. *Note:* With oscilloscopes having calibrated gain controls it should only be necessary to set both to the same sensitivity, choosing the value of this latter to fill the screen as described above. Check that the resulting straight line is indeed at 45° by

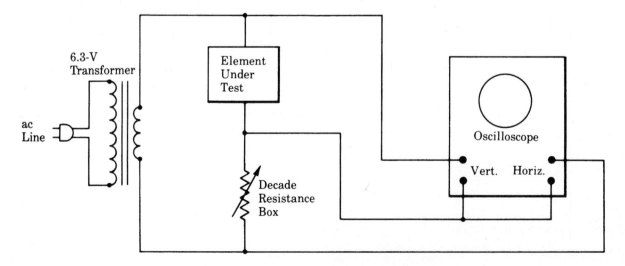

Figure 34.6 *Circuit for impedance measurement*

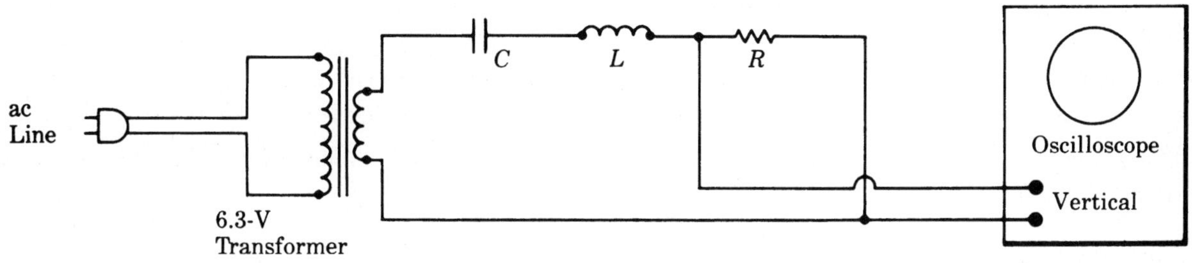

Figure 34.7 *Circuit for observing resonance*

seeing that it passes exactly diagonally through the squares of the *graticule*, that is, the grid of 1-cm squares usually found superposed on the screen of instruments of this quality.

3. Disconnect the horizontal input lead from the vertical input terminal and reconnect it to the "low" side of the resistance box so that your circuit is again as shown in Fig. 34.6. Adjust the resistance box until a line at exactly 45° to the horizontal is displayed. *Note:* This line will extend up to the left rather than up to the right like the line obtained in Procedure 2 because of the polarity reversal discussed in Procedure 1. Record the final setting of the decade box.

4. Remove the carbon resistor from your circuit and measure its resistance directly with the multimeter switched to be an ohmmeter. Have your instructor tell you how to do this if he or she has not already done so. Also note that this resistor has its nominal value marked on it by means of a color code. Your instructor will explain this code to you. The colors associated with the ten digits (0–9) are listed in Table XV at the end of the book. Record the value of your carbon resistor as measured by the ohmmeter.

5. Connect the inductor in your circuit as the element under test. Because the inductor is not a pure inductance but presents an impedance consisting of a resistance and its inductive reactance in series, the pattern will not be a circle but an ellipse similar to that shown in Fig. 33.4 except for the left-right reversal. Adjust the resistance box so that the horizontal and vertical pattern amplitudes defined in that figure are exactly equal. Record the resistance box setting and the values of the vertical amplitude and vertical intercept.

6. Remove the inductor from your circuit and measure its dc resistance with the multimeter set up as an ohmmeter. Record your result.

7. Connect the capacitor decade in your circuit as the element under test and set it for a capacitance of 2 μF. Adjust the resistance box so that equal horizontal and vertical amplitudes are obtained. Practical capacitors approach the ideal rather closely, so that their impedance is usually very nearly a pure capacitative reactance. You should therefore get a good circular pattern. In this case, the vertical amplitude and vertical intercept are the same, and Equation 33.9 of Experiment 33 gives $\phi = 90°$ as it should for a pure capacitance. Observe this fact, and record the setting of the resistance box. *Note:* The transformer may disturb the sinusoidal nature of the voltage enough so that some distortion may be apparent in the displayed circle, but if a reasonably good transformer is used, this effect will be negligible.

8. Repeat Procedure 7 with the capacitor decade switched to 4 μF.

9. Rewire your components in the circuit of Fig. 34.7 in order to observe series resonance. Use the decade capacitor at C, the iron core inductor at L, and the resistance box set to 5 Ω at R. Notice that R serves as a current transducer; that is, the current through it produces a proportional voltage across it according to Ohm's law and this voltage is then displayed on the oscilloscope. The vertical oscilloscope deflection is thus proportional to the current flowing through the circuit. Set the oscilloscope to view the 60-Hz alternating current and look for series resonance by varying the decade capacitor until the sine curve appearing on the screen has maximum amplitude. This corresponds to resonance, for the circuit impedance is a minimum when $X = 0$; hence a fixed voltage source such as the transformer secondary coil will produce maximum current. Record the value to which C is set when maximum current amplitude is observed.

DATA

Resistance Measurement

Resistance box setting _____ Percent error _____

Measured value of carbon resistor _____

Inductance Measurement

Resistance box setting   _____   Percent discrepancy   _____

Pattern amplitude   _____   Reactance of inductor   _____

Pattern intercept   _____   Inductance of inductor   _____

Phase angle ϕ _____   Given value of the inductance   _____

Resistance of inductor from oscilloscope   _____   Percent discrepancy   _____

Resistance of inductor from ohmmeter   _____

Capacitance Measurement

Value of capacitance	Resistance box setting	Capacitance		Percent discrepancy
		Reactance	Measured value	
2 μF				
4 μF				

Series Resonance

Value of C _____   Calculated value of the frequency   _____

Value of L _____   Percent error   _____

CALCULATIONS

1. Refer to Fig. 33.4 of the last experiment and, by writing out an appropriate derivation in the space below, satisfy yourself that Equation 33.9 is correct.

2. The resistance box setting obtained in Procedure 3 should be equal to the resistance of the carbon resistor. Compare it with the value measured in Procedure 4 by calculating the percent error.

3. Calculate the phase angle between the current through and the voltage across the inductor from the data of Procedure 5. The resistance box setting obtained in this step is the inductor's impedance. Using Equations 34.12 and 34.13, compute the resistance and reactance of the inductor. Compare the resistance so obtained with that measured with the ohmmeter by calculating the percent error.

4. From the reactance of the inductor found in Calculation 3 and the known line frequency, compute the inductance of the inductor. Compare your result with the known value by calculating the percent discrepancy. *Note:* The value of the inductance in henrys marked on the inductor is nominal and will in general be smaller than your measured value because it is usually given as the effective value when a superposed dc current is flowing. Thus close agreement between the marked value and the result of your measurement cannot be expected. If a better value of the inductance is available, your instructor will give it to you when you have finished the experiment so that a more meaningful comparison can be made.

5. Calculate the capacitance from your data in Procedure 7 and compare it with the known value by calculating the percent discrepancy. The settings of the decade capacitor should be accurate to 3% or better.

6. Repeat Calculation 5 using the data of Procedure 8.

7. Using the resonant value of C found in Procedure 9 and the inductance found in Calculation 4, compute the resonant frequency of your series *RLC* circuit. Compare your result with the known value of the ac line frequency, 60 Hz. *Note:* The value 60 Hz for this frequency is not nominal but very precise (about 1 part in 10,000), since the accuracy of certain kinds of electric clocks depends on it. Thus, if you have good values for L and C, your calculated frequency should be very nearly equal to 60 Hz. Estimate the precision with which you could determine C and the error to be attached to your value of the resonant frequency, and see if the "true" value of 60 Hz lies within this range.

QUESTIONS

1. Discuss the precision with which you can read the amplitudes and intercepts on the oscilloscope screen in Procedures 5, 7, and 8 and estimate the resulting errors in your measurements of the inductance and resistance of the inductor and the capacitance of the capacitor. Then see if the known values of these parameters fall within the ranges of these errors.

2. Show that if the henry were written in terms of the dimensions of fundamental quantities in the SI system, it would be a newton-meter-second^2/coulomb^2.

3. The magnetic field of a very long solenoid (a single-layer cylindrical coil) is, to a good approximation, concentrated in the region inside the solenoid, where it is uniform, directed parallel to the solenoid axis, and given by

$$B = \frac{\mu N i}{h}$$

where B is the field strength in teslas, μ is the permeability of the core material (the material on which the solenoid is wound), N is the number of turns in the coil, i is the current through the coil in amperes, and h is the length of the solenoid in meters. Using the definitions of Φ and L, derive an expression for the inductance of a long solenoid.

4. Prove by any method you choose that the average of $(\sin 2\pi ft)(\cos 2\pi ft)$ is zero over one period and hence over any integral number of periods.

5. Derive Equations 34.14 and 34.15 as outlined in the Theory section.

6. Derive Equation 34.16.

7. Explain why the decade box resistance obtained in Procedure 5 is equal to the impedance of the inductor being tested. In particular, point out why it is equal to the *impedance*, not the resistance or reactance of the inductor.

8. Explain why no vector representing Z can ever be drawn in the left half-plane (the region to the left of the vertical axis) in diagrams such as those shown in Fig. 34.5 as long as only passive circuit elements (elements like resistors, capacitors, and inductors that do not include a source of power) are considered.

9. In high-quality oscilloscopes having the sweep frequency control calibrated directly in seconds per centimeter, the horizontal gain control is automatically switched out of the circuit when the sweep system is being used so that the pattern width is permanently preset rather than being adjustable. Why is this so?

10. What serves as the time standard in electric clocks that do *not* depend on the 60-Hz line for this purpose?

11. The AM broadcast band runs from 540 to 1600 kHz (thousands of cycles per second). One common tuning method used in older radio receivers was to have an inductor connected to a variable capacitor driven by the tuning dial. A station was selected by adjusting C so that resonance was obtained with the fixed L at the station frequency. (a) The variable capacitor used usually had a maximum capacity of 365 pF (10^{-12}F). Find the value of the inductance needed to tune in the low-frequency end of the broadcast band. (b) What must the variable capacitor's minimum value be in order that the high-frequency end of the broadcast band may be tuned in?

Geometrical Optics 35

Although light is an electromagnetic wave, its wavelength is so short that when dealing with it in equipment whose dimensions are large by comparison (such as lenses, mirrors, etc.), we may ignore its wave nature. Light can then be discussed in terms of beams or rays that travel in straight lines (the law of rectilinear propagation) unless bent by reflection or refraction. When a light ray strikes the boundary between two media of different densities, all or part of it is reflected according to the law of reflection. If the medium into which the incident ray is directed is transparent, most of the light enters this sec-ond medium, but the ray is bent where it passes through the boundary if it does not strike normally (that is, at right angles to the boundary surface). This bending of light rays is called *refraction* and follows a third law. The purpose of this experiment is to study these three laws by tracing the pattern of light rays. We will find that a great deal of this work is geometrical in nature; hence the study of light rays in experimental setups in which their wave nature may be ignored is called *geometrical optics* and the laws of reflection, refraction, and rectilinear propagation are known as the laws of geometrical optics.

THEORY

A light ray travels in a straight line if it is in a transparent medium of uniform density. This simple fact bears the somewhat grandiose name of *the law of rectilinear propagation.* If, however, a light ray strikes a boundary surface separating two transparent media of different density, some of the light will be reflected and some will proceed into the new medium but in an altered direction. The situation is illustrated in Fig. 35.1, which shows a ray in Medium 1 incident on a boundary separating Mediums 1 and 2 and producing a reflected ray coming back in Medium 1 and a refracted ray proceeding into Medium 2. Regular reflection takes place at a smooth surface such as the boundary shown in this figure. Such a surface might be the surface of a mirror (the boundary between glass and air) or of a lake (a boundary between water and air). The law of reflection states that a single ray of light will proceed, after reflection, in a direction such that the angle of incidence equals the angle of reflection. These angles are the angles made by the respective rays with the *normal,* that is, the perpendicular drawn to the boundary surface at the point of incidence. In addition, the incident ray, the normal, and the reflected ray must all lie in the same plane. Fig. 35.1 shows a ray incident on a boundary between two media, the normal, the angle of incidence i, the reflected ray, and the angle of reflection s. The law of reflection requires $i = s$.

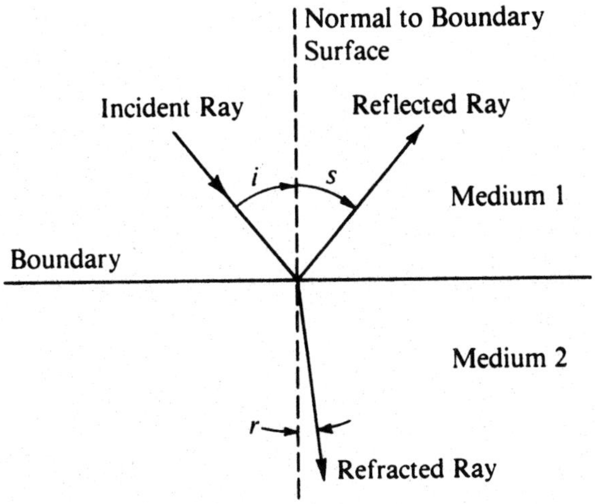

The intermedium boundary does not, of course, have to be plane. Consider, for example, a spherical mirror, that is, a mirror whose surface is a portion of a sphere. It can be shown fairly easily that if a concave spherical mirror is held facing a beam of parallel light rays, the reflected rays will (to a close approximation) converge to a point. This point is called the mirror's *principal focus, F* (see Fig. 35.2). Another important point is the *center of curvature, C,* which is the center of the spherical surface of which the mirror is a part. The *principal axis* of a mirror is a straight line drawn from the center of curvature through the focus to the mirror's surface. The distance from the center of curvature to the mirror along the principal axis is obviously a radius R of the reflecting surface. The distance from the focus to the mirror along the principal axis is the mirror's *focal length, f,* and can be shown to be $\frac{1}{2}R$.

A spherical mirror can also be convex instead of concave, in which case it will diverge incoming paraxial rays instead of converging them, as shown in Fig. 35.3. In this case, the reflected ray *seems* to come from a point *behind* the mirror. This point is therefore taken as the principal focus F, but since no rays actually go through this point, it is called a *virtual* focus. Moreover, because

Figure 35.1 *Reflection and refraction at a plane surface*

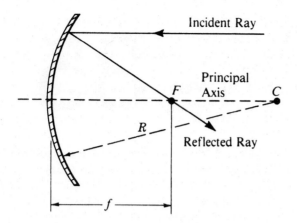

Figure 35.2 *The concave spherical mirror*

where
i = angle of incidence
r = angle of refraction
v_1 = velocity of light in Medium 1
v_2 = velocity of light in Medium 2
n_1 = index of refraction of Medium 1
n_2 = index of refraction of Medium 2

The index of refraction of a particular medium is the ratio of the velocity of light in vacuum to its velocity in the medium. Thus $n_1 = c/v_1$ and $n_2 = c/v_2$, where c represents the velocity of light in vacuum. Hence $v_1/v_2 = n_2/n_1$, as indicated in Equation 35.1. Note that the velocity of light in air is very nearly the same as its velocity in vacuum, so that the index of refraction of air is very nearly 1. It may be taken as equal to 1 for most purposes.

A lens is a piece of glass or other transparent material bounded in its simplest form by spherical surfaces and used to converge or diverge rays of light passing through it. It is similar to a spherical mirror in this respect but is used to converge or diverge *transmitted* rays rather than rays that are reflected back on themselves, and it does this by refraction rather than reflection. Fig. 35.4 illustrates this situation for a converging lens bounded by convex spherical surfaces. These surfaces have centers of curvature C_1 and C_2, and the principal axis of the lens is the line joining these points. The principal focus F is the point on the left of the lens to which rays coming in from the right parallel to the principal axis are converged by the lens's action. F lies on the principal axis at a distance f, the focal length, from the lens. As long as the same medium (for example, air) lies on both sides of the lens, it makes no difference which way the light passes through it, and parallel rays incident from the left will be converged to focal point F' located on the principal axis at the same focal length f to the right of the lens that F was to the left. The focal length should properly be measured from the center of the lens, but we commonly deal with so-called thin lenses, whose thickness is so small that measuring f from the point

in this case the center of curvature C lies behind the mirror, the mirror's radius of curvature R is considered negative, and for similar reasons the focal length f (which remains equal to $\frac{1}{2}R$) is considered negative also. It should be noted that the occurrence of a negative focal length indicates a diverging optical element, mirror or lens.

Refraction is the bending of light rays when they pass obliquely from one medium to another (see Fig. 35.1). The direction of a particular ray after refraction is given by the law of refraction, often called Snell's law after its discoverer, Willebrord Snell (1591–1626). This law states that the ratio of the sine of the angle of incidence to the sine of the angle of refraction is equal to the ratio of the velocity of light in the first medium (the medium of incidence) to that in the second. As in the case of reflection, the incident ray, the refracted ray, and the normal to the boundary all lie in the same plane, and the angle of refraction is the angle between the refracted ray and the normal, as shown in Fig. 35.1. Snell's law may be stated mathematically by the expression

$$\frac{\sin i}{\sin r} = \frac{v_1}{v_2} = \frac{n_2}{n_1} \qquad (35.1)$$

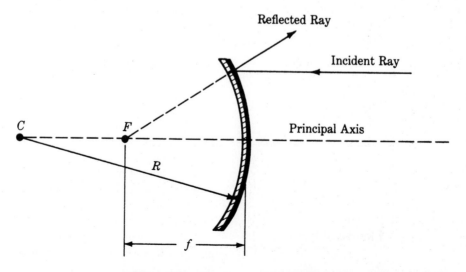

Figure 35.3 *The convex spherical mirror*

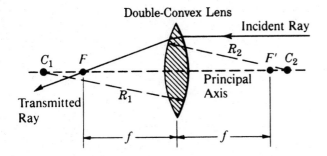

Figure 35.4 *The double-convex converging lens*

where the principal axis intersects the lens surface introduces negligible error.

The focal length of a thin lens must obviously be determined not just by the radii of its two surfaces but also by the index of refraction of the material of which the lens is made. The expression for f in this case must therefore be much more complicated than the simple one already mentioned for the spherical mirror, both because the more complicated law of refraction is involved and because, as seen in Fig. 35.4, refraction takes place at two surfaces. Detailed analysis shows that

$$\frac{1}{f} = (n-1)\left(\frac{1}{R_1} - \frac{1}{R_2}\right) \qquad (35.2)$$

where f is the focal length in meters, n is the index of refraction of the lens material, and R_1 and R_2 are the radii of the two surfaces as defined in Fig. 35.4. Note that if f is in meters, R_1 and R_2 must be in meters also.

This formula is known as *the lens maker's equation* because it is mostly of interest to someone who is handed a piece of glass or plastic having a certain index of refraction and is told to make a lens of a specified focal length. It will not concern us much in the present experiment, in which we will measure the focal length of a given lens, but the role of the radii of curvature R_1 and R_2 is worth some discussion.

The lens of Fig. 35.4 is called *double convex* because both surfaces are convex, and it should be noted that both contribute to converging the incident ray. In this situation, the radius R_1 of the right-hand surface, whose

center C_1 is in the space to the left of the lens, is considered positive, whereas the radius R_2 of the left-hand surface, with center C_2 on the right, is considered negative. The term $1/R_2$ in Equation 35.2 is therefore negative, with the result that the quantities $1/R_1$ and $1/R_2$ will add in this equation, thus expressing the fact that both surfaces contribute to the convergence of rays by increasing the quantity $1/f$, which means shortening the focal length. Notice that a *shorter* focal length implies *stronger* convergence, and the reciprocal $1/f$ is therefore called the *power* of the lens, as it is a direct measure of the lens's ability to converge rays. If f is given in meters, the power $1/f$ is measured in *diopters,* whose dimensions are m^{-1}.

A lens can also diverge incoming rays by being made with concave surfaces. This situation is illustrated in Fig. 35.5, which shows a paraxial ray coming in from the right and being diverged by a thin, double-concave diverging lens. The dotted line indicates that the refracted ray *appears* to be coming from a point on the axis to the right of the lens, so that in the present case this point is considered the principal focus F, and its distance from the lens is the focal length f. A glance at Fig. 35.4 shows that the positions of F and F' are reversed as compared to their positions with the converging lens, and the focal length of a diverging lens is therefore taken to be negative just as it was with the diverging mirror. Note also that no ray actually comes from F so that it is a virtual focus. Thus the diverging lens, like the diverging mirror, has virtual focal points and negative focal lengths. And just as the focal length of the diverging mirror can be considered negative because it is one half of the negative radius of curvature, so the negative focal length of the diverging lens can be obtained from Equation 35.2. Thus, if we apply our convention about signs to R_1 and R_2 in Fig. 35.5, we find that now R_1 is negative and R_2 positive. Thus both the $1/R_1$ and $1/R_2$ terms in this equation will be negative, giving a negative power and hence a negative focal length for the lens. Note that R_1 and R_2 need not be equal, and in fact you will encounter plano-concave and plano-convex lenses, for which the radius pertaining to the flat surface is infinite so that its reciprocal in the lens maker's equation is zero. There are also concave-convex lenses with

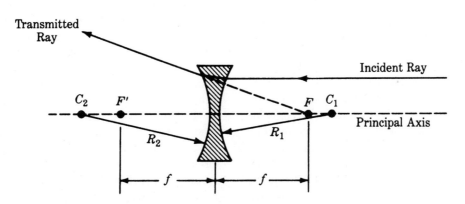

Figure 35.5 *The double-concave diverging lens*

one convex surface and one concave, an arrangement that seems contradictory, as one surface then wants to diverge incoming rays and the other to converge them. Such lenses have certain specialized applications that we will not consider.

Snell's law also shows that if a ray of light passes from one medium into another whose index of refraction is less than that of the first (for example, a ray from an underwater source passing from water into air), the angle of refraction r is greater than the angle of incidence i and the ray is bent away from the normal. As the angle of incidence is increased, eventually the refracted ray will graze the boundary surface. In this case $r = 90°$ and Equation 35.1 becomes $\sin i_c = n_2/n_1$, where i_c, the angle of incidence for which the refracted ray grazes the boundary, is called the *critical angle*. If the angle of incidence is made greater than the critical value, Equation 35.1 requires $\sin r$ to be greater than 1, which is impossible. It's mathematics' way of telling us that there can be no $\sin r$, and hence that there is no refracted ray. In fact, in this case no light proceeds into the new medium and total reflection occurs at the bounding surface.

A prism is a solid piece of transparent material having flat surfaces. When a ray of light impinges on a prism, it is refracted on entry into the prism material and again when it exits through one of the other surfaces. Prisms of various shapes are used to deflect light rays in predetermined desired ways, but a common form is the triangular prism whose principal use is in optical spectrometers. Used this way, it deviates the incident beam through an angle D, called the *angle of deviation*, as shown in Fig.

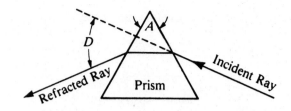

Figure 35.6   *Deviation by a triangular prism*

35.6. D depends on the index of refraction of the prism material, and because the velocity of light in transparent media varies with the light's wavelength, the index of refraction and hence D will be different for light of different colors. A prism spectrometer can thus be used to separate a beam of light into its component wavelengths. For light of a fixed wavelength, D depends on the direction of the incident ray (the initial angle of incidence), but for one particular direction of incidence the angle of deviation becomes a minimum. Minimum deviation occurs when the path of the ray inside the prism is symmetrical. This means that the initial angle of incidence is equal to the final angle of refraction, and the path of the ray inside the prism is parallel to the face opposite the prism's apex angle A (see Fig. 35.6). Under these conditions the index of refraction of the prism material is given by

$$n = \frac{\sin \frac{1}{2}(A + D)}{\sin \frac{1}{2}A} \tag{35.3}$$

This constitutes one method of measuring index of refraction.

APPARATUS

1. Low-power helium-neon laser
2. Drawing board
3. Plane mirror
4. Large cylindrical concave mirror
5. Large cylindrical convex mirror
6. Plano-convex lens with cylindrical curved surface
7. Semicircular lens

8. Large 60° triangular prism
9. Meter stick
10. Protractor
11. Drawing compass
12. White paper sheets for drawing board
13. White cards, masking tape, pins

PROCEDURE

1. Tape a sheet of white paper to the drawing board. Place the plane mirror on the sheet near the center with the reflecting surface facing your right. Plug in and turn on the laser and allow a few minutes for it to warm up. *Note:* There are two important features of laser light. First, a laser emits a narrow beam that spreads out very little and is thus ideal for use as a "ray" in this experiment. Second, laser light is highly monochromatic, meaning that it contains, to a close approximation, just one wavelength. For a helium-neon laser this wavelength is 6328 angstrom units. An angstrom unit (abbreviated Å) is 10^{-10} m or 0.1 nm; thus the wavelength of the helium-neon laser may also be quoted as 6.328×10^{-7} m,

6.328×10^{-5} cm, or 632.8 nm. Although knowledge of the wavelength of the light used in this experiment is not necessary because we are doing geometrical optics, you should realize that the refractive index you will encounter for the lens and prism material will have the value corresponding to 6328 Å.

2. Place the laser on the laboratory bench just beyond the right edge of the drawing board and direct it so that the beam passes over the board just above the paper surface and strikes the mirror at an angle of incidence of about 30°. CAUTION: Although a low-power laser is used in this experiment so that the beam will not hurt your skin, *be very careful not to look into the beam.* Keep it down on the bench and take care not to shine it in anyone's face, including your own.

3. Using the mirror as a straightedge, draw a heavy line with a sharp pencil to represent the reflecting surface on the paper. Make a mark on the paper directly under the point where the laser beam strikes the mirror. Also make marks near the right-hand edge of the paper directly under the incoming (incident) beam and under the outgoing (reflected) beam. *Note:* In clear air the laser beam will be invisible because no part of the light travels to your eye. It can be made visible by blowing chalk dust or cigarette smoke into the beam path, in which case the small particles will scatter some of the light to your eye so that you can see it. There will be enough scattering at the mirror surface so that you will be able to see a red spot where the beam strikes and so be able to mark the paper immediately below it. To place a mark under the incident and reflected rays, hold your pencil vertical with the point down and move it sideways through the beam. You will see the beam spot on the pencil and will easily be able to tell when the pencil is centered in the beam path. You can then lower the point onto the paper to make the desired marks.

4. Remove the mirror from the paper and point the laser toward the wall or other backstop behind the bench to stop the beam. Do this whenever you are not actually using the laser. With the drawing compass and meter stick, construct the normal to the line representing the mirror at the point of incidence. Draw this normal as a broken line. Draw solid lines representing the incident and reflected rays. Use the meter stick as a straightedge and connect the point you marked under the incoming beam in Procedure 3 and the point of incidence to represent the incident ray. Connect the point under the reflected ray and the point of incidence to represent the reflected ray. Place arrowheads on these lines to indicate the direction of propagation in each case. With the protractor, measure and record the angles of incidence and reflection.

5. Remove the sheet of paper from the drawing board and set it aside for inclusion in your report. Put down a new sheet and secure it with tape. Place the concave mirror on this sheet near its left edge and facing to the right. Note that for the two-dimensional work we are doing here the mirror is cylindrical rather than spherical. Trace the reflecting surface on the paper with a sharp pencil so that you get a heavy curved line representing the mirror. Now remove the mirror and locate its center of curvature on the paper as follows: draw any two chords to your curved line and construct their perpendicular bisectors. The intersection of these two lines is the mirror's center of curvature. Pick a point near the middle of the curved line and, using the meter stick as a straightedge, draw a straight line from this point through the center of curvature. This line will serve as the mirror's axis. Now construct two lines, one on each side of the axis, parallel to it, and separated from it by 2 in. These

lines should extend from the right-hand edge of the paper to the curved line representing the mirror and will represent incoming rays. Put arrowheads on them to show that they are directed toward the mirror.

6. Replace the concave mirror on the paper so that its edge matches the curved line traced from it in Procedure 5, thus ensuring that the mirror is back in its original position. Direct the laser beam at the mirror so that it lies right above one of the incoming ray lines drawn in Procedure 5. Do this by making the spot where the beam strikes the mirror come right above the point where the ray line intersects the curved mirror tracing, while at the same time a pencil held vertically in the beam path near the right edge of the paper points to the line below it. Observe the reflected ray and mark a point directly under it near the right-hand edge of the paper as described in Procedure 3.

7. Repeat Procedure 6 with the laser beam directed along the other ray line.

8. Remove the mirror from the paper and direct the laser beam at the wall to stop it. Draw lines representing the reflected rays from the respective points of incidence on the mirror through the points marked in Procedures 6 and 7. These lines should intersect on the axis, the point of intersection being the mirror's principal focus. If they do not, mark the best average position on the axis and call this the focus. With the meter stick, measure and record the focal length and radius of curvature of the concave mirror. Put arrowheads on your reflected ray lines to indicate that they are directed away from the mirror.

9. Repeat Procedures 5–8 for the convex mirror. Note, however, that this mirror diverges the reflected rays so that, instead of *coming to* a focus, they appear to *come from* a focus behind the mirror. To handle this situation, modify your procedure as follows:

   a. Place the mirror near the center of the paper rather than near the left edge to leave room to draw projections of the rays and axis behind it.

   b. Extend the axis line across the paper behind the mirror.

   c. To locate the principal focus, draw the projections of the reflected rays as broken lines behind the mirror. These projections are called *virtual rays*, since there is really no light there. They should intersect at a point on the axis. This is the principal focus. As described in the Theory section, it is a virtual focus, and the focal length and radius of curvature are considered to be negative.

10. Remove the sheet of paper on the board for inclusion in your report and replace it with a clean sheet. Place the plano-convex lens on the paper near its center with the convex side facing to the right. With a sharp pencil trace the outline of the lens. Then remove the lens and construct the perpendicular bisector to the straight line representing the lens's flat side. Extend this perpendicular bisector all the way across the paper. It will represent the lens axis. Also construct two lines, one on each side of the axis, parallel to it and spaced away from it by 2 in. These lines should be drawn from the right edge of the paper to the convex surface of the lens and will represent incoming rays. Put arrowheads on them to show that they are directed toward the lens.

11. Replace the lens on the paper, centering it carefully in the tracing to ensure its return to its original position. Direct the laser beam at the lens, adjusting the laser so that the beam lies above one of the incoming ray lines drawn in Procedure 10. Do this as described in Procedure 6. Observe the refracted ray issuing from the left (plane) side of the lens. Mark the point where this ray crosses the axis. Also mark the paper directly below the point where the beam exits the lens.

12. Repeat Procedure 11 with the laser beam directed along the other ray line.

13. Remove the lens from the paper and point the laser beam at the wall to stop it. Draw lines representing the refracted rays through the points marked in Procedures 11 and 12. Note that the points where these rays intersect the axis should coincide, this being the lens's principal focus. If they do not, mark the best average position on the axis and call this the focus. Establish a point on the axis in the center of the lens tracing and measure the focal length as the distance from this point to the focus. Record your result.

14. Replace the sheet of paper on the board with a new sheet. Lay the semicircular lens on this sheet near its center with the flat side facing to the left. Aim the laser beam in from the right so that it strikes "normally" in the center of the curved surface of the lens, passes through the lens material without being refracted, and exits normally through the middle of the flat surface, again without refraction because the ray is normal to the boundary between the two media involved.

15. Stick a pin into the drawing board at the midpoint of the flat surface of the lens, to serve as a pivot about which the lens may be conveniently rotated. Note that because this is a semicircular lens, the pin is at the center of the curved surface. Carefully rotate the lens about the pin as an axis, noting that in this way the laser beam is kept normal to the curved surface at the point of incidence so that no refraction occurs as the beam enters the lens material. At the point of *exit*, however, the angle of incidence is being increased from zero, and the beam is bent *away* from the normal because it is passing from a *denser* medium (the lens material) to a *less dense* medium (air). Observe the refracted beam by holding a small white card in its path and noting the spot where it strikes. Continue rotating the lens until the refracted beam just grazes the flat face. You will now notice two interesting effects: (1) Just as you get to an angle of refraction of 90°, the refracted beam will disappear (the spot will vanish from the card) and (2) a strong reflected beam emerging normally from the curved side will appear and can be detected by observing the spot on your card when it is held in the appro-

priate position. This is total reflection occurring as you set the angle of incidence of the ray *inside the lens* on the flat surface to its critical value. With the lens in this critical position, trace its outline on the paper with a sharp pencil. Mark points under the incoming beam where it comes over the right edge of the paper and where it strikes the curved surface of the lens. Also mark points under the totally reflected beam where it exits the curved surface and where it passes over the paper's right edge.

16. Remove the lens from the drawing board and direct the laser beam toward the wall to stop it. Construct the perpendicular to the line representing the flat face of the lens at the point of incidence. This point should be the center of this line and should be marked for you by the pin, which may now be removed. The perpendicular should be drawn as a broken line and represents the normal to the lens's flat surface. Also draw the incident and totally reflected rays through the appropriate points marked in Procedure 15 and place arrowheads on them to show their directions. Note that these rays are radii of the curved lens surface inside the lens. Measure and record the angles of incidence and reflection at the flat surface.

17. Remove the sheet of paper on the board for inclusion in your report and replace it with a fresh sheet. Place the triangular prism on the paper near the center with one side toward you. Direct the laser beam at the prism as illustrated in Fig. 35.6 and observe the refracted beam by holding the white card in its path and noting the spot where it strikes. Carefully rotate the prism and watch how the spot moves on the card. You should be able to find an arrangement in which rotation of the prism in one direction causes the spot to move to an extreme position and then start to move back. Set the prism so that the spot is in its extreme position. This is minimum deviation.

18. Mark points under the incident beam where it comes over the right-hand edge of the paper on the drawing board and where it strikes the prism. Similarly, mark points under the refracted beam where it emerges from the prism and where it passes over the left edge of the paper. With a sharp pencil, and being careful not to let the prism move, trace the prism's outline on the paper.

19. Remove the prism and shut off the laser. Draw the incident and refracted rays through the points established in Procedure 18 and place arrowheads on them to show their directions. Draw a line from the point of incidence of the incoming ray on the prism to the point at which the refracted ray exits the prism to represent the ray inside the prism material. See if this ray is parallel to the prism base (the face opposite the apex angle). Extend the incident ray line through the prism as a broken line so that you can measure the angle of deviation D as shown in Fig. 35.6. With the protractor, measure and record this angle and the prism's apex angle A.

DATA _____

Plane Mirror

Angle of incidence   _____   Percent error   _____

Angle of reflection   _____

Curved Mirrors

Mirror	Focal length	Radius of curvature	$\frac{1}{2}$ Radius of curvature	Percent error
Concave				
Convex				

Lens

Focal length   _____

Critical Angle

Angle of incidence   _____   Index of refraction of lens material   _____

Angle of reflection   _____

Triangular Prism

Apex angle A _____   Index of refraction of prism material   _____

Angle of deviation D _____

CALCULATIONS _____

1. Compare the angles of incidence and reflection measured in Procedure 4 by finding the difference between them, treating this difference as an error, and calculating the percent error.

2. From your data in Procedure 8, calculate one half the radius of curvature of the concave mirror. This should be the mirror's focal length. Compare it with your measured focal length by finding the percent error.

3. Repeat Calculation 2 for the convex mirror using your data from Procedure 9.

4. Assuming the index of refraction of air to be 1, calculate the index of refraction of the semicircular lens material from the value of the critical angle measured in Procedure 16. *Note:* The critical angle is the angle of incidence measured in this procedure. It should be equal to the angle of reflection. Is it?

5. Use Equation 35.3 to obtain the index of refraction of the triangular prism material using your data from Procedure 19.

QUESTIONS

1. What is meant by the chromatic aberration of a lens?

2. (a) What is meant by the spherical aberration of a mirror? (b) Why are the concave mirrors used in automobile headlights parabolic instead of spherical?

3. What is wrong with a lens that has astigmatism?

4. Under what conditions is the angle of refraction greater than the angle of incidence?

5. A ray of light shines from under water through the surface into the air above. Find the critical angle of incidence of this ray on the water-air interface. Take $n = \frac{4}{3}$ for water and 1 for air.

6. Show that the angle of deviation D produced by a triangular prism is minimum when the prism is positioned so that the ray inside it is parallel to the prism's base.

7. (a) Compute the velocity of light in the triangular prism from your value of the index of refraction for this prism and the known value of the velocity of light in vacuum. (b) Repeat for the semicircular lens if its material has a different index of refraction.

8. A ray is incident on the plane surface of a transparent medium having index of refraction n. Derive an expression for the angle of incidence that will cause the reflected ray and the refracted ray (the ray transmitted into the medium) to be perpendicular to each other. The incident and reflected rays are in air, whose index of refraction can be taken to be 1. *Note:* The angle of incidence found in this question is called *Brewster's angle*.

9. Show that for a spherical concave mirror that is small enough so that the angle of incidence of any ray coming in parallel to the mirror's axis and intercepted by the mirror is small, the rays are converged to a point (the principal focus) that is one half the mirror's radius of curvature away from the mirror.

10. A ray of light impinges at an angle of incidence i on the surface of a transparent slab of thickness d. The material of the slab has index of refraction n. (a) Show that the ray that emerges on the other side of the slab is parallel to the incident ray. (b) Derive an expression giving the lateral displacement of the emergent ray from the incident one in terms of i, d, and n.

11. A woman looks at herself in a plane mirror. (a) Why does her image appear to be behind the mirror? (b) Is this a virtual or real image? (c) Why is it reversed left-to-right but not top-to-bottom?

12. A six-foot man's eyes are 4 inches below the top of his head. (a) How tall must a flat mirror mounted on the wall be so that the man can just see all of himself when he looks in it? (b) How high above the floor should the bottom edge of this mirror be placed?

13. A straight stick is held so that it extends through the surface and down into a pool of clear, still water. Explain why the stick appears bent at the surface to an observer viewing it from above. *Hint:* Remember that the human eye knows nothing about reflection or refraction but judges an object's location on the assumption that rays coming to it from the object travel in straight lines.

14. Find out if your plano-convex lens is made of the same material as either the semicircular lens or the prism. If it is, determine the radius of the convex surface and use this radius and the appropriate index of refraction (from either Calculation 4 or 5) in the lens maker's equation to obtain a theoretical value of the plano-convex lens's focal length. Compare this result with the experimental value found in Procedure 13 and explain any significant difference.

Image Formation with Lenses: The Telescope and Microscope **36**

The formation of images by lenses is one of the most important studies in geometrical optics. The purpose of this experiment is to observe the real images formed by various lenses and to verify the lens equation; in particular, to measure the focal length of some positive and negative lenses and the equivalent focal length of a combination of thin lenses. Combinations of lenses are used extensively in many kinds of optical instruments, and it is instructive to see how such a combination works in actual practice. As examples, this experiment also presents methods of constructing an astronomical telescope, an opera glass, and a microscope, and of determining the magnifying power of each.

THEORY

When a beam of rays parallel to the principal axis of a lens impinges upon a converging lens, it is brought together at a point called the principal focus of the lens. The distance from the principal focus to the center of the lens is the focal length of the lens; the focal length is positive for a converging lens and negative for a diverging one. The relation between the object distance p, the image distance q, and the focal length f of a thin lens may be determined by the application of some plane geometry to the diagram of Fig. 36.1, which shows a converging lens of focal length f forming a real image at a distance q to the right of the lens from an object placed at distance p to the lens's left. Note that in this arrangement the rays are traveling left to right so that it is the right-hand focal point that is labeled as the principal focus F, although a lens in air is completely reversible, exhibiting the same focal length in either direction.

Any object is made visible by light reflected in all directions from each part of it. Of the rays coming from the tip of the arrow representing the object in Fig. 36.1, we have chosen three whose behavior after passing through the lens is obvious. Ray No. 1 starts out parallel to the axis and must therefore pass through focal point F; ray No. 2 passes through the center of the lens and is therefore undeflected; and ray No. 3 passes through focal point F' and must therefore exit the lens parallel to the axis. These three rays (and hence all others leaving the object's tip in such a direction as to be intercepted by the lens) intersect at a point that is the image of that tip. Rays from other points on the object will come together at corresponding points on the image, so that a complete real image of the object is formed as shown.

Now the triangles ABF' and CGF' are both right triangles and the angles at F' in each one are vertical. Hence these triangles are similar, so that their corresponding parts are in proportion, and we can write

$$\frac{AB}{CG} = \frac{AF'}{CF'} = \frac{p-f}{f} \qquad (36.1)$$

Moreover, triangles DEF and CHF are also similar so that

$$\frac{DE}{CH} = \frac{DF}{CF} = \frac{q-f}{f} \qquad (36.2)$$

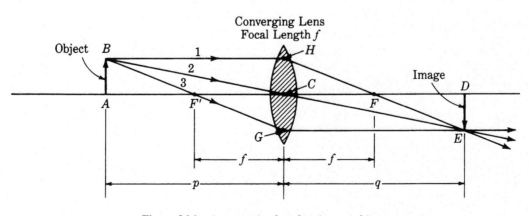

Figure 36.1 *A converging lens forming a real image*

But because *AB* and *CH* are opposite ends of the rectangle *ABHC*, they are equal, and *DE* = *CG* for the same reason. Hence *DE/CH* = *CG/AB*, and substitution in Equations 36.1 and 36.2 yields

$$\frac{f}{p-f} = \frac{q-f}{f}$$

or
$$(p-f)(q-f) = f^2 \qquad (36.3)$$

Equation 36.3, in which the object and image positions appear as their distances $p-f$ and $q-f$ from the two focal points, is called *Newton's form of the lens equation*. A little algebra readily transforms it to the more usual form

$$\frac{1}{p} + \frac{1}{q} = \frac{1}{f} \qquad (36.4)$$

The linear magnification produced by a lens is defined as the ratio of the length of the image to the length of the object. In Fig. 36.1 this is seen to be $-DE/AB$, where the minus sign has been inserted to indicate that the image is inverted with respect to the object. Because the triangles *DEC* and *ABC* in Fig. 36.1 are another similar pair, $DE/AB = DC/AC = q/p$, and so the linear magnification can be stated as the ratio of the image distance to the object distance. Thus,

$$\text{Magnification} = -\frac{q}{p} \qquad (36.5)$$

Notice that the magnification will be negative with a real, inverted image, but in the case of a virtual image the image distance q will be negative. Hence, the magnification will be positive as it should be with the erect image obtained under these circumstances.

An interesting example of this is the magnifying glass, which is simply a converging lens of the sort shown in Fig. 36.1 but is placed so that the object falls inside the focal length f. Thus p is less than f and the lens equation (36.4) shows that q will then be negative and hence the image erect and virtual. Note that this is simply a consequence of the fact that the lens, although converging, does not have sufficient power (short enough focal length, i.e., $f > p$) to overcome the strong divergence

of the rays coming from the nearby object. Note also that we can then write

$$\frac{1}{q} = \frac{1}{f} - \frac{1}{p} = \frac{p-f}{fp} = -\frac{f-p}{fp}$$

and hence from Equation 36.5

$$\text{Magnification} = -\left(-\frac{fp}{f-p}\right)\Big/p = \frac{f}{f-p}$$

The denominator in this result is clearly less than the numerator and can be made quite small if p is just a little less than f, so that a large magnification is obtainable.

The principal focal length of a converging lens may be determined by forming an image of a very distant object on a screen and measuring the distance from the screen to the lens. This distance will be the focal length, since the rays of light from a very distant object are very nearly parallel. In other words, p in Equation 36.4 is infinite so that $1/p$ is zero and $q = f$. A better method is to place an object at a known distance p from the lens, measure the resulting image distance q, and calculate the focal length from Equation 36.4.

Another method, which demonstrates the meaning of this equation very clearly, makes use of the following arrangement: An object and a screen on which an image of the object is to be focused are placed a known distance D apart, and a converging lens of focal length f is placed between them. Since q is the distance from the lens to the image (that is, the screen), the sum $p + q$ must be equal to D. Hence $q = D - p$, and substitution in Equation 36.4 to eliminate q produces a quadratic equation in p whose solutions are

$$p_a = \frac{D}{2}\left(1 + \sqrt{1 - 4\frac{f}{D}}\right) \qquad (36.6)$$

and

$$p_b = \frac{D}{2}\left(1 - \sqrt{1 - 4\frac{f}{D}}\right) \qquad (36.7)$$

There are thus two positions of the lens between the object and the screen for which an image on the screen will

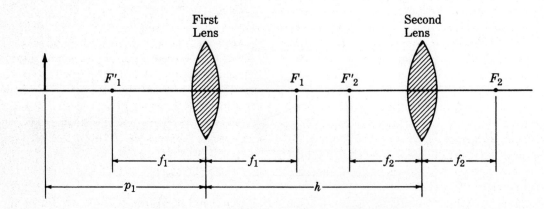

Figure 36.2   *An optical system of two converging lenses*

be in focus *provided D is at least four times the lens's focal length*. If D is just equal to $4f$, the square root in Equations 36.6 and 36.7 disappears and $p_a = p_b = D/2$; that is, there is only one lens position, namely, the midpoint of the object-screen distance, for which a focused image is obtained. This condition is easily observed and the corresponding value of D measured. The focal length of the lens is then found by dividing this result by 4.

The lens equation (36.4) may also be used to handle the situation where an optical system is formed in which light rays pass through two or more lenses in sequence. Fig. 36.2 illustrates the sort of arrangement that might be encountered. The procedure is to consider the first lens in front of the object first and apply the lens equation to it as if it were acting alone. An image distance q_1 is found, and the image appearing at q_1 is then treated as the object of the second lens. In Fig. 36.2, q_1 will be positive, the real image formed by the first lens (which in this illustration is converging) appearing on the right of that lens. The object distance p_2 for the second lens is then $h - q_1$, where h is the space between the lenses, and the lens equation may then again be applied with $p_2 = h - q_1$ to find an image distance q_2, which gives the position of the final image *measured from the location of the second lens*. Notice that in the present arrangement, positive object distances are measured to the left of the pertinent lens in each case, positive image distances to the right, and converging lenses have positive focal lengths. If these sign conventions are adhered to, the above-described procedure will handle any optical system regardless of the lens spacing h or of whether the lenses are converging or diverging. Note in particular that if q_1 is larger than h, the second lens's object distance p_2 will be negative.

A useful example of a two-lens system is the arrangement of two thin lenses that touch each other. In this case, $h = 0$, $p_2 = -q_1$, and two successive applications of the lens equation will show that the combination has an equivalent focal length f given by

$$\frac{1}{f} = \frac{1}{f_1} + \frac{1}{f_2} \qquad (36.8)$$

where f_1 is the focal length of the first lens and f_2 is the focal length of the second lens. This arrangement is useful because a concave lens by itself cannot form a real image, since it is a diverging lens; hence a different method must be used for measuring its focal length. This may be done by placing the negative lens in contact with a positive lens of shorter and known focal length, measuring the equivalent focal length of the combination experimentally, and then using Equation 36.8 to solve for the focal length of the negative lens.

Other well-known examples of multilens optical systems are the telescope and the microscope. There are three main types of telescopes: the astronomical telescope, the terrestrial telescope, and the opera glass. In the astronomical telescope, a so-called objective lens, which

is a positive lens of long focal length, forms a real inverted image of a distant object in its focal plane. Note that p is really infinite for the astronomical telescope's objective, so that $q = f$ for this lens. The second lens or *eyepiece* is a positive lens of short focal length used as a magnifying glass to produce an enlarged virtual image of the real image formed by the objective. For this purpose, the eyepiece is placed so that the focal point of the objective lies between the eyepiece and its focal point. In other words, the arrangement is similar to that of Fig. 36.2 but with h less than $f_1 + f_2$ so that F_2' lies to the left of F_1 to make $p_2 = h - q_1 < f_2$. This will result in a negative q_2 and hence a virtual image as already described in the discussion of the magnifying glass. The eye looks directly at the magnified virtual image produced by the eyepiece.

A disadvantage of the astronomical telescope, which is of little consequence when it is used to look at the stars, is that the image is inverted. This will be understood by noting that the eyepiece forms a virtual image that is not reinverted. In the terrestrial telescope, a third lens is introduced between the objective and the eyepiece. It is a converging lens and is placed beyond the image formed by the objective at such a distance as to form a second real image that is the inversion of the first. The two inversions thus produce a real, erect image that the eyepiece then views as in the astronomical telescope.

Because of its three lenses and the spacing required between them, the terrestrial telescope tends to be very long. This disadvantage is overcome without sacrificing the advantage of an erect final image in the opera glass, a telescope in which the eyepiece is a divergent lens. It is placed so that the rays from the objective strike it before they converge to form an image. Thus, if Fig. 36.2 is to represent an opera glass, the second lens is diverging and is spaced from the first lens by a distance h less than q_1. In fact, h is made equal to the difference between the absolute values of the two focal lengths f_1 and f_2. For the divergent eyepiece used in this case, f_2 is shorter than f_1 and negative; hence $h = f_1 - |f_2|$ so that the spacing will not only be less than q_1 but less than f_1. The second lens therefore lies within the focal length of the first. Thus the image that would have been formed by the objective serves as the virtual object (p_2 negative) for the diverging eyepiece, from which the rays diverge as if they came from an enlarged virtual image. This type of telescope has a small field of view, but it is clearly shorter than either of the other two types and, as already noted, has the advantage of giving an erect image of a distant object.

Like the astronomical telescope, the microscope uses two convergent lenses, one as an objective and one as an eyepiece, but in this case the objective is the lens with the shorter focal length. The object is placed at a distance only slightly greater than the focal length of this lens, so that a highly enlarged real image is formed. The longer focal length lens is used as the eyepiece and is placed as in the telescope so that the real image produced by the objective falls just inside its focal length. A further magnified virtual image is thus obtained. The microscope gives a much greater magnification than does the eyepiece used alone as a magnifier because in the microscope the eyepiece is used to view the enlarged real image that the objective forms from the object, rather than viewing the object directly.

The magnifying power of an optical instrument is defined as the ratio of the angle subtended at the eye by the image of an object viewed through the instrument to the angle subtended at the eye by the object viewed directly. This definition is somewhat different from that of linear magnification noted earlier (Equation 36.5), but it is an important characteristic of a telescope or a microscope. Thus much of the experimental work on these instruments will be devoted to measuring their magnifying power.

APPARATUS

1. Optical bench
2. Illuminated object
3. Lens holders
4. Screen
5. Two converging lenses, A and B (of about 20-cm and 10-cm focal length, respectively)
6. One diverging lens, C (of about −15-cm focal length)
7. Vertically mounted metric scale
8. Telescope magnification scale. This scale is made up of a strip of white paper 2 ft long with a series of thick black lines drawn horizontally across it at regular intervals of about 2 in. The lines are numbered to facilitate counting.
9. Vernier caliper
10. Desk lamp

PROCEDURE

1. Measure the focal length of lens A directly by obtaining the image of a very distant object on the screen and measuring the image distance. The object may be a tree or a house about a block away. Record the focal length so determined.

2. Repeat Procedure 1 for lens B.

3. Determine the focal length of lens A by the use of the lens equation. Place the illuminated object at one end of the optical bench and the screen at a distance of about five times the focal length of the lens. With the object and screen fixed, find the position of the lens for which a sharp, enlarged image is produced on the screen. Make sure that the object, lens, and screen all lie along the same straight line (the principal axis of the lens) and that they are all perpendicular to the axis. Measure and record to a precision of 1 mm the object distance p from the illuminated object to the lens and the image distance q from the lens to the screen. Then use either the vernier caliper or the metric scale to measure the sizes of the object and image. Record these measurements to 0.5 mm or, if you can, to 0.1 mm.

4. Using the arrangement of Procedure 3, move the lens back and forth to find two positions for which the image on the screen is in focus. Move the screen a few centimeters closer to the object and again observe that a focused image is obtained for two different lens positions. Notice that these positions are closer together than they were before the screen was moved. Continue to move the screen closer to the object and to observe that there are two lens positions that give a focused image until these two positions coincide at the midpoint between the object and the screen. Measure and record the value of the object-screen distance D corresponding to this condition.

5. Repeat Procedure 3 using lens B.

6. Repeat Procedure 4 using lens B.

7. Repeat Procedure 3 using the combination of lenses A and B in contact.

8. Repeat Procedure 3 using the combination of lenses B and C in contact.

9. Construct a simple astronomical telescope using the long-focus converging lens A as the objective and the short-focus converging lens B as the eyepiece. Special care must be taken to have the various optical parts at the same height and well lined up. In addition, the lenses must be perpendicular to the optic axis of the system. Mount the eyepiece near one end of the optical bench and the screen in front of it at a distance of its focal length away. Place the illuminated object at the far end of the laboratory table and point the optical bench at it. Mount the objective lens on the optical bench on the other side of the screen from the eyepiece and move it until it focuses a real inverted image of the object on the screen. Then remove the screen and adjust the eyepiece so that on looking through it you can see a sharp inverted image of the illuminated object.

10. Measure the magnifying power of your telescope by taping the telescope magnification scale to a distant wall and pointing your telescope at it. Illuminate the scale with the desk lamp if necessary and readjust the eyepiece to bring the scale into sharp focus. Look through the telescope with one eye and directly at the scale with the other. The magnified image of the scale will be superimposed on the unmagnified scale. Count the number of divisions of the scale viewed directly that cover exactly one division of the magnified scale. This number gives the magnifying power of the telescope.

11. Construct an opera glass using the same lens for the objective as before and the diverging lens as the eyepiece. This time place the screen at one end of the optical bench, point the optical bench at the illuminated object at the far end of the laboratory table, and mount the objective lens so that a real inverted image of the object is formed on the screen. Mount the eyepiece on the same side of the screen as the objective lens and spaced from this lens by a distance equal to the difference of their focal lengths. Remove the screen and adjust the eyepiece so that on looking through it you can see the largest clear and erect image that can be formed.

12. Determine the magnifying power of the opera glass as in Procedure 10.

13. Construct a microscope using lens B as the objective and lens A as the eyepiece. Place the illuminated object at one end of the optical bench and mount the short-focus lens in front of it at a distance of a little more than this lens's focal length. Mount the screen on the other side of this lens and slide it along the optical bench until a real inverted image is sharply focused on it. Then mount the eyepiece on the other side of the screen from the objective and at a distance from the screen equal to the eyepiece's focal length. Finally remove the screen and adjust the eyepiece so that when you look through it you see a sharp inverted image of the illuminated object.

14. Replace the illuminated object with the vertical metric scale; position the latter so that it is the same distance from your microscope objective as the illuminated object was. Illuminate the scale with the desk lamp. Look through your microscope and make small adjustments to the eyepiece position to bring the scale divisions into sharp focus. Place the vernier caliper against the eyepiece and adjust the jaws until they span exactly five millimeter divisions as seen in the microscope. Remove both lenses and mount the caliper on the eyepiece support without disturbing the setting of its jaws. Look at the scale directly through the space between the jaws, keeping your eye in the same position it was in when looking through the microscope, and note the length of scale bracketed by these jaws. You will have trouble distinguishing the millimeter divisions at this distance, but you should be able to see the centimeter divisions quite clearly. Read the length of the scale seen between the caliper jaws, estimating to 0.1 cm. This is the length subtended at the eye by the same angle that subtended 5 mm (0.5 cm) in the image. This length in centimeters divided by 0.5 cm is therefore the magnifying power of the microscope.

DATA

Lenses	Focal length measured directly	Object distance p	Image distance q	Distance D for single lens position	Size of object	Size of image
A						
B						
A and B						
B and C						

Lenses	Focal length from lens equation	Focal length from Equation 36.8	Focal length from D for single lens position	Magnification	
				Size of image / Size of object	$-\dfrac{q}{p}$
A					
B					
A and B					
B and C					
C					

Magnifying power of the astronomical telescope _____

Magnifying power of the opera glass _____

Magnifying power of the microscope _____

CALCULATIONS

1. Calculate the focal length of lens A from the data of Procedure 3 and the focal length of lens B from the data of Procedure 5. Compute the magnification produced in each case from the ratio of the size of the image to the size of the object and from the ratio of the image distance to the object distance. Enter these results in the boxes provided for them in the second data table.

2. Calculate the focal lengths of lenses A and B from the values of the object-screen distance D found in Procedures 4 and 6, respectively.

3. Calculate the focal length of the combination of lenses A and B from the data of Procedure 7. Compare the result with the value obtained from theory, using Equation 36.8.

4. Calculate the focal length of the combination of lenses B and C from the data of Procedure 8. By using this value of the focal length of the combination, compute the focal length of the negative lens C, using Equation 36.8.

QUESTIONS _____

1. Explain in your own words what is meant by a real image and a virtual image.

2. Derive Equation 36.4 from Equation 36.3.

3. (a) Draw a ray diagram similar to Fig. 36.1 but with the object distance p less than the focal length f so that the lens functions as a magnifying glass. (b) Show from your diagram that the image is erect and virtual and that the magnification is $f/(f-p)$.

4. (a) What is meant by the spherical aberration of a lens? (b) How can this aberration be reduced in a simple lens?

5. Derive Equation 36.8. Assume that the two lenses in contact with each other are so thin that the combination is thin also.

6. State what is meant by the *power* of a lens and write Equation 36.8 in the form it would take if powers were used instead of focal lengths. Does your result remind you of something you encountered in your study of electricity?

7. In a film projection apparatus, it is desired to produce pictures 12 ft wide on a screen 50 ft from the lens. The size of the picture on the film is 1 in. wide. What must be the focal length of the projection lens used?

8. Substitute $q = D - p$ in Equation 36.4 and show how the solutions 36.6 and 36.7 are obtained. What happens mathematically if D is less than $4f$ in these equations, and what does this tell you about what you would observe physically in this case?

9. If a telescope is accurately focused on a distant object, in what direction must the eyepiece be moved to focus on a nearby object? Explain.

10. What is the basic difference between the astronomical telescope and the microscope?

11. Draw a ray diagram, approximately to scale, showing the illuminated object, the real image formed by the objective, and the virtual image formed by the eyepiece of your astronomical telescope. Represent the object by a small arrow.

12. The magnifying power of an astronomical telescope 24 in. long is equal to 15. Determine the focal length of the objective lens and that of the eyepiece. Assume that the telescope is focused on a distant object.

13. Redraw Fig. 36.2 for the case of an opera glass.

14. Explain how the method of Procedure 14 measures the magnifying power of the microscope in accordance with the definition of magnifying power.

15. For both the astronomical telescope and the opera glass, the distance between the lenses is approximately equal to the *algebraic* sum of their focal lengths. Explain why this is so, remembering that the focal length of a diverging lens is negative, so that in the opera glass the lens separation is the *difference* between the *values* of the focal lengths.

Analysis of Light by a Prism Spectrometer 37

Light is made up of many different colors, each of which has a definite wavelength. When a beam of light is passed through a glass prism, it is broken up into a spectrum. Each chemical element produces its own characteristic spectrum, so that by analyzing the light given off by a substance it is possible to identify the elements it contains. The object of this experiment is to introduce you to the prism spectrometer and to use this important optical instrument to study the bright-line and continuous spectra obtained from certain light sources.

THEORY

In general, light is made up of many component wavelengths, each of which corresponds to a definite color. When a beam of light is passed through a glass prism, refraction takes place when the beam enters the prism and again when it leaves, with the result that the beam is bent through an angle D, called the angle of deviation. This angle is a function of the index of refraction n of the glass, as discussed in the Theory section of Experiment 35 (Fig. 35.6 and Equation 35.3). Since n is different for different wavelengths (because they propagate with different velocities in transparent materials such as glass), each component wavelength in the incident beam is bent through a different angle. The incident beam is thus separated into beams of its component colors (wavelengths) that may then be displayed as a *spectrum*. We say that the incoming light has been *dispersed* into a spectrum, and the property exhibited by the prism of separating this light into its constituents is called *dispersion*.

Spectrum analysis is the decomposition of a beam of light into its constituent wavelengths and the examination of the image so formed. A spectrometer (or spectroscope, as instruments not provided with means for recording the analyzed light are often called) is an optical instrument for producing and analyzing spectra. A diagram of a prism spectrometer is shown in Fig. 37.1, and a typical laboratory instrument appears in Fig. 37.2. It consists of three essential parts: a collimator, a prism, and a telescope. The collimator is a tube with a converging lens at one end and a fine slit of adjustable width at the other. The length of the tube is equal to the focal length of the lens so that the collimator renders the rays of light from the slit parallel. The prism deviates these rays and disperses them into a spectrum. The telescope contains an objective lens, which brings the rays of light to a focus in its focal plane, and an eyepiece, through which the image of the spectrum is viewed. If the light is all of one color, or monochromatic, a single image of the slit appears. If a source contains several colors, several images of different color will appear side by side, each being an image of the slit formed by one component of the light. The telescope can be rotated about the prism so that light emerging at different angles can be viewed, and the angular position of each slit image can be read off from the divided circle under the telescope mount.

There are three main types of spectra: continuous, bright-line, and absorption. A continuous spectrum is produced by light from an incandescent solid or liquid. It contains all wavelengths and hence appears as a continuous gradation of colors, merging with one another with

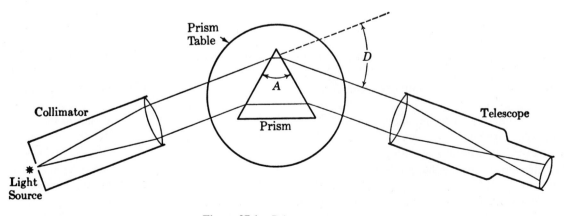

Figure 37.1 *Prism spectrometer*

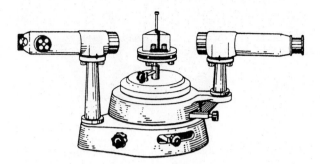

Figure 37.2 *A typical laboratory prism spectrometer*

TABLE 37-1   HELIUM LINES

Wavelength, Å	Color
7065	red
6678	red
5876	yellow
5048	green
5015	green
4922	blue-green
4713	blue
4471	deep blue
4388	blue-violet
4026	violet
3964	violet
3889	violet

TABLE 37-2   MERCURY LINES

Wavelength, Å	Color
6907	red
5790	yellow
5770	yellow
5461	green
4916	blue-green
4358	blue
4078	violet
4047	violet

very gradual changes from red to violet. The bright-line spectrum is produced by exciting an element in the gaseous state by means of an electric discharge or by heating. Such a spectrum consists of bright lines, each an image of the spectrometer slit formed by light of a definite color, separated by dark spaces. The number and relative positions of the lines depend on the element used and the method of excitation employed. An absorption spectrum is produced when light from a source furnishing a continuous spectrum is passed through some cooler absorbing medium. The resulting spectrum is crossed by dark spaces or lines that form the absorption spectrum of the material.

Wavelengths of visible light are usually expressed in angstrom units (abbreviated Å) or, in more recent literature, in nanometers. An angstrom unit is defined as 10^{-10} m and thus equals 10^{-8} cm. A nanometer is 10^{-9} m. Hence a nanometer is 10 Å.

The principal lines in the helium spectrum are given in Table 37-1. The principal lines in the mercury spectrum are given in Table 37-2. In these spectra other fainter lines are also present, but the ones listed are the brightest and easiest to find.

APPARATUS

1. Spectroscope
2. Mercury discharge tube
3. Helium discharge tube
4. Hydrogen discharge tube
5. Sodium light source
6. Desk lamp
7. Discharge-tube power supply
8. Stand and clamp for holding the discharge tubes (unless part of the power supply)

PROCEDURE

1. The spectrometer should be already adjusted when you come to the laboratory, with the collimator and telescope properly focused and aligned and the prism correctly mounted and leveled. CAUTION: Do not handle the spectrometer until the instructor has demonstrated its operation to you. This instrument is easily thrown out of adjustment, and needless delay or even permanent damage can be caused by careless handling. Be extremely careful when using it.

2. Mount the helium discharge tube in its holder and place it immediately in front of the slit at the end of the spectrometer's collimator. The spectrometer should be positioned so that the slit faces directly into the center of the discharge tube. Connect the leads from the power supply to the discharge-tube terminals and turn the supply on. CAUTION: The discharge tube operates at a high voltage, and its power supply can deliver enough current to give a painful shock. Do *not* touch either the tube or the wires leading to it while the power supply is in operation.

3. Adjust the angular position of the telescope so that the helium spectrum is visible when you look into the eyepiece. A small adjustment of the eyepiece position may be necessary to bring the spectrum into sharp focus. Note that the spectrum consists of a number of vertical lines of different colors. Each of these lines is an image of the collimator slit formed by the pertinent wavelength, but because most optical spectrometers use such slits, the word "line" has come to mean a particular wavelength in a spectrum. Now adjust the spectrometer for minimum deviation. To do this, move the telescope so that the bright yellow line at 5876 Å is roughly in the center of the field of view. Slowly turn the prism table in such a direction as to diminish the angle of deviation while watching this line. It should move first one way and then the other as the prism is rotated. If it moves out of the field of view, the telescope position must be readjusted to follow it. Lock the prism table in the position corresponding to the extreme position of the yellow line. This is the orientation giving minimum deviation for light of this wavelength. The prism also gives maximum dispersion when set this way, which is why it is so adjusted for use in a spectrometer. *Note:* When you rotate the prism table, every line in the spectrum will move through a minimum deviation position, but the table setting for minimum deviation will be slightly different for different lines, that is, different wavelengths. The yellow line is chosen because it is at about the middle of the visible spectrum. Therefore the prism setting for minimum deviation of this line represents a good average for the entire spectrum. Do not change the prism table position after the minimum deviation setting just described has been found.

4. Adjust the position of the telescope so that the first red line of the helium spectrum falls exactly on the intersection of the eyepiece cross hairs. Note that directly under the arm supporting the telescope is a clamping device that rotates with the arm and permits the telescope to be clamped to the base by means of a radial screw. A tangent screw on the arm allows a fine adjustment of the telescope setting; however, the tangent screw can be used only after the telescope has been clamped with the radial screw. Rotate the telescope until the first red line is near the vertical cross hair and then clamp it by means of the radial screw. Use the tangent screw to set the intersection of the cross hairs exactly on the red line. Record the angular setting of the telescope to the nearest minute of arc, reading the divided circle with the help of the vernier. Also record the first red line's color and known wavelength (see Table 37-1).

5. Repeat Procedure 4 for the other lines of the helium spectrum, taking readings of the telescope setting for each of the lines that can be recognized.

6. Turn off the discharge tube power supply and replace the helium discharge tube with the mercury tube. Turn the supply on and repeat Procedures 4 and 5 for the mercury spectrum.

7. Repeat Procedure 6 using the hydrogen tube. Four lines should be observed and telescope settings obtained for them. Record the longest wavelength (red) line as H_α, the next as H_β, and so on.

8. Turn off the power supply, remove the discharge tube, and set up a sodium light source in front of the spectroscope slit. Turn on the source and wait for it to warm up so that a brilliant yellow light is produced. Look through the telescope and adjust it so that the yellow sodium doublet falls exactly on the intersection of the cross hairs. Record the telescope setting.

9. Examine the spectrum of the light from a tungsten lamp. Record the setting of the telescope for the upper and lower limits of the visible spectrum.

DATA _____

Spectrum	Color	Known wavelength	Setting of telescope	Wavelength from calibration curve
Helium				
Mercury				
Sodium				
Visible				

Hydrogen Spectrum

Line	Color	Setting of telescope	Wavelength from calibration curve	Reciprocal of wavelength	$1/n^2$
H_α; $n = 3$					
H_β; $n = 4$					
H_γ; $n = 5$					
H_δ; $n = 6$					

Measured value of Rydberg constant from slope _____

Measured value of Rydberg constant from intercept _____

Balmer series limit _____

Accepted value of Rydberg constant _____

CALCULATIONS _____

1. Draw a graph with wavelengths in angstrom units as abscissas and the settings of the telescope as ordinates for the lines of the helium spectrum from the data of Procedures 4 and 5 and the wavelengths given in Table 37-1. The smooth curve so obtained is the calibration curve of the instrument. The wavelengths of the lines observed in other spectra may be determined from it.

2. From your calibration curve, read off the wavelengths of the mercury lines corresponding to the telescope readings obtained in Procedure 6. Tabulate the results and see how they compare with the known values given in Table 37-2.

3. Repeat Calculation 2 for the hydrogen readings obtained in Procedure 7. No table of known wavelengths is given for the hydrogen lines in this experiment. However, the hydrogen spectrum observed with a glass discharge tube and prism is part of the so-called *Balmer series* in hydrogen. The wavelengths of lines in this series are given by the Balmer formula

$$\frac{1}{\lambda} = R_H \left(\frac{1}{2^2} - \frac{1}{n^2} \right) = \frac{R_H}{4} - \frac{R_H}{n^2} \tag{37.1}$$

where λ is the wavelength in angstrom units, R_H is the famous Rydberg constant, and n is an integer having the value 3 for the longest wavelength (H_α) line in the series, 4 for the next longest (H_β), and so on. Calculate $1/\lambda$ for the four lines you observed and plot the results as ordinates against the corresponding values of $1/n^2$ as abscissas. To obtain more accurate results, construct the graph using as large a scale as possible; that is, utilize the entire sheet of graph paper. Extend the line to $1/n^2 = 0$.

4. According to Equation 37.1, your graph in Calculation 3 should be a straight line of slope $-R_H$. Obtain a measured value of the Rydberg constant in this way.

5. Equation 37.1 also indicates that $1/\lambda$ tends to a maximum value $R_H/4$ and hence λ to a minimum value $4/R_H$ as n gets very large. This minimum value of λ is called the *series limit*. Find it from the $1/n^2 = 0$ intercept on the graph in Calculation 3. Obtain a second measured value of the Rydberg constant from this intercept and compare it with that found in Calculation 4 and with the accepted value listed in Table I at the end of the book. Enter the accepted value in the space provided in the data sheet so that it is displayed along with your two measured values. Watch your units.

6. Measure the wavelength of the sodium doublet by using the setting of the telescope in Procedure 8 and reading off the wavelength from the calibration curve. Compare your results with the known value (use 5893 Å, the mean of 5890 Å and 5896 Å, which are the wavelengths of the two lines forming the so-called doublet).

7. From the data of Procedure 9 and the calibration curve determine the range of the visible spectrum in angstrom units.

QUESTIONS _____

1. Compare the appearance of the bright-line spectrum with that of the continuous spectrum and explain the difference.

2. Of what use is the calibration curve and why is it needed?

3. Compare your results for the range of the visible spectrum with the known value, which you should look up and record in the spaces provided in the data table.

4. Does the index of refraction of glass increase or decrease with increasing wavelength? Explain your answer.

5. The Balmer formula, Equation 37.1, predicts spectrum lines in hydrogen corresponding to all integer values of n equal to or greater than 3. Why, then, did you see only four lines in Procedure 7?

6. (a) What are the frequencies in hertz of the limits of the visible spectrum whose wavelengths you found in Calculation 7? (b) How many times the lower limit frequency is the upper limit frequency? Is it as much as twice (an octave)?

7. (a) Find out what the Doppler effect is and explain it in your own words. (b) Show how this effect can be applied to light waves in determining the speed of stars toward or away from Earth.

8. Calculation 6 indicates that the yellow light from your sodium lamp consists of two closely spaced lines (a "doublet") with respective wavelengths of 5890 Å and 5896 Å rather than a single line, even though you could only see one line in your spectrometer. Explain why this is so.

Interference of Light 38

One of the important methods of measuring the wavelength of light is to allow monochromatic light to be reflected from the two surfaces of a very thin film of varying thickness. The light reflected from one surface will *interfere* with that reflected from the other to produce interference fringes. The light's wavelength can then be calculated from the dimensions of the film and the number of fringes. The object of this experiment is to study the interference of light, to measure the wavelength of the yellow light from a sodium lamp, and to determine the radius of curvature of a convex lens from an observation of the interference pattern called Newton's rings.

THEORY

Interference is the effect produced when two or more wave trains are superimposed so as to either reinforce or cancel each other; the effect's magnitude depends on the relative phases of the two trains. With light, interference occurs when two monochromatic beams having identical frequencies are superimposed, darkness (destructive interference) being produced when the two beams are united so that the positive vibration peaks of one match the negative ones of the other (a 180° phase difference). Conversely, a bright spot (constructive interference) is seen when the two waves are matched peak to peak (zero phase difference). In most interference experiments, the two beams are produced by splitting a single beam to ensure that the frequencies will be the same. The phase difference is obtained by making the two beams traverse paths of different length before being recombined. A path difference of half a wavelength produces a relative phase shift of 180°; hence a path difference of a whole wavelength or any integer number of wavelengths produces no phase shift at all. Therefore, a path difference of any integer number of wavelengths (an even number of half wavelengths) gives rise to constructive interference, whereas a difference of an integer number of wavelengths plus a half wavelength (an odd number of half wavelengths) gives rise to destructive interference.

One method of producing interference is to allow monochromatic light to fall upon two pieces of plate glass placed at a very small angle to each other (see Fig. 38.1). Light will be reflected from each of the glass surfaces. The light reflected from the lower surface of the upper plate and the light reflected from the upper surface of the lower plate will interfere, the second reflected beam traveling a longer distance, namely, across the air film and back. Since the air film varies regularly in thickness, the interference pattern produced will consist of a series of parallel bright and dark lines that are called *interference fringes*. However, one point should be noted in connection with this method: reflection of the light at the upper surface of the lower glass plate (where the light strikes a denser medium) causes a 180° phase shift, which is equivalent to an extra half wavelength of path for this beam. Consequently, with this setup the conditions for constructive and destructive interference are reversed; an actual path difference of an even number of half wavelengths results in destructive interference, while an odd number gives the constructive effect.

The wavelength of the yellow light from a sodium lamp may be determined by using interference fringes. The experimental arrangement for doing this with two glass plates that make a small angle with each other is shown in Fig. 38.2. Note that the plates touch along one edge and are separated by a thin steel strip of thickness T at the other. If L is the distance from the touching edge to the edge of the strip and y is the thickness of the air wedge along a line parallel to and at a distance x from the touching edge, then by similar triangles

$$\frac{y}{x} = \frac{T}{L} \quad \text{or} \quad y = \frac{xT}{L} \quad (38.1)$$

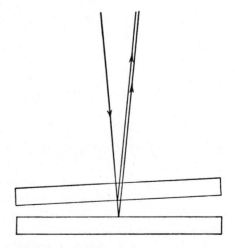

Figure 38.1 *Interference from two plane surfaces*

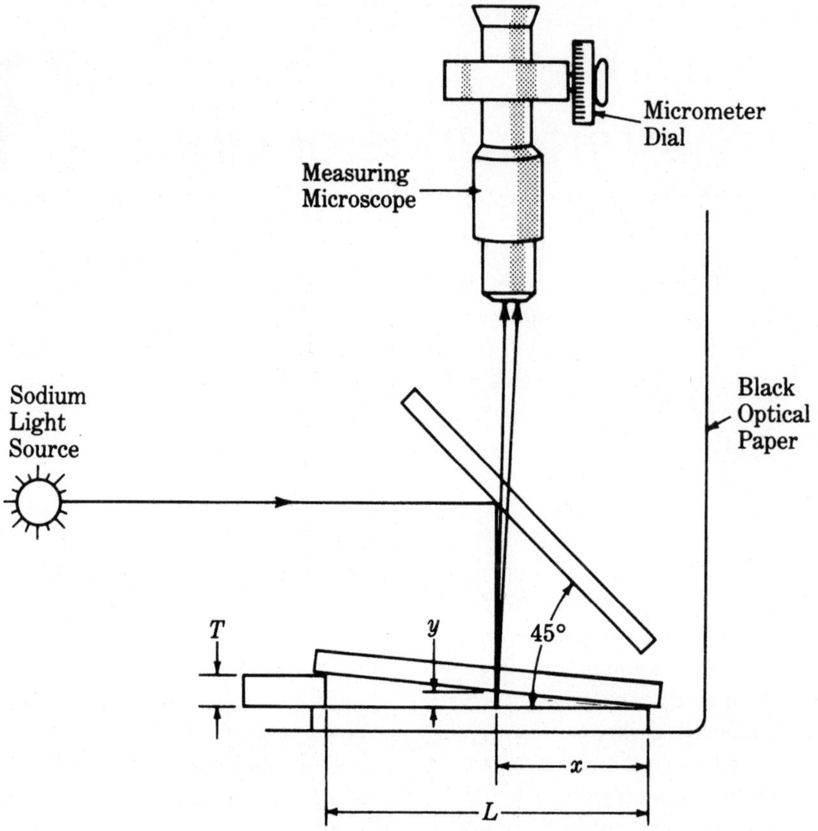

Figure 38.2 *Apparatus for viewing interference fringes*

Now at any place where the path difference is an odd number of half wavelengths, the two reflected rays will be in phase because of the phase reversal at the lower plate, and a bright line or fringe will appear at this value of x. Conversely, where the path difference is an even number of half wavelengths, a dark line will be seen. Hence, for a dark line, $2y = m\lambda$ where m is an integer and $2y$ is the round-trip distance across the air wedge. Substitution in Equation 38.1 then yields

$$x_m = \frac{mL\lambda}{2T} \qquad (38.2)$$

where x_m is the distance of the mth dark line from the touching edge. The distance between adjacent dark lines is clearly

$$x_{m+1} - x_m = \frac{(m+1)L\lambda}{2T} - \frac{mL\lambda}{2T} = \frac{L\lambda}{2T} \qquad (38.3)$$

from which we get

$$\lambda = \frac{2T}{L} \text{ (spacing between adjacent fringes)} \quad (38.4)$$

Note that attention must be paid to the units in this relation. If T, L, and the fringe spacing are all in centimeters, λ will be in centimeters also and must be multiplied by 10^7 for nanometers or 10^8 for angstrom units.

Another very interesting example of interference produced by reflection from the two sides of an air

wedge bounded by glass surfaces is called *Newton's rings*. The arrangement, which appears in Fig. 38.3, is seen to be similar to that of Fig. 38.2 except that the upper glass plate and steel strip have been replaced by a convex lens whose spherical lower surface touches the lower glass plate at its central point and forms an air wedge whose thickness increases as we go radially outward from that point. Again, destructive interference will occur when $2y = m\lambda$ and will be seen as a dark ring of radius r_m about the central point, where r_m is the radial distance from this point to where the lower lens surface has curved up to the distance $m\lambda/2$ above the upper surface of the glass plate. To find r_m we need a relation between r and y as defined in Fig. 38.3 analogous to Equation 38.1 for the arrangement of Fig. 38.2. A study of Fig. 38.3 shows that

$$(R - y)^2 + r^2 = R^2 \qquad (38.5)$$

which, when multiplied out, reduces to

$$r^2 = 2Ry - y^2 \qquad (38.6)$$

where r is the radial distance outward from the point where the lens contacts the glass plate and R is the lens's radius of curvature. Because $y \ll R$, the term y^2 in Equation 38.6 is negligible compared to $2Ry$, and we can write $r^2 \approx 2Ry$ from which there follows

$$r_m^2 = m\lambda R \qquad (38.7)$$

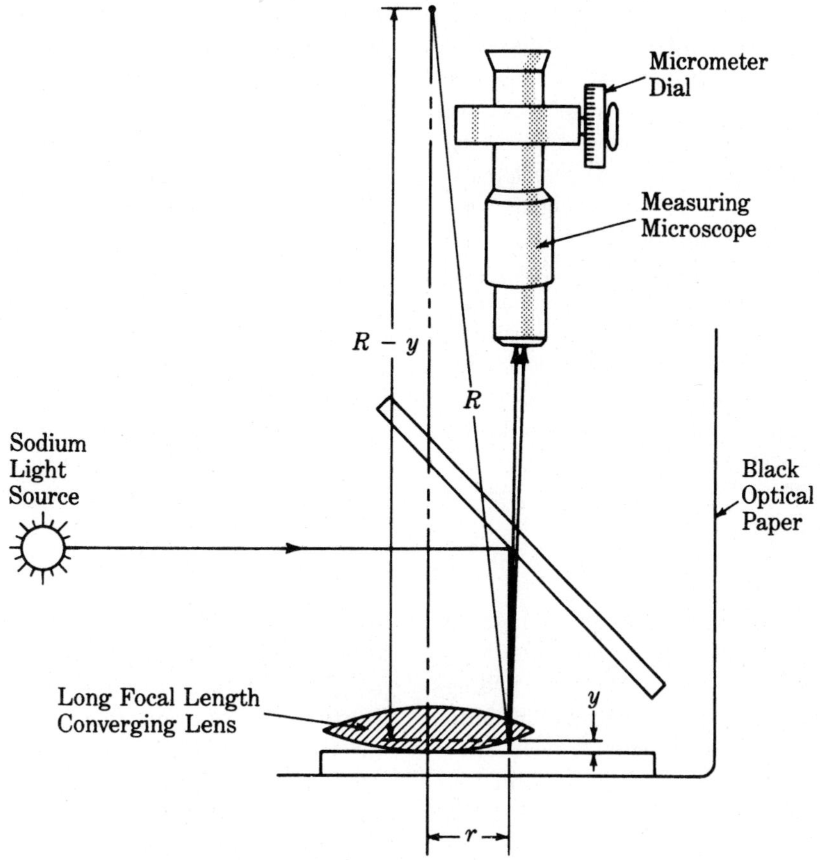

Figure 38.3 *Apparatus for viewing Newton's rings*

Here, as already noted, r_m is the radius of the mth dark ring. This result shows that a plot of the squares of the radii of Newton's rings against the corresponding values of the index m should be a straight line of slope λR. The wavelength of light can then be found if the radius of curvature of the lens is known. Conversely, if λ is known, the value of R for the lens could be determined (as, for example, if a source of light of a known wavelength were used).

APPARATUS

1. Sodium light source
2. Three optically flat glass plates, about 7 cm square
3. Measuring microscope
4. Thin strip of steel about 0.0075 cm thick
5. Converging lens, 20- to 25-cm focal length
6. Centimeter scale
7. Micrometer caliper
8. Stand and clamps
9. Sheet of black optical paper

PROCEDURE

1. Measure the thickness of the thin strip of steel with the micrometer caliper and record the reading.

2. Be sure that the glass plates are clean. Be careful not to get finger marks on the surfaces that are to be placed together. Lay the sheet of optical paper on the microscope base and bend it up to form a backstop for the incident light as shown in Fig. 38.2. Hold the paper down on the base by laying one of the glass plates on it. Then carefully lay the other plate down flat upon the first with the best surfaces in contact, so that the upper plate projects slightly at the thin end of the air film, again as shown in Fig. 38.2. Separate the plates at the other end by the thin steel strip, placing the edge of the strip parallel to the ends of the glass plates.

3. Measure the length L of the air film (the distance between the edge of the steel strip and the line along which the two plates rest on each other) and record this reading.

4. Place the support stand beside the measuring microscope, mount the third glass plate in the clamp, and attach this clamp to the stand with a right-angle clamp so that the plate is held under the microscope objective but above the other two plates at an angle of 45° to the horizontal. See Fig. 38.2, which shows the complete experimental arrangement. Set up the sodium light source so that it shines horizontally into the 45°-plate, which serves as a partial mirror to reflect the light down onto the two plates forming the air film but allows the light reflected from the film to pass through to the microscope. Note that much of the light from the source passes through the 45°-plate and is absorbed by the optical paper backstop, and some of the light reflected from the air film is reflected back to the source instead of being transmitted. These losses result in a decrease in the intensity of the observed pattern, but as the light source is very strong, there is plenty of intensity remaining for observation.

5. Turn the source on and wait for it to warm up so that a brilliant yellow light is produced. Then look through the microscope and adjust its focus (by sliding the microscope barrel up and down in its mount) until you see a sharp image of the interference fringes. They consist of a series of dark parallel lines across the microscope's field of view and are perpendicular to the length of the air film (parallel to the edge of the steel strip). Adjust the eyepiece focus until you also see a sharp image of the microscope cross hairs, and rotate the microscope barrel until one of the cross hairs (the double one if your microscope is so equipped) is parallel to the dark fringes.

6. Your measuring microscope has a calibrated micrometer screw that allows you to traverse the cross hairs across the fringe pattern through distances that can be measured to high precision. Begin by setting the parallel cross hair right on top of one of the dark fringes toward the left side of the image. If your microscope has a double cross hair, set it so that the dark fringe falls in the center of the space between the pair. Double cross hairs are sometimes provided because this setting can then be made with higher precision. Observe and record the reading of your microscope's micrometer dial. Then traverse the cross hairs until you come to the tenth dark fringe to the right of the one you started from. Set the parallel cross hair exactly on this fringe and again read the micrometer dial setting. Note that you will have turned the micrometer dial through many complete revolutions in traversing ten fringes. The best way to take account of this is to count the number of times you turned the dial through zero, multiply this number by the number of dial divisions in one revolution (usually 10 large or 100 small divisions),

and add on the final dial setting to obtain the final reading to be recorded.

7. Repeat Procedure 6 twice to obtain a total of three sets of measurements of the distance spanned by ten dark fringes.

8. Carefully remove the upper one of the two horizontal glass plates and the steel strip. Wipe off the lens with a damp cloth or lens paper to be sure it is free of lint, dust, and finger marks, and place it on the glass plate that is still under the microscope. If the lens is plano-convex instead of double convex, be sure the convex side is down, that is, in contact with the plate surface, and position the lens directly below the microscope. Look through the microscope and adjust its focus until you see a sharp image of the series of dark circles that constitute Newton's rings. Be careful about this observation: If you did not get the lens in the right position under the microscope, you may miss the ring pattern entirely, and in any case you may have to move the lens so that the center of the pattern is in the center of your field of view. Moreover, when you have the pattern in focus, there should be a dark spot at the center. If there is not, it means that either the lens or the glass plate is still dirty, even though the dirt may be just a speck of dust. Again, clean them carefully until you get a sharp pattern of Newton's rings with a dark spot in the middle.

9. Measure the diameter of the first Newton's ring (the first dark circle around the central dark spot) by moving the cross hairs in the measuring microscope until the vertical one is tangent to the circle on the left. This cross hair is the double one if your microscope is so equipped, and in this case you should position the pair so that the dark, rather thick circular line just comes into the space between them. In the case of a single cross hair, position it so that it is tangent to the inner line of the circle rather than to its outer edge. Read and record the micrometer dial setting. Then traverse the cross hairs to the right until the vertical one is tangent to the circle on the right in the same manner as on the left. Count the number of times you turned the micrometer dial through zero, note the final dial setting, and record the final reading as the number of turns multiplied by the divisions per turn plus the final setting, as you did in Procedure 6.

10. Repeat Procedure 9 twice to obtain a total of three sets of measurements of the diameter of the first dark ring.

11. Repeat Procedures 9 and 10 for the second, third, fourth, and fifth rings.

12. Carefully remove the lens from your apparatus and measure its focal length by the method of Procedure 1 in Experiment 36.

DATA _____

Thickness T of the thin steel strip   _____   Length L of the air film   _____

Micrometer Dial Readings for the Width Spanned by Ten Fringes

Trial	Initial setting	Final setting	Difference Image	Difference Actual	Deviation
1					
2					
3					
Average					

Measured wavelength of
sodium light   _____   Percent error   _____

Diameters of Newton's Rings

Ring No. m	Trial	Initial setting	Final setting	Diameter Image	Diameter Actual	Actual radius	Actual radius^2
1	1						
	2						
	3						
	Average						
2	1						
	2						
	3						
	Average						
3	1						
	2						
	3						
	Average						
4	1						
	2						
	3						
	Average						
5	1						
	2						
	3						
	Average						

Value of R from slope of graph   _____   Measured focal length of lens   _____
Calculated focal length of lens   _____   Percent error   _____

CALCULATIONS

1. Subtract the initial settings of the micrometer dial from the final settings recorded in Procedures 6 and 7. Note that the final setting is the dial reading plus the total number of divisions turned through from the zero *before* the initial dial setting, so that subtraction of this initial setting gives you the total number of dial divisions turned through in traversing the ten fringes. Be careful of your units. For example, for the Gaertner M101 or Ml01A microscope equipped with the M202 eyepiece, each *small* division on the micrometer dial represents a cross-hair motion of 0.0025 mm. If you are using a different microscope, your instructor will give you its calibration. Enter your differences between initial and final dial settings along with the proper units in the *image* column in the pertinent table in the data sheet.

2. The measurements made in Procedures 6 and 7 and hence the width spanned by ten fringes found in Calculation 1 are all distances along the *image* of the interference pattern formed by the microscope's objective lens. They are therefore all magnified by the magnification produced by that lens and must be divided by that magnification to get the actual distances along the interference pattern itself. Divide each of your results from Calculation 1 by this factor to get the actual distances, and record them in the appropriate column. For the Gaertner M101 and Ml01A microscopes, the magnification factor is 3.2; if you are using a different microscope, this factor will be supplied by the instructor.

3. Compute the average of your three actual distances spanned by ten fringes. Then compute the deviation of each distance from the average and find the average deviation (a.d.).

4. Find the spacing between adjacent fringes by dividing your result from Calculation 3 by 10, and use this spacing and your data from Procedures 1 and 3 in Equation 38.4 to determine the wavelength of the light from your sodium lamp.

5. Use the a.d. found in Calculation 3 to find the error to be attached to your result in Calculation 4. Now compare this result with the known value, 5893 Å, for the average wavelength of the sodium doublet by noting whether this known value falls within the limits of error you just determined. Also calculate the percent error by which your measured value differs from the known or accepted value.

6. Subtract the initial micrometer dial settings from the final settings obtained in Procedures 9–11 to obtain the diameters of the five Newton's rings you measured just as you did for the set of ten fringes in Calculation 1. Enter your results as the diameters of the images.

7. Divide your results from Calculation 6 by the objective magnification as you did in Calculation 2 to obtain the actual ring diameters. Then divide each actual ring diameter by 2 to get the actual radius r_m and square each of these radii to get r_m^2. Enter all these results in the appropriate data table boxes.

8. Calculate the average radius and an average of the radius squared for each ring from $m = 1$ to $m = 5$.

9. Plot your values of r_m^2 as ordinates against the corresponding values of the index m as abscissas. According to Equation 38.7, this graph should be a straight line of slope λR, where λ is the wavelength of the light used and R is the radius of curvature of the lens. Note that the origin should be a point on your graph.

10. Find the slope of the graph you made in Calculation 9 and from it determine R, using 5893 Å for λ. Then use the lens maker's equation (Equation 35.2) to find the focal length of your lens and compare this result with the focal length measured in Procedure 12 by calculating the percent error. *Note:* If your lens is plano-convex, the radius of the flat side is infinite and its reciprocal zero. If your lens is double convex, assume that both sides have the same radius of curvature. Also assume that your lens is made of crown glass for which the index of refraction is given in Table XIII at the end of the book unless your instructor gives you a better value. Note that you should not be surprised if you find an appreciable difference between your calculated and measured values of the lens's focal length, especially if you are uncertain about the index of refraction of the lens material.

QUESTIONS _____

1. Explain how the method used in this experiment could be employed to test whether a surface is optically plane or not.

2. Explain why the fringes sometimes run diagonally across the plates.

3. Suppose that white light were used to produce the interference fringes in this experiment. What would be the appearance of the fringes produced?

4. Two plane glass plates touch at one edge and are separated at the opposite edge by a strip of tinfoil. The air wedge is examined with sodium light reflected normally from its two surfaces, and a total of 35 dark interference fringes are observed. Compute the thickness of the tinfoil.

5. How is this experiment evidence for a wave theory of light? Explain.

6. Explain why interference patterns caused by light reflected from the top surface of the upper plate or from the bottom surface of the lower plate in Fig. 38.1 are not observed.

7. In the setup to observe Newton's rings, why should you see a black spot in the center of the pattern?

8. Did you get good agreement between the value of your lens's focal length obtained from observing Newton's rings and the directly measured value? If not, discuss why you might expect a disagreement.

9. Prove that the area between two adjacent Newton's rings is a constant regardless of which pair of adjacent rings is chosen.

10. A certain double convex, crown-glass lens has a focal length of 30 cm. Find the radius of the tenth dark Newton's ring viewed with sodium light ($\lambda = 5893$ Å).

The diffraction grating is a most important and interesting optical device. It is among the simplest devices for producing spectra so that the wavelengths contained in a given light beam may be measured, and it has the advantage of making calibration with a known spectrum unnecessary. The object of this experiment is to study the operation of a diffraction grating and, in particular, to see how it may be used to measure the wavelength of light from a helium-neon laser. The wavelength produced by such a laser is known to great accuracy, so that an excellent check on a grating's ability to measure wavelength can be made.

THEORY

The diffraction grating provides the simplest and most accurate method for measuring wavelengths of light. It consists of a very large number of fine, equally spaced parallel slits. There are two types of diffraction gratings: the reflecting type and the transmitting type. The lines of the reflection grating are ruled on a polished metal surface, and the incident light is reflected from the unruled portions. The lines of the transmission grating are ruled on glass; the unruled portions of the glass act as slits. Gratings usually have about 10,000 to 20,000 lines per inch. The grating used in this experiment is a transmission grating replica.

The principles of diffraction and interference are applied to the measurement of wavelength in the diffraction grating. Let the broken line in Fig. 39.1 represent the edge view of a magnified portion of a diffraction grating, and let a beam of parallel rays (a plane wave) of monochromatic light impinge on the grating from the left. By Huygens's principle, the light spreads out in every direction from the apertures of the grating, each of which acts as a separate new source of light. Because the grating slits are very narrow, they act as line sources, and the emitted secondary wavelets are, as Fig. 39.1 shows, cylindrical waves. The envelope of these wavelets determines the position of the complete advancing wave. The figure shows the instantaneous positions of several successive wavelets after they have advanced beyond the grating. Lines drawn tangent to these wavelets connect points in phase, hence they represent the new wave fronts. One of these wave fronts is tangent to wavelets that have all advanced the same distance from the slits; this wave front is parallel to the original one. A converging lens placed in the path of these rays would form the central image. Another wave front is tangent to wavelets whose distances from adjacent slits differ by one wavelength. This wave front advances in direction 1 and forms the first-order spectrum. The next wave front is tangent to wavelets whose distances from adjacent slits differ by two wavelengths. This wave front advances in direction 2 and forms the second-order spectrum. Spectra of higher orders will be formed at correspondingly greater angles.

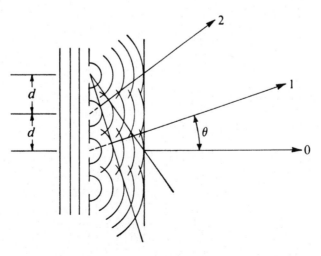

Figure 39.1 *The diffraction grating*

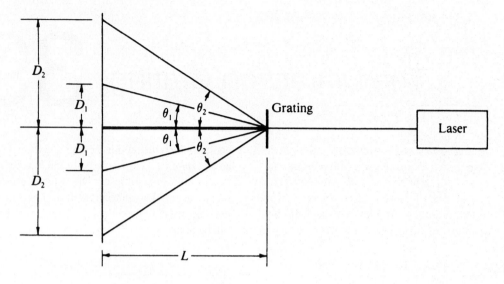

Figure 39.2 *Experimental arrangement for the diffraction grating*

Looking at Fig. 39.1, we note that ray 1 has been drawn perpendicular to its wave front at the wave front's point of tangency with a Huygens wavelet that is coming from the third slit down from the top. A projection of this ray backward will pass through the center of the slit, because a line drawn perpendicular to a tangent at the point of tangency is a radius of the circle in question. In the right triangle of ray 1, the associated wave front, and the line representing the grating, we see that the angle between the wave front and the grating is the same as θ, the angle between rays 1 and 0. Consideration of this triangle reveals that $\sin \theta = 2\lambda/2d = \lambda/d$, where d is the distance between slits. Similarly, for ray 2, which, for convenience, has been drawn to the center of the second slit down, we note that $\sin \theta = 2\lambda/d$ where now θ is the angle between rays 2 and 0. In general, we will find that $\sin \theta = n\lambda/d$, or

$$n\lambda = d \sin \theta \qquad (39.1)$$

where n is an integer (0, 1, 2, 3, etc., not to be confused with the index of refraction) and is called the *order* of the spectrum. In this equation, λ is of course the wavelength of the light and will be in the same units as d, the spacing between the grating lines. The wavelength of visible light is usually expressed in angstrom units; one angstrom unit is 10^{-8} cm. However, the nanometer (10^{-9} m, hence 10^{-7} cm or 10 Å) is becoming increasingly popular in expressing optical wavelengths. The grating constant d is ob-

tained by noting the number of lines per inch or per centimeter marked on the grating and taking the reciprocal to find the spacing between lines. This spacing will then be in inches or centimeters, so that a conversion must be made to obtain λ in appropriate units.

In this experiment, a helium-neon laser will be used to provide a narrow beam of highly monochromatic light ($\lambda = 6328$ Å). A plan view of the experimental arrangement is shown in Fig. 39.2. The laser beam strikes the grating normally and the direct ray proceeds in a straight line to the screen. The first-order ($n = 1$) diffracted rays are deflected through angles θ_1 on either side of the central ray and strike the screen at points spaced D_1 on either side of the central spot. Similarly, the second-order ($n = 2$) diffracted rays each make angle θ_2 with the central ray and strike the screen at points separated from the central spot by D_2. These angles are determined by measuring D_1, D_2, and L, the distance between the grating and the screen, and noting that $\tan \theta_1 = D_1/L$ and $\tan \theta_2 = D_2/L$.

Because the laser light contains only one wavelength, only one pair of spots will appear on the screen in each order. However, if the experiment is run with light that is not monochromatic, there will be as many pairs of spots in each order as there are different wavelengths in the light from the source, the diffracting angle θ for each wavelength being given by Equation 39.1. We will observe these results when white light from a tungsten lamp is substituted for the laser light.

APPARATUS

1. Low-power helium-neon laser
2. Diffraction grating
3. Optical bench
4. Long-focus converging lens (about 25 cm)
5. Ground-glass screen

6. Lens holders and optical bench clamps
7. Meter stick
8. Masking tape
9. Tungsten lamp
10. Stand and clamps for mounting the light source

PROCEDURE _____

1. Set up the optical bench and mount the ground-glass screen at one end. Mount the grating about 30 cm in front of it using one of the lens holders and optical bench clamps. The laser is then mounted so that the beam is directed normally at the grating, as shown in Fig. 39.2.

2. Turn the laser on and observe the central spot on the screen. CAUTION: The low-power laser used in this experiment cannot hurt you even if the beam is allowed to fall on your skin for a prolonged period. However, *you must not look into it or allow anyone else to do so.* Keep the laser beam down below eye level and do not let it shine in anyone's face including your own.

3. With the usual size ground-glass screen and a grating of about 15,000 lines per inch placed 30 cm away, the first-order spots will appear on the screen, but the second-order ones will miss it on either side. To display both orders and facilitate measurement of the separations D, tape the meter stick to the screen so that the central spot falls approximately at the 50-cm mark and the diffracted spots are displayed right on the stick. Adjust the orientation of the screen so that it and the meter stick are perpendicular to the central ray. Do this by rotating the screen in its mount until the two second-order spots are equidistant from the central spot.

4. Read and record the positions of the two first-order spots on the meter stick.

5. Similarly, read and record the positions of the two second-order spots. These spots will be fainter than the first-order ones; you may need to darken the room in order to see them clearly.

6. Measure and record the distance L from the grating to the surface of the meter stick. If the length graduations on the optical bench are used for this purpose, allowance must be made for the thickness of the meter stick and one half the thickness of the ground-glass screen.

7. Shut off the laser, replace it with the tungsten lamp, and mount the converging lens on the optical bench between the lamp and the grating. Align the lamp and the lens so that the light passing through the lens falls on the middle of the grating. The lens should be moved back and forth along the optical bench until the central spot observed on the screen is a focused image of the lamp filament. The lamp must be mounted at a distance greater than the lens's focal length behind the grating to make this adjustment possible.

8. Examine the continuous spectrum displayed on the meter stick in first order, noting the relative positions of the different colors. The second-order spectrum will probably be too faint to be seen. Record the positions of the inner ends of the first-order spectra observed on either side of the central white spot.

9. Record the positions of the outer ends of the displayed spectra.

DATA _____

Number of lines per inch on
   the grating   _____

Grating constant d in centimeters   _____

Distance L from grating to meter stick
   in centimeters   _____

Source	Meter stick reading		Calculated values						
	Right	Left	Difference	D	tan θ	θ	sin θ	λ	% error
Laser (first-order)									
Laser (second-order)									
Lamp (violet end)									
Lamp (red end)									

Range in wavelength of visible
   spectrum   _____

CALCULATIONS

1. Compute the grating constant (that is, the distance between the lines on the grating) in centimeters from the number of lines per inch stated on the grating.

2. Find D_1, the displacement of the first-order spot from the central spot, by computing the difference between the two spot positions recorded in Procedure 4 and dividing by 2.

3. Calculate θ_1, the deviation of the first-order diffracted laser beam, from your values of D_1 and L.

4. Calculate the laser light wavelength by using Equation 39.1 with $n = 1$, $\theta = \theta_1$ (from Calculation 3), and the grating constant d found in Calculation 1. Note that with d in centimeters, λ will come out in centimeters and must be multiplied by 10^8 to give angstrom units.

5. Find the percent deviation of your computed value of the laser light wavelength from the known value (6328 Å).

6. Repeat Calculations 2–5 for the second-order diffracted laser light, using the data of Procedure 5.

7. Repeat Calculations 2–4 for the violet end of the continuous spectrum obtained with the tungsten lamp, using the data of Procedures 7 and 8.

8. Repeat Calculations 2–4 for the red end of the continuous spectrum, using the data from Procedure 9.

9. Calculate and record the range in wavelengths of the visible spectrum as measured in this experiment.

QUESTIONS _____

1. Why does a grating produce several spectra, while a prism produces only one?

2. (a) Discuss the difference between the continuous spectrum obtained with a grating and one obtained with a prism. (b) Which color is deviated most in each case? Why?

3. A monochromatic beam of light impinges normally on a grating and forms an image of the source slit 10 cm from the central image on a screen. The screen is 50 cm from the grating and is set parallel to it. The grating has 10,000 lines per inch. What is the wavelength of the light?

4. Why are the higher-order spectra more accurate than the first-order spectrum in determining wavelengths of light?

5. A grating having 15,000 lines per inch is used to produce a spectrum of the light from a mercury arc shining through a slit. What is the angle between the first-order green line and the second-order green line? See Experiment 37 for the wavelength of the mercury green line.

6. Consider two rays coming from adjacent slits in a grating of grating constant d and intersecting at the point where they strike a distant screen. Use the rule for constructive interference (see Experiment 38) to derive Equation 39.1. *Note:* Because the distance from the grating to the screen is very much larger than the separation between the slits, the two rays are almost parallel.

7. (a) Why in the present experiment did you not observe a third-order diffracted beam of laser light? (b) What is the highest order in which you would expect to be able to view the mercury green line (see Experiment 37 for λ) when using a grating having 15,000 lines per inch?

8. Is it possible for the red end of the visible spectrum viewed in first order to overlap the violet end viewed in second order? Assume that the visible spectrum runs from 7000 to 4000 Å.

9. What would be the effect of the plane of the grating not being perpendicular to the beam of light coming from the laser? If you can, show how Equation 39.1 should be modified to take account of an angle ϕ between this beam and the normal to the grating plane.

10. What is the effect on the observed spectrum of using a larger grating, that is, one having the same grating constant but a greater total number of lines? Assume an incident beam whose cross section covers the grating's entire active surface. *Hint:* How does the pattern produced in a Young's experiment, which, in effect, uses a grating with only two slits, differ from the pattern you observed with your grating of many thousands of slits?

Polarized Light 40

Experiments on interference and diffraction show that light is composed of waves, but they do not offer any evidence as to whether these waves are transverse or longitudinal. However, experiments on polarization show that light waves must be transverse. Polarization, which refers to the direction assumed by the wave vector in the plane perpendicular to the direction of propagation, is only possible with transverse waves. The purpose of this experiment is to study some of the fundamental phenomena of polarized light.

THEORY

Light waves consist of transverse electromagnetic vibrations that travel with finite velocity. The electric vibrations and the magnetic vibrations are perpendicular to the direction of propagation of the light and also perpendicular to each other. In ordinary light, the electric vibrations occur in all directions in a plane perpendicular to the direction of propagation as shown in Fig. 40.1 (a). This light is unpolarized. It is possible to produce *plane-polarized* light, that is, light in which the electric vibrations all occur in some one direction perpendicular to the light ray, that is, to the direction of propagation. Vertical plane polarization is illustrated in Fig. 40.1 (b) and horizontal plane polarization in Fig. 40.1 (c). Notice that by definition of a transverse wave, in all these cases the direction of the electric vibrations (the electric vector **E**) lies in a plane perpendicular to the ray. In the two cases of plane polarization, however, another plane, perpendicular to the one just mentioned, is defined by the ray and the one direction occupied by the electric vector. This plane, which appears only in Fig. 40.1, parts (b) and (c), is called the *plane of polarization*—hence, the term "plane polarized" for the light represented by these two figures.

Light may be plane polarized by reflection from a nonconducting reflector or by passing through a doubly refracting crystal. When light is reflected from the surface of a transparent dielectric medium such as glass or water, it is polarized to some extent if the angle of incidence is greater than 0° but less than 90°. The *reflected* light is *completely* polarized if the ray reflected from the medium is at right angles to the refracted ray proceeding into the medium. Calculation of the angle of incidence i_p for which this condition is satisfied may be carried out by reference to Fig. 40.2. Clearly, $r_1 + r_2 + 90° = 180°$ or $r_1 + r_2 = 90°$. But by the law of reflection, $r_1 = i_p$, and by the law of refraction (Snell's law, Equation 35.1), $\sin i_p = n \sin r_2$. Then, since $r_2 = 90° - r_1 = 90° - i_p$, $\sin i_p = n \sin (90° - i_p) = n \cos i_p$, and there results

$$n = \frac{\sin i_p}{\cos i_p} = \tan i_p \qquad (40.1)$$

This is *Brewster's law* and the angle of incidence i_p whose tangent is equal to the index of refraction of the medium and for which the reflected beam is completely polarized is called *Brewster's angle*. For a piece of flat, unsilvered glass it is about 57°. Note that the direction of polarization of the reflected beam is parallel to the reflecting surface (perpendicular to the plane of the paper

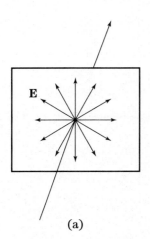
(a)

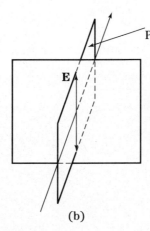
(b)

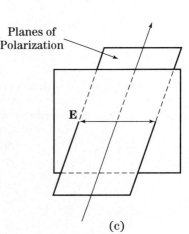
(c)

Figure 40.1 *Polarization of light*

in Fig. 40.2). Note also that although this beam is completely polarized, the refracted beam in the dielectric medium is not. This situation may be understood by considering that the incident (unpolarized) light is made up of two components, one polarized perpendicular to the plane of the paper in Fig. 40.2 and the other lying in that plane. The former is largely reflected, although a part proceeds into the new medium, but at Brewster's angle *all* of the component polarized in the plane of the paper is transmitted into the medium. Thus, the transmitted beam is "enriched" in this latter component, while the reflected beam contains none of it.

In a doubly refracting crystal such as calcite, light propagates at two different velocities depending on whether it is polarized along or perpendicular to the so-called optic axis. The electric vibrations of a light ray propagating along the optic axis are perpendicular to this axis even if the light is unpolarized, because light waves are transverse. Such a ray travels with a speed $v_\perp$ regardless of its polarization. However, if a ray is directed perpendicular to the optic axis in a doubly refracting crystal, we must consider two components. One is polarized perpendicular to the optic axis and travels with the same speed $v_\perp$ as does a wave traveling along the optic axis and thus is called the *ordinary ray*. The other component, polarized along the optic axis, travels with a different speed $v_\parallel$ and is called the *extraordinary ray*. The crystal then exhibits two indices of refraction, $n_\perp = c/v_\perp$ and $n_\parallel = c/v_\parallel$. If a ray of unpolarized light travels through a doubly refracting crystal in a direction making some angle (other than 0° or 90°) with the optic axis, there will still be an ordinary ray polarized perpendicular to this axis that obeys Snell's law in the usual manner. In this case, however, the extraordinary ray's electric vibrations are neither parallel nor perpendicular to the optic axis

and thus involve both indices of refraction. Therefore, this ray does not obey the simple form of Snell's law and is consequently bent away from the ordinary ray. Hence, a doubly refracting (also called a birefringent) crystal has the property of dividing a beam of unpolarized light into two beams plane polarized at right angles to each other.

Some crystals have the property of transmitting a light beam polarized along the optic axis and absorbing one polarized at right angles to this axis. Unpolarized light incident perpendicular to the optic axis can be plane polarized in this manner. This property is called *dichroism.* Tourmaline is the most common example of a dichroic crystal.

Natural crystals are usually too small to be used in an experiment where a wide beam of polarized light is needed; for this purpose, Polaroid sheets are used. In a Polaroid, very minute crystals are distributed uniformly and densely in a cellulose film mounted between two glass plates. The individual crystals are needle-shaped and submicroscopic in size and are turned so that their axes are parallel by stretching the film before it is mounted on glass. Like tourmaline, Polaroids are dichroic and produce polarized light in the same manner.

A Polaroid has an optic axis in the plane of the sheet along which the electric vibrations of a normally incident light ray must lie in order for the ray to get through without absorption. If there is an angle θ between the optic axis and the direction of the electric vibrations, as shown in Fig. 40.3, then the electric vector **E** in this direction may be broken up into a component $E_\perp = E \sin \theta$, which is perpendicular to the optic axis and gets absorbed by the Polaroid, and a component $E_\parallel = E \cos \theta$, which is parallel to the optic axis and is therefore transmitted with little or no absorption. Since the intensity of a light beam is

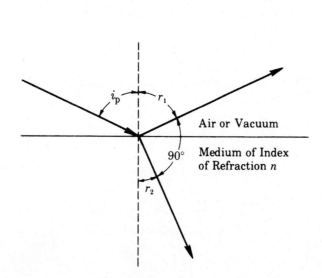

Figure 40.2 *Brewster's angle*

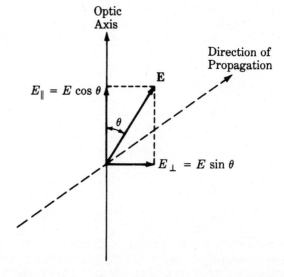

Figure 40.3 *Components of **E** with respect to the optic axis*

proportional to the square of the amplitude of the electric vibrations, the ratio of the intensity of the polarized light transmitted by the Polaroid to the incident intensity is

$$\left(\frac{E_{\parallel}}{E}\right)^2 = \cos^2\theta \qquad (40.2)$$

This result is called the law of Malus after its discoverer, E. L. Malus (1775–1812), and states that if plane-polarized light is normally incident on a Polaroid or other dichroic material, the transmitted intensity is proportional to the square of the cosine of the angle between the optic axis and the direction of polarization. Note that the intensity becomes zero when this angle is 90°.

APPARATUS

1. Two Polaroid disks, each mounted in a holder in which it can rotate
2. Optical bench with clamps and holders
3. Light source with lens to produce a parallel beam of light
4. Desk lamp
5. Photometer with optic probe
6. Calcite crystal
7. Pile of thin glass plates
8. Black glass mirror
9. Piece of cellophane
10. Thin piece of mica
11. Unannealed glass plate
12. Annealed glass plate
13. Piece of celluloid
14. Iris diaphragm

PROCEDURE

1. Place the light source at one end of the optical bench and, if a separate lens is used, mount it one focal length from the source. Place the photometer's optic probe at the other end of the bench and position it to receive the light. Mount one of the Polaroid disks on the bench in front of it, positioned so that the light must pass through the disk to get to the probe. This Polaroid is the analyzer. Rotate it and notice the photometer reading. Adjust the photometer's sensitivity control to give a nearly full-scale reading. If your instrument has no such control, mount a diaphragm in front of the lens and adjust it to obtain the same result. The light coming directly from the lamp is unpolarized. The intensity of the light transmitted by the analyzer should be nearly independent of the angle that its axis makes with the vertical.

2. Place the second Polaroid disk on the optical bench in front of the lens and about 4 in. away from it, in such a position that all the light passes through it. This is the polarizer. Place it with its axis in the vertical direction.

3. Now rotate the analyzer and take readings of the photometer for different angular settings. Begin with the axis vertical (and hence parallel to the polarizer's axis), and take readings for angles 0°, 10°, 20°, 30°, etc., up to 180°. You will observe that the intensity of the light transmitted by the analyzer depends upon the angle between the axis of the analyzer and that of the polarizer.

4. Remove the optic probe. Allow the light to pass through both Polaroids, leaving a space of about 6 in. between them. Set the two Polaroids for extinction, that is, so that the minimum amount of light passes through them. Now place a piece of cellophane in the space between the Polaroids, look at the cellophane through the analyzer, and observe the field of view. Fold the cellophane into several thicknesses and observe the effect on the color produced. Describe what you see in the space below.

5. Place a thin piece of mica in the space between the Polaroids, look at the mica through the analyzer, and observe the field of view. Again describe what you see in the space below.

6. Repeat Procedure 5 using the unannealed glass plate.

7. Repeat Procedure 5 using the annealed glass plate.

8. Repeat Procedure 5 using a piece of celluloid. Twist the celluloid slightly out of shape and observe the regions where strains are produced.

9. Remove the polarizer, light source, and lens from the optical bench. Place the black glass mirror in a holder on the optical bench and allow the unpolarized light from the desk lamp to fall on it. Rotate the mirror holder so as to make the angle of incidence about 55°. Observe the light reflected from the mirror by looking at it through the analyzer Polaroid. By rotating this Polaroid, notice whether the reflected light is polarized. Report your results in the space below.

10. Repeat Procedure 9 using a pile of thin glass plates as the mirror. Determine whether the reflected light is polarized and also whether the transmitted light is polarized. State your findings in the space below.

11. Make a dot on a piece of paper. Place a calcite crystal over the dot and look through it. You will observe two images. Rotate the crystal and notice what happens to the two dots. Now look at the dots through the calcite and a Polaroid. Rotate the Polaroid and notice the result. The two dots represent two beams of light coming from the calcite crystal. Determine whether these two beams are polarized and find the relative direction of vibration of the two beams. Report your results in the space below.

12. If the sun is shining, look at the blue light from the sky through the analyzer. Notice particularly the light coming from a direction roughly perpendicular to the sun's direction. Determine whether this light is polarized and report your results in the space below.

Analyzer angle θ	$\cos^2 \theta$	Photometer reading	Analyzer angle θ	$\cos^2 \theta$	Photometer reading
0°			100°		
10°			110°		
20°			120°		
30°			130°		
40°			140°		
50°			150°		
60°			160°		
70°			170°		
80°			180°		
90°					

CALCULATIONS

1. The percent polarization is given by the expression

$$P = 100 \frac{I_{max} - I_{min}}{I_{max} + I_{min}}$$

where P is the percentage polarization, I_{max} is the largest value of the photometer reading observed, and I_{min} is the smallest value observed. Calculate the percent polarization from your readings in Procedure 3.

2. Calculate and record the value of $\cos^2 \theta$ for each of the angles between the polarizer and analyzer optic axes used in Procedure 3.

3. Plot a graph using the photometer readings obtained in Procedure 3 as ordinates and the values of $\cos^2 \theta$ as abscissas. Note that $\cos^2 \theta$ will be the same for $\theta = 80°$ and $100°$, for $\theta = 70°$ and $110°$, etc., and you should therefore get the same photometer readings for the two angles in each of these pairs. Differences in these readings represent errors in the experiment. Plot all points and draw the best straight line through them.

QUESTIONS

1. Mention some of the uses of polarized light.

2. Explain the difference between transverse waves and longitudinal waves.

3. What is meant by a plane-polarized beam of light?

4. (a) Describe the construction of a nicol prism. (b) Explain how it produces a polarized beam of light.

5. A so-called quarter-wave plate is a slice cut from a calcite or other birefringent crystal in such a way that the optic axis lies in the plane of the slice. The name comes from the fact that the thickness of the slice is made equal to $\lambda/4$, where λ is the wavelength of the light to be used. Look up and learn what is meant by *circular polarization* and how a quarter-wave plate can be used to produce it. Then explain in your own words how a quarter-wave plate works.

6. (a) Could one have elliptical polarization? (b) If so, discuss ways in which elliptically polarized light might be produced. *Hint:* Review the discussion of lissajous figures in Experiment 33 and make the analogy.

7. Two Polaroids are set for extinction, that is, with their optic axes at right angles, and a third Polaroid is then placed between them. (a) Explain how the third Polaroid can be adjusted so that light can pass through the set of three even though no light can get through the original pair. (b) What setting of the third (middle) Polaroid results in maximum transmitted intensity?

Radioactivity: Its Source and Its Detection 41

In 1896 the French physicist Henri Becquerel (1852–1908) discovered that minerals containing uranium emitted a radiation that resembled X-rays in that it could expose a photographic plate inside a covering envelope and ionize the surrounding air. Later Marie Curie (1867–1934) and her husband Pierre (1859–1906), after two years of hard labor during which nearly a ton of uranium ore was processed by hand, discovered two new chemical elements: polonium and the one that made them world famous, the intensely radioactive element radium. From then on, through the work of Ernest Rutherford (1871–1937), James Chadwick (1891–1974), and many others, there was continuous progress in the understanding of the atomic nucleus and how it could emit the various forms of so-called "emanations" that were observed.

In this experiment we shall study the nucleus to see under what circumstances it can be radioactive (that is, emit radiation) and look at the kinds of radiation that are emitted. To this end, we will consider various means for detecting this radiation and will, in particular, gain experience with one of the most frequently used detectors available today—the Geiger-Müller counter. In this way we shall gain an understanding not only of the nucleus and of nuclear radiation but of how on the one hand it can provide a vast source of useful energy and on the other the weapons we have come to dread in less than a century following Becquerel's first observation. Hopefully we can then regard nuclear energy in the proper light without letting the term "radiation" inspire the terror associated with an unknown bogey.

THEORY

Although Becquerel and the Curies showed us that there were radioactive (that is, radiation-emitting) elements, it was Rutherford who gave us the model of the atom that is still in use today albeit much modified by the later developments in quantum mechanics. He it was whose experiments showed that the atom seemed to consist of a heavy, positively charged core around which electrons circulated under the influence of the electrostatic attraction between their negative charge and the positive core. Because both the gravitational and the electrostatic forces depend inversely on the square of the distance separating the bodies involved, the Rutherford atom resembles a tiny solar system and, interestingly enough, like the solar system, consists mostly of empty space. The core, which occupies a tiny fraction of the whole atomic volume despite its embodying most of the atom's mass, is thus properly called the nucleus, just as the sun could be called the nucleus of our solar system.

The lightest atom, hydrogen, appears to have a single particle as its nucleus. It has a mass of about 1836 times that of an electron and carries a positive charge equal to the electron's charge in magnitude. The hydrogen atom, like all atoms, is thus electrically neutral as a whole and may be thought of as consisting of this single particle, called a *proton*, with a balancing negative electron in orbit around it. Heavier atoms may then be considered to have nuclei made up of several protons surrounded by an equal number of electrons. The *atomic number Z* of an atom is defined as the number of protons making up the nucleus or the number of electrons orbiting around it.

This picture is fine as far as it goes, but we must remember that the nucleus occupies a very small volume (about that of a sphere 10^{-13} cm in diameter), which means that the repulsive force between adjacent protons would, because they're so close together, be large enough to disrupt it. There appears to be a short-range but very powerful force called, for that very reason, the *strong force*, which binds the protons forming the nucleus together, but it would not be strong enough to oppose the electrostatic repulsion successfully were it not for the existence of the *neutron*. Discovered by Chadwick in 1932, the neutron has about the same mass as the proton and is similar to it in many ways but is uncharged. It therefore contributes to the strong force holding the nucleus together without contributing to the electrostatic repulsion that wants to blow it apart. Thus, in general, a nucleus contains both protons and neutrons, and the *atomic weight A* gives the total number of particles present. The number of neutrons is thus $A - Z$. Note that an element's chemical identity is determined by its atomic number Z, which gives the number of electrons in the atom, on which all the atom's chemical characteristics depend. Naming the element and giving its atomic number are thus completely synonymous statements. However, different atoms of a particular element may have different numbers of neutrons in their nuclei and therefore different atomic weights, although they are all identical chemically. Atoms differing only in the number of neutrons they contain are called *isotopes* of the particular element. Because these atoms all have the same chemical properties, they cannot

351

be separated by the usual chemical means, and special methods have had to be developed to separate them. The highly secret and, to this day, mysterious Manhattan Project of World War II was simply a big research program aimed at separating appreciable quantities of the rare isotope U^{235} of uranium ($Z = 92$, $A = 235$) from the more abundant isotope U^{238} ($Z = 92$, $A = 238$), U^{235} being the *nuclide* (by which we mean a particular isotope of a particular element) needed for the first atomic bomb.

From the foregoing we conclude that a given element may have different numbers of neutrons in its nucleus, but there are limits on how many or how few can be present and still allow that nucleus to be stable, that is, not radioactive. We have seen that if there are too few neutrons there will not be enough "nuclear glue" (as the neutron is sometimes called) to oppose the electrostatic repulsion of the protons. The nucleus does not blow up under these circumstances, however, but is found to stabilize itself by having one of its protons change into a neutron by emitting a *positron,* a particle similar to the electron but charged positively rather than negatively. The atom is then *transmuted* into a new element whose atomic number Z is reduced by 1. Similarly, if there are too many neutrons (an isolated neutron being itself unstable and decaying into a proton and an electron), one of them changes into a proton, and an electron is ejected from the nucleus. Here the resulting transmuted element has an atomic number one greater than its parent's.

In general, we find that, at least for the lighter elements, the most stable isotope in each case is the one having the same number of neutrons as protons. Thus O^{16}, the most common isotope of oxygen, whose atomic number is 8, also has eight neutrons to give an atomic weight $A = 16$. We note that O^{17} and O^{18} exist and are stable, although they constitute only a tiny fraction of the atoms in naturally occurring oxygen. Similarly, C^{12} is the common isotope of carbon with six protons ($Z = 6$) and six neutrons, although C^{13} with seven neutrons is a rare but stable isotope. Note that the name *atomic weight* for A stems from the fact that on a mass scale in which both the neutron and the proton are taken to have mass 1 and the mass of the electrons in an atom is disregarded as being negligibly small, the number A of nuclear particles also gives the mass of the particular nuclide. This is not quite correct, as the proton and neutron masses are not quite equal and the *binding energy* holding the nucleus together, which is sufficient to affect the overall mass through the Einstein relation $E = mc^2$, must be taken into account. Today the *atomic mass unit* (amu) is defined so that the carbon 12 isotope has a mass of *exactly* 12 amu, which leads to an atomic mass of 1.008 for hydrogen. Thus, the actual atomic masses of the various nuclides, even when given in atomic mass units, are not integers such as A but are fairly close. Moreover, chemists often quote the atomic mass of a particular element as a weighted average of the masses of that element's naturally occurring isotopes, which can lead to results bearing little relation to the value of A for any of the isotopes involved. This will be especially apparent with elements having several isotopes present with appreciable percentages in the naturally occurring form. The classic example is natural chlorine, which contains 75% Cl^{35} and 25% Cl^{37}, so that chemists quote the atomic mass of chlorine as 35.5 amu, obtained by taking the properly weighted average of 35 and 37.

Finally, we note that as we go to heavier elements, more and more neutrons are needed to hold the nucleus together, until, as we have seen in the case of uranium, their number becomes much larger than Z. Ultimately, in the very heavy elements, we find that within the nucleus there is a chance that two neutrons and two protons get together to form a helium nucleus ($Z = 2$, $A = 4$), the most tightly bound of all nuclei, which then separates itself from the larger nucleus and is forced out by electrostatic repulsion, the "nuclear glue" no longer being effective between the newly formed helium nucleus and the remaining protons and neutrons. The probability of this happening is not very large, which is why we still find heavy elements like uranium in nature, and therefore why the spontaneous emission of radiation was first discovered with such elements. The "radiation" thus consists primarily of a stream of helium nuclei, but the remaining atoms, which now have two fewer protons and two fewer neutrons, are often left in unstable states requiring the change of a neutron into a proton with the emission of an electron and the disposal of extra energy, which is emitted as electromagnetic waves. One thus expects three different types of "radiation" from a naturally radioactive sample, only one of which could strictly speaking be called radiation, the other two being streams of particles. The three types were identified by Rutherford and his collaborators, who placed a radioactive sample at the bottom of a small hole drilled in a lead block as shown in Fig. 41.1. A photographic plate was positioned above the block, as shown, the whole apparatus was highly evacuated, and a strong magnetic field was applied perpendicular to the plane of the figure. When the

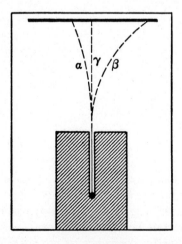

Figure 41.1 *Separation of alpha, beta, and gamma rays by a magnetic field*

plate was developed, three distinct spots were observed, two due to emanations that followed a curved path in the magnetic field and the third due to an undeflected emanation. At the time, Rutherford was unaware of the nature of these three components, and so he simply named them α rays, β rays, and γ rays. There was no deep reason for doing this; a physicist calls anything that proceeds along a particular path a ray, and Rutherford could perfectly well have labeled his three rays a, b, and c. Being a classically trained scholar, however, he chose to use the Greek alphabet. In 1907 Rutherford himself found that the alpha rays were streams of helium nuclei, which, being relatively heavy and positively charged, were deflected by a relatively small amount as illustrated. Later the beta rays were identified as electrons. Because of their relatively small mass and negative charge, they were deflected the other way in the magnetic field and to a much greater extent than the alpha rays. Finally, the gamma rays turned out to be just electromagnetic radiation, which is not deflected by a magnetic field. It is an interesting comment on the progress of understanding in physics that the terms alpha, beta, and gamma remain in use to this day even though they are of only historical significance—the legacy of a great researcher's first steps in the investigation of radioactivity. Helium nuclei are still often referred to as alpha particles and electrons as beta particles, and the term gamma rays usually means electromagnetic waves whose wavelength is shorter than that commonly associated with X-rays, occupying the range from about 0.005 to 1.40 A.

The occurrence of radioactive disintegration (or decay, meaning the spontaneous emission of the rays described above with the consequent transmutation of the nucleus) is statistical in nature; that is, for any individual atom of a radioactive substance, there is a definite probability that it will decay within a certain time interval. The probability that such disintegration will occur per unit time is called that atom's *decay constant*, usually represented by the Greek letter λ, not to be confused with the usual symbol for wavelength.* Because λdt then expresses the probability that the given atom will decay in the short time interval dt so that the decrease in the total number of atoms in the sample must at any instant be λdt times the number present at that instant, this situation is similar to that discussed in connection with the exponential function in Experiment 31. Indeed, because we again have a decrease proportional to the number present, the number N of atoms present at time t in a radioactive sample is given by

$$N = N_0 \, e^{-\lambda t} \qquad (41.1)$$

where N_0 is the number present initially (at $t = 0$) and e is the base of natural logarithms. As we saw in Experiment 31, an exponential of this kind never goes to zero, but in the present case we see that when $t = 1/\lambda$ the number N has fallen to $1/e$ times its initial value and becomes negligibly small after a period of several times $1/\lambda$ has elapsed. Thus, just as in Experiment 31 we spoke of the time constant RC as a measure of how long it takes the charge on the capacitor to fade away, so here we can treat the reciprocal of the decay constant as a measure of how long the sample will last. However, it is more meaningful to speak of the atom's *half-life* $t_{1/2}$, which simply means the time required for the number of atoms initially present to decrease by a factor of 2. Putting $N = N_0/2$ and $t = t_{1/2}$ in Equation 41.1 leads immediately to $0.5 = e^{-\lambda t_{1/2}}$, $\ln 0.5 = -\lambda t_{1/2}$, and hence $t_{1/2} = 0.693/\lambda$. Obviously, the greater the probability of decay, the shorter the half-life. Half-lives for different radioactive isotopes vary over a wide range, some being extremely short—for instance, less than a millionth of a second for polonium 212 (Po^{212}, sometimes called thorium C′)—and some quite long, such as 1620 years for radium. The half-life of uranium is even longer, which is why there was still some around for the Curies to discover, whereas polonium 212 is not a naturally occurring nuclide.

The intensity of the rays emitted by a radioactive substance decreases as the distance from the source is increased. Because the rays spread out in all directions, their intensity follows the *inverse square law of radiation*. This law states that the radiation received on a small surface held perpendicular to the rays varies directly with the source's intensity and inversely as the square of the surface's distance from the source. This law applies strictly for a point source or, in practice, when the distance is very great compared to the source size.

Experimentally, the radiations given off by radioactive substances may be detected and measured in several ways. These include the use of photographic plates, the ionization chamber, the Wilson cloud chamber, and the Geiger-Müller counter. With a photographic plate, the rays that penetrate the sensitive emulsion produce a latent image that can be developed. The density of the film is a measure of the intensity of radiation. The Wilson cloud chamber is another instrument used to study penetrating radiation. The cloud chamber works on the principle that when air saturated with water vapor is expanded suddenly, the cooling effect causes the formation of a cloud of tiny water drops. Drops of water form around dust particles or any ionized particles present in the chamber. When a ray of penetrating radiation passes through the chamber, it produces ions along its path. The air is then suddenly expanded, and liquid droplets form on the ions. These water drops appear momentarily as a white line that may be seen or photographed.

Another method of measurement is to use an ionization chamber. When a gas is exposed to radioactive radiations, it becomes ionized, and the amount of ionization

---

*Physicists love to represent things with Greek letters, possibly because they often run out of our ordinary letters, both upper and lower case. There have been instances where so many symbols were needed that the entire English and Greek alphabets were exhausted and the Phoenician alphabet had to be pressed into service!

produced is a measure of the intensity of the rays. The ionization chamber consists of a closed metal box provided with a thin aluminum or plastic window and containing an insulated metal electrode maintained at a fairly high potential (either negative or positive) with respect to the box itself. If some penetrating radiation enters the chamber, it will ionize the gas therein, producing a large number of electron-ion pairs. Under the influence of the applied voltage, the positive ions go one way and electrons the other, so that an electric current flows through the chamber. The amount of ionization, and hence the magnitude of the current, is proportional to the intensity of the radiation, which can thus be measured by measuring the current.

The Geiger-Müller counter is one of the most important instruments of modern physics, giving us the most convenient method of measuring penetrating radiation. It consists of a metal tube fitted inside a thin-walled glass cylinder, with a fine tungsten wire stretched along the axis of the tube. The arrangement is illustrated in Fig. 41.2. The cylinder contains a gas such as air or argon at a pressure of about 10 cm of mercury. A difference of potential of about 1000 V is applied between the wire and the metal cylinder, the positive voltage being applied to the wire and the negative to the cylinder. This difference of potential is slightly less than that necessary to produce a discharge. When any ionizing radiation, such as a high-speed particle from a radioactive source, enters the tube, ions will be produced by the freeing of electrons from the gas molecules. These ions and electrons are accelerated by the electric field and produce more ions by collision. The process repeats itself, creating an avalanche of electron-ion pairs that are immediately attracted to the two electrodes. Thus a large current rapidly builds up through the tube. However, this current must pass through resistor R shown in Fig. 41.2, producing a potential drop that causes the net voltage across the tube to fall below the extinction value, that is, the value needed to maintain the discharge. The current thus stops almost as soon as it has appeared, but this fairly intense current pulse gives rise to a large voltage pulse across R, which, with a little amplification, can operate an electronic or mechanical counter. In this manner, the passage of a single particle through the Geiger tube can be registered. Because of its ability to give a pulse for each particle passing through it, the tube itself is often referred to as the counter.

The electrical pulses from the Geiger tube are usually fed to a counting circuit. In some instruments the individual pulses are actually counted. In others, a count-rate meter is used, which integrates the pulses and gives a reading in terms of counts per minute (cpm). In the present experiment, however, some of the counting rates will be too slow to give a good reading on such a meter. Therefore, a counter that counts the incoming pulses over a set period of time, such as 1 minute, will be used.

Different types of Geiger counter tubes are used for the detection of different kinds of particles. When the walls of the tube are made comparatively thick, the tube is used for the more penetrating radiations, such as X rays or gamma rays. If the tube is to be used for less penetrating radiations, such as alpha or beta particles, then very thin aluminum or glass bubble windows are used to permit the particles to enter the tube. One type of Geiger counter uses a thin-walled stainless steel tube for the detection of beta and gamma rays.

The proper operation of the tube as a counter is dependent upon the voltage applied to it. If the voltage is too low, the passage of radiation through the tube will not cause a current pulse. On the other hand, if the voltage is too high, there will be a discharge within the tube even without any radiation. Hence it is very important to determine the characteristic curve of the Geiger tube that is to be used in a set of measurements. This curve shows the effect of the voltage applied to the tube on its operation as a counter and clearly indicates the voltage needed for proper results. Each tube has its own characteristic curve and its own best operating voltage. The curve also provides an understanding of how the Geiger counter works.

The *characteristic curve* of a Geiger tube is obtained by measuring its counting rate when placed near a radioactive source of moderate and constant intensity. The distance between the tube and the source is kept fixed. The counting rate is then determined for each of a series of increasing applied voltages. At low voltages, no counts are produced. As the voltage is increased, a particular voltage is reached at which the tube just begins to produce counts, at a slow rate. This is the *threshold voltage,* or the starting voltage. As the applied voltage is increased, the counting rate also increases, until a region is reached where the counting rate is practically constant for an increase in voltage of 100 to 200 V. This region is

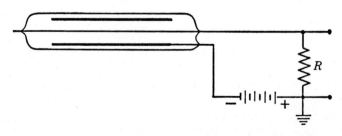

Figure 41.2   *The Geiger counter (schematic diagram)*

known as the *plateau,* or the flat portion of the curve. As the applied voltage is increased still further, the counting rate increases rapidly because of spurious counts. The *operating voltage* for the Geiger tube is usually taken at approximately the center of the plateau region.

When the intensity of radiation produced by a beta-gamma source is to be measured with a Geiger counter, the counter tube is placed at a definite distance from the source and the count rate observed. For a study of the variations in intensity with distance, the source is first placed a short distance from the tube, and this distance is then increased in regular steps. The corresponding intensity of radiation reaching the tube at each distance is recorded. A typical setup is shown in Fig. 41.4.

It is also instructive and useful to study the so-called *penetrating power* of nuclear radiation, that is, its ability to pass through various materials. Alpha rays, for example, are completely absorbed by a few centimeters of air, by an aluminum foil only 0.006 cm thick, or by a sheet of ordinary writing paper. Beta rays are about 100 times more penetrating than alpha rays; it takes a sheet of aluminum about 0.3 cm thick to stop them. Gamma rays are the most penetrating of all; they can still be detected after passing through a block of iron 30 cm thick. The ionizing power of these kinds of radiation is roughly inversely proportional to their penetrating power.

In this experiment, we shall measure the penetrating power of gamma rays from a particular source by interposing successively thicker layers of an absorbing material between that source and a Geiger counter. The intensity of radiation as indicated by the counter is to be recorded for each absorber thickness beginning with no absorber at all.

The absorbing power of any given material is expressed by a parameter μ called the material's *linear absorption coefficient.* μ is a constant characteristic of that material also depends on the wavelength of the gamma rays used. It is defined as the fractional decrease in the intensity I of the radiation per unit length of path in the absorber. The situation is thus similar to that already described for radioactive decay, where the decay constant λ could be considered as the fractional decrease per unit time in the number of atoms present, and leads to a similar equation, although we are now dealing with a decrease in intensity with distance rather than a decrease in number of atoms with time. We therefore have

$$I = I_0 e^{-\mu d} \qquad (41.2)$$

where I_0 is the intensity of the radiation before it enters the absorber and I is the intensity after passing through thickness d.

The similarity between Equations 41.1 and 41.2 is obvious, and just as we could speak of a half-life in the former case, we can now conveniently represent the penetrating power of the rays in a particular absorber by introducing that absorber's *half-value thickness,* defined as

the thickness required to reduce the intensity to half its initial value. Then, as in the case of Equation 41.1, we can write Equation 41.2 as

$$\frac{I}{I_0} = e^{-\mu d} \qquad (41.3)$$

and, taking $I/I_0 = 0.5$ and $d = d_{1/2}$, where $d_{1/2}$ stands for the half-value thickness, obtain $d_{1/2} = 0.693/\mu$. Moreover, as we have already seen in Experiment 31, and as explained in the section in the Introduction on plotting curves, equations such as 41.1 and 41.2 will plot as straight lines on semilogarithmic graph paper. Fig. 41.3(a) shows what happens when Equation 41.2 is plotted on standard linear paper for an absorber with a half-value thickness of 2 cm. If, on the other hand, semi-

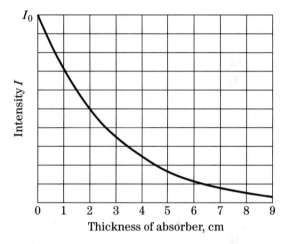

(a) Linear Scale

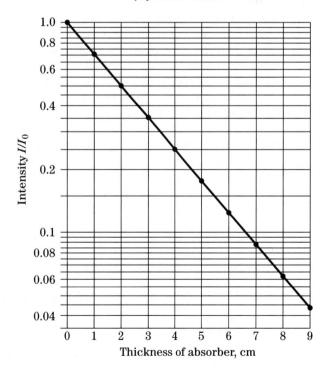

(b) Logarithmic Scale

Figure 41.3   *Absorption of gamma rays*

logarithmic paper is used and Equation 41.2 is put in the form of Equation 41.3, points taken from the curve of Fig. 41.3(a) plot as a straight line in Fig. 41.3(b). We shall thus have occasion to use semilogarithmic graph paper in this experiment just as we did in Experiment 31.

When a sensitive instrument is used to measure radiation, it is found that it will show a small reading even when no radioactive sources are in the vicinity. This small reading, known as the background, is due to cos-mic rays coming from outer space and to minute traces of radioactive materials present in the laboratory. To obtain an accurate measurement of the intensity of radiation produced by a sample, a measurement is first made with the sample in position, then with the sample removed. The intensity of radiation due to the sample is obtained by subtracting the second reading, which is the background measurement, from the first.

APPARATUS

1. Geiger counter tube in holder (probe)
2. Electronic counter (scaler)
3. Calibrated mounting board
4. Stand for Geiger tube
5. Source of beta-gamma radiation
6. Set of 12 aluminum plates, each about 6 mm thick and about 10 cm square
7. Vernier caliper
8. Stopwatch or stop clock (not needed if a timing circuit is built into the scaler)

PROCEDURE

1. Make sure that the scaler unit is turned off and that its high-voltage controls are turned fully counter-clockwise to the zero position. The instrument's line cord may then be plugged into a 115-V, 60-Hz ac outlet.

2. Place the holder containing the Geiger tube on the mounting board so that the window faces along the board toward where the radiation source will be placed. The Geiger tube cable should be plugged into the appropriate jack on the rear of the scaler.

3. Turn the power switch on and observe that the numerical display lights up reading some random number. Press the "stop" and "reset" switches to reset the display to zero.

4. Place the source of beta-gamma radiation in its holder about 20 cm from the Geiger tube so that it faces the tube's window (see Fig. 41.4). If your scaler has a built-in timer, set it to the manual position so that the timing circuit is inoperative. Then press the "count" switch to put the scaler into operation.

5. Slowly increase the high-voltage controls until counting begins as indicated on the numerical display. Raise the voltage another 50 volts and observe that counting is proceeding at a good rate. Then stop the count by pressing the "stop" switch and reset the display to zero with the "reset" switch.

6. If your scaler has a built-in timer, set it to count for 1 minute. Otherwise, prepare the stop clock to time the counting for this period. Press the "count" switch and count the Geiger tube pulses for 1 minute. Repeat this procedure, adjusting the distance between the source and the tube until about 2000 counts per minute are obtained. Keep the positions of the source and the tube fixed for Procedures 7–10.

7. Return the high-voltage controls to zero, reset the display to zero, and disable the built-in timer by setting its control to "manual." Press the "count" switch and slowly turn up the high voltage until counting just starts. The voltage at which these pulses first appear is called

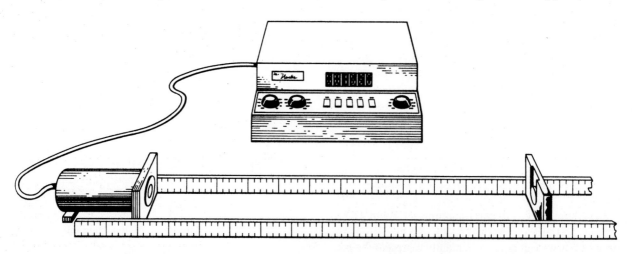

Figure 41.4 *A typical Geiger counter tube and scaler-timer*

the *threshold voltage* for your Geiger tube. Record its value.

8. Stop the count and reset the display to zero. Set the timer or arrange your stop clock to count pulses for one minute. Press the "count" switch and thus determine the counts per minute obtained at the threshold voltage.

9. Increase the high voltage by 20 volts and record this new value. Reset the display to zero, press the "count" switch, and again observe the number of counts obtained in one minute. Record this count rate.

10. Repeat Procedure 9 several times, increasing the voltage by 20 V each time and recording the counts per minute obtained with each voltage. Notice that after the first big increase in the count rate, the readings remain quite constant over a fairly wide range of voltage. This is the Geiger tube's plateau region. Then, with a further increase in voltage, there may be a considerable increase in the counting rate. Do *not* increase the voltage any further but lower it quickly to avoid damaging the tube and stop taking data.

11. From the data of Procedures 7–10, choose a voltage in the middle of the plateau region. Record this as the operating voltage for your Geiger tube. Set the scaler's high-voltage controls to this value.

12. Move the beta-gamma source in its holder so that it faces the Geiger tube window and is exactly 5 cm away. Reset the scaler to zero and then count pulses for 1 minute. Record the number of counts per minute so obtained.

13. Repeat Procedure 12 for tube-to-source distances of 6, 7, 8, 9, 10, 12, 14, 16, 19, 22, and 25 cm. If the number of counts registered in 1 minute gets small, increase the counting time to 2 or 5 minutes as required, but don't forget to divide the count by 2 or 5 to obtain the counts per minute. Record the count rate for each distance.

14. Measure the background radiation. To do this, remove the radioactive source from the vicinity of the Geiger tube. It should be placed at least 10 ft away so

that the intensity of the radiation produced by it at the Geiger tube will be negligible. You will notice that the tube still continues to count at a slow rate. These pulses are produced by the background radiation due mostly to cosmic rays. Count the number of pulses obtained in a 5-minute interval and record this value.

15. Measure the thickness of a stack of 10 of the aluminum plates with a vernier caliper and record the value.

16. Place the holder containing the Geiger tube on the stand. A typical setup for absorption measurement is shown in Fig. 41.5.

17. Press the "stop" and "reset" switches on your scaler to zero the display. Then place the source of beta-gamma radiation at the bottom of the stand so that it faces the Geiger tube window. If your scaler has a built-in timing circuit, set it to count for 1 minute. Otherwise, get the stop clock ready to time the counting period for this interval. Press the "count" switch, count the pulses from the Geiger tube for 1 minute, and record this result as the count rate (counts per minute) obtained with no aluminum absorbing plates inserted.

18. Insert one aluminum plate in the stand between the Geiger tube and the beta-gamma source so as to shield the tube from the source. Zero the display on the scaler and again count the pulses coming from the Geiger tube for 1 minute. Record your result as the count rate with one plate as the absorber.

19. Repeat Procedure 18 with 2, 3, 4, etc., plates used as the absorber, until you have inserted all 12 plates. If the number of counts collected in a 1-minute interval gets small (less than 200) as the number of plates is increased, use a 5-minute counting period and divide the result by 5. Record the count rate (counts per minute) obtained with each number of plates. Switch off the scaler and tube power supply and return the radioactive source to your instructor when you have finished the experiment.

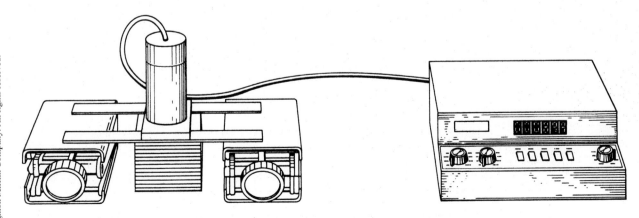

Figure 41.5　*Typical apparatus assembly showing the Geiger tube on a stand above the absorbing plates*

DATA _____

Threshold voltage _____   Operating voltage _____

   Plateau region from _____ to _____

Geiger tube voltage	Count rate, cpm		Geiger tube voltage	Count rate, cpm

Cosmic ray background:

   Counts in 5 min _____   Counts per min _____

Distance, cm	Intensity of radiation, cpm		Distance, cm	Intensity of radiation, cpm	
	Observed	Corrected		Observed	Corrected
5			12		
6			14		
7			16		
8			19		
9			22		
10			25		

Value of slope from the graph _____   Percent deviation from the theoretical value _____

Number of plates	Intensity of radiation, cpm		Fraction I/I_0 of intensity with no plates
	Observed	Corrected	
0			1.000
1			
2			
3			
4			
5			
6			
7			
8			
9			
10			
11			
12			

Thickness of 10 aluminum plates _____   Linear absorption coefficient _____

Thickness of 1 plate _____   Half-value thickness:

From absorption curve _____

From Equation 41.3 _____

Percent difference _____

CALCULATIONS _____

1. Plot a graph showing the relation between the count rate and the voltage applied to the Geiger tube. Plot the voltage along the x axis and the corresponding values of the count rate along the y axis. Choose the scales on the coordinate axes in such a way that the curve will extend over almost all of a full sheet of graph paper. Indicate the position of each point on the graph by a small pencil dot surrounded by a small circle. Then draw a *smooth* curve through the plotted points.

2. Observe the plateau region of your curve; that is, notice the region where the count rate remains nearly the same for a wide range of voltage. This is the region where the curve is nearly horizontal. Record the width of this plateau region; that is, record the value of the voltage at the beginning and at the end of the flat portion of the curve.

3. Mark the operating voltage selected in Procedure 11 on your curve. Confirm that it is located at or near the middle of the plateau region.

4. From your 5-minute count of the cosmic ray background obtained in Procedure 14, calculate the background count rate in counts per minute.

5. From each observed value of the intensity of radiation, both in Procedures 12 and 13 and in Procedures 17–19, subtract the value of the cosmic ray background obtained in Calculation 4 to get the corrected values of the radiation intensity.

6. Using the data of Procedures 12 and 13 as corrected in Calculation 5, plot the values of the distance of the source of radiation from the Geiger counter against the corresponding values of the radiation intensity in counts per minute on a sheet of 2-cycle full logarithmic graph paper. Read the instructions given in the Introduction for plotting such a graph. Indicate the position of each point on the graph by a small pencil dot surrounded by a small circle. Then draw the best-fitting straight line through the plotted points. This line does not necessarily have to pass through all the points, but it should have, on the average, as many points on one side of it as on the other.

7. Theoretically, if the radiation intensity is inversely proportional to the square of the distance between the source and the Geiger tube, then the logarithm of the intensity should be a linear function of the logarithm of the distance with slope –2. Find the slope of the graph drawn in Calculation 6 and compare it with this theoretical value by computing the percent deviation.

8. Find the fractional intensity observed with each number of absorbing plates by dividing the intensity observed with the plates in position by the intensity observed with no plates. Use the corrected values of intensity. Record your results.

9. Plot an absorption curve for gamma rays on semilogarithmic graph paper. Read the instructions given in the Introduction for plotting a graph of this kind. Plot the absorber thickness (the number of plates) along the linear axis and the fractional intensity along the logarithmic axis. Indicate the position of each point on the graph by a small pencil dot surrounded by a small circle. Then draw the best-fitting straight line through the plotted points. *Note:* If the gamma rays used are made up of different wavelengths, the absorption curve will not be a straight line because each wavelength will have its own characteristic value of μ. In this case, the best-fitting straight line will give an average value of μ.

10. From the slope of your absorption curve, determine the linear absorption coefficient of aluminum for the gamma rays used. Be careful with your units.

11. Determine the half-value thickness of aluminum for the gamma rays used. Read the number of plates corresponding to $I/I_0 = 0.5$ from the absorption curve plotted in Calculation 9. Then multiply this number (not necessarily an integer) by the thickness of one plate to obtain the half-value thickness in centimeters. Express your result in centimeters of aluminum.

12. From Equation 41.3 and the subsequent discussion, calculate the half-value thickness of aluminum for the gamma rays used in this experiment. Use the value of the linear absorption coefficient obtained in Calculation 10. Compare this half-value thickness with the one obtained in Calculation 11 by finding the percent difference.

QUESTIONS _____

1. Discuss the differences between the ionization chamber and the Geiger-Müller tube, and the advantages and disadvantages of each.

2. Suppose that the voltage applied to the Geiger tube fluctuated by ±20 V about the value of the operating voltage that you determined. What percent fluctuation in the count rate would this produce?

3. A source of gamma radiation is of such intensity as to produce 30,000 cpm on a Geiger counter placed 5 cm from the source. At what distance would the source produce 30 cpm?

4. An example of radioactive decay is given by the radioactive gas radon, a product of radium. Radon loses its activity quite rapidly, its half-life being 3.82 days. A practical use is made of this fact in that thin-walled glass or metal tubes filled with radon are used for therapeutic work. These tubes are enclosed in hollow steel needles that may be inserted into a diseased tissue, which is thus exposed to the radiation given off by the radon. The short life of the radon is a safeguard against overexposure. Calculate the activity of the radon in these tubes after 19.1 days as compared with the initial activity.

5. Your instructor will tell you if your Geiger tube has an operating voltage recommended by its manufacturer. If so, compare this voltage with the one you chose in Procedure 11.

6. Bismuth-214 (Bi^{214}, sometimes called radium C) sends out a group of beta rays whose linear absorption coefficient in aluminum is 13.2 cm^{-1}. A beam of these beta rays is passed through an aluminum plate 3.03 mm thick. Find what fraction of the beam is absorbed by the plate. For the absorption of beta rays, use the same relation as was used for the absorption of gamma rays.

7. A beam of gamma rays of energy 1.5 MeV (million electron volts) is passed through an iron plate 10 cm thick. If the linear absorption coefficient of iron for these rays is 0.40 cm^{-1}, find what fraction of the beam is transmitted by the plate.

1.0
0.9
0.8
0.7
0.6
0.5
0.4
0.3
0.2
0.1

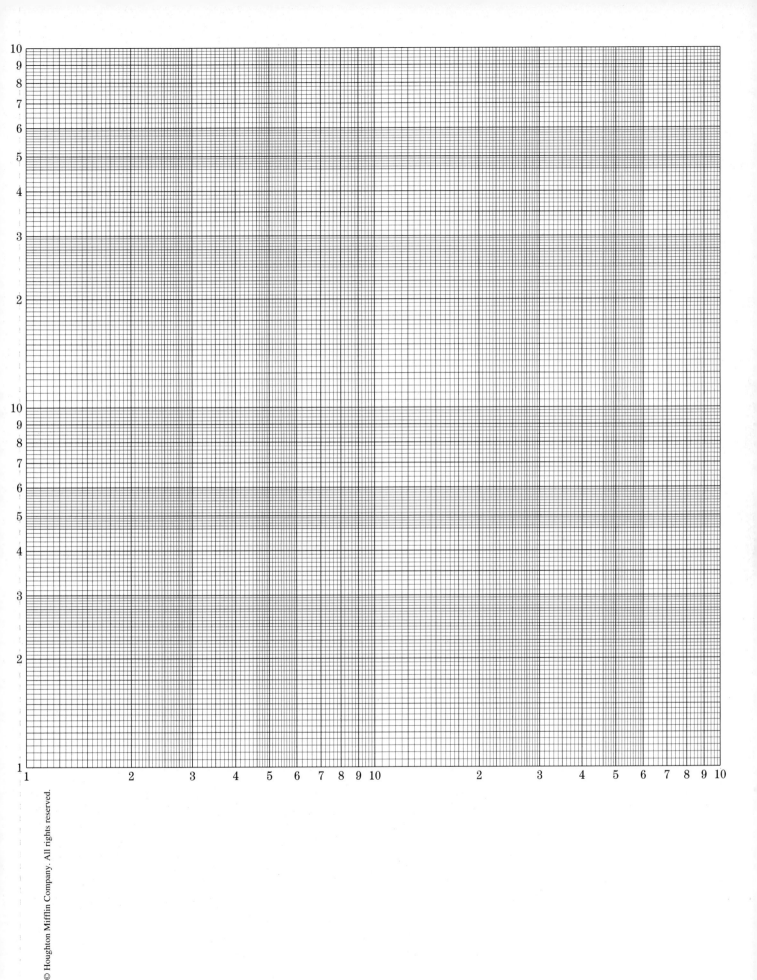

The Hydrogen Spectrum and The Rydberg Constant $\large 42$

The research work done in the field of spectroscopy, both experimental and theoretical, has contributed a great deal to our knowledge of the physical nature of things, a knowledge not only of Earth, but of the sun, the stars, and interstellar space. This experiment contains a discussion of the Bohr theory of the hydrogen atom, which marked the beginning of a new era in spectroscopy and atomic structure associated with the original introduction of quantum theory. The object of this experiment is to measure the wavelengths of the first four lines of the Balmer series of the hydrogen spectrum by means of a diffraction grating spectrometer, and from these measurements to substantiate Bohr's early theory and calculate the value of the Rydberg constant.

THEORY

The *bright-line* spectrum of an element is produced by exciting that element in the gaseous state with an electric discharge or by heating. Such a spectrum consists of bright lines, each being an image of the spectrometer slit formed by light of a particular color and hence wavelength. The number and relative positions of the lines depend on the element and on the method of excitation employed. Wavelengths of visible light are usually expressed in angstrom units, although in recent years there has been an attempt to make the nanometer the standard unit of wavelength for infrared, visible, and ultraviolet electromagnetic radiation. One angstrom unit (abbreviated Å) is defined as 10^{-10} m and is therefore 10^{-8} cm. A nanometer is 10^{-9} m and hence 10^{-7} cm. Clearly there are 10 Å in a nanometer.

The frequencies of the light emitted by an excited atom of a gas or vapor at low pressure are characteristic of that particular atom, and the spectrum produced is the bright-line spectrum of that element. When an atom absorbs radiant energy, it absorbs the same frequencies of light it would emit if it were excited. The sun's spectrum is an absorption spectrum. The main body of the sun emits a continuous spectrum, but the relatively cooler gases surrounding the sun absorb the energy corresponding to the line spectra of the elements present in the sun's atmosphere. These absorption lines appear as dark lines or missing lines in the continuous spectrum of the sun. They were first observed by Joseph von Fraunhofer (1787–1826) and are called *Fraunhofer lines.*

It is found that among the many lines appearing in the spectrum of an element, certain ones may be picked out that are clearly related to each other. Such a group forms a series following a simple law, and there may be many different series for the same element. Over a period of years, many attempts were made to explain the origin of spectral lines and the various relationships found. It was thought at first that the frequencies of the light emitted by a particular element might be related like the fundamental and harmonics of a vibrating system. But this explanation turned out to be wrong. Even the one visible series in hydrogen, the simplest element and therefore the element with the simplest spectrum, defied explanation until finally, in 1885, Johann Jakob Balmer (1825–1898), a Swiss schoolteacher and lecturer, found an empirical formula that correctly gave the wavelengths of the lines of this series. The Balmer formula may be written

$$\frac{1}{\lambda} = R_{\mathrm{H}}\left(\frac{1}{2^2} - \frac{1}{n^2}\right) \tag{42.1}$$

where λ is the wavelength of a visible line in the hydrogen spectrum, R_{H} is a constant called the Rydberg constant, and n takes on the integral values 3, 4, 5, etc., corresponding to the various lines. Thus setting $n = 3$ gives the wavelength of the longest wavelength (red) line H_α in the visible spectrum of hydrogen; $n = 4$ the next longest (H_β); $n = 5$ the next (H_γ); etc. For very large n, the wavelength approaches a limiting short value $4/R_{\mathrm{H}}$ called the *series limit*. Note that in Equation 42.1 the Rydberg constant R_{H} must have the dimensions of reciprocal length and that the unit of length chosen dictates the units in which the wavelength λ will come out. In particular, if R_{H} is given in centimeters^{-1}, $1/\lambda$ will be in *wave numbers,* a quantity proportional to the frequency of the emitted light. Wave numbers are popular with spectroscopists who want to talk about the frequency of the radiation with which they are dealing but dislike handling the large numbers involved, the frequency of light waves being of the order of 10^{14} Hz. If the wave number of a particular line is known, the frequency can be found simply by multiplying by the velocity of light *in centimeters per second,* this choice of units being left over from days before the SI system was generally accepted.

A theoretical derivation of the Balmer formula was introduced in 1913 by Niels Bohr (1885–1962) and proved to be the beginning of a vast development in atomic theory. Bohr took as his starting point the nuclear atomic model already proposed by Ernest Rutherford (1871–1937). This model of atomic structure postulates a heavy nucleus containing most of the atomic mass and all of the atom's positive charge and around which one or more negative electrons circulate in orbits under the influence of the electrostatic attraction between the unlike charges. The situation is very simple for the case of one-electron atoms such as hydrogen, singly ionized helium, doubly ionized lithium, etc. The attractive force between the electron and the nucleus is then simply

$$F = \frac{1}{4\pi\epsilon_0} \cdot \frac{(Ze)e}{r^2} = \frac{Ze^2}{4\pi\epsilon_0 r^2}$$

where F is the force in newtons, Z is the atomic number (1 for hydrogen, 2 for helium, etc.), e is the charge on the electron in coulombs, ϵ_0 is the permittivity of free space, and r is the distance between the electron and the nucleus in meters. Assuming a circular orbit, we note that r is the radius of this orbit and F provides the centripetal force mv^2/r necessary to maintain it. We therefore require

$$\frac{Ze^2}{4\pi\epsilon_0 r^2} = \frac{mv^2}{r}$$

or

$$mv^2 = \frac{Ze^2}{4\pi\epsilon_0 r} \qquad (42.2)$$

where m is the mass of the electron in kilograms and v its speed in meters per second.

Equation 42.2 is a necessary relation between the speed of the electron and the radius of its orbit in the Rutherford model, but it does not predict that any particular orbit will be preferred. Moreover, the model itself contains a serious flaw in that, according to classical electromagnetic theory, the radial acceleration of the charged electron should cause electromagnetic energy to be radiated continuously until the electron falls into the nucleus and the atom collapses.

Since actual hydrogen atoms neither radiate continuously nor collapse, Bohr postulated the existence of certain *stationary orbits* in which the electron did not radiate classically. These orbits are specified by the *quantum condition*

$$mvr = n\frac{h}{2\pi} \qquad (42.3)$$

which states that the electron's angular momentum mvr cannot have any value but only certain allowed ones. These values must all be integral multiples of a basic package or *quantum* $h/2\pi$ of angular momentum that cannot be subdivided. Here h is Planck's constant, and the fact that the angular momentum must be an integral multiple of $h/2\pi$ is expressed in Equation 42.3 by requiring that n be an integer (1, 2, 3, etc.). An integer that tells how many quanta are to be taken is called a *quantum*

number, and *n,* the first such number in atomic theory, is called the *principal quantum number.* Notice that h must have the same dimensions as angular momentum—namely kilogram-meter²-seconds⁻¹ or joule-seconds.

Equations 42.2 and 42.3 are two simultaneous equations in v and r and may be solved simultaneously for these quantities. The result is

$$r = n^2\frac{\epsilon_0 h^2}{\pi m Z e^2} \qquad \text{and} \qquad v = \frac{1}{n} \cdot \frac{Ze^2}{2h\epsilon_0} \qquad (42.4)$$

These relations show that only certain values of r and v are allowed—namely those obtained when the positive integers are successively substituted for n.

Following in the footsteps of the German physicist Max Planck (1858–1947), Bohr further assumed that in transferring from one allowed orbit to another, an electron had to make the change instantaneously (in order not to be "caught" in a forbidden position) and accomplished this by either absorbing or giving up the energy difference between these two states (orbits) in a single package, that is, an energy quantum. To determine the size of this package, we must investigate the energy the electron has when moving in any given orbit. Its kinetic energy is $mv^2/2$, and Equation 42.2 shows that this is just $Ze^2/8\pi\epsilon_0 r$. The electron also has potential energy $-Ze^2/4\pi\epsilon_0 r$ by reason of its position in the electric field of the nucleus considered as a point charge Ze. Note that the negative sign comes from assuming that the zero of potential energy is at infinity and that since the electron and the nucleus attract each other, work is done *by* the electric field on the electron rather than *on* the field when the electron is brought from infinity into its orbit.

The total energy of the electron is the sum of its kinetic and potential energies and is thus given by

$$E = \frac{Ze^2}{8\pi\epsilon_0 r} - \frac{Ze^2}{4\pi\epsilon_0 r} = -\frac{Ze^2}{8\pi\epsilon_0 r}$$

Notice that it is zero when the electron is stopped at infinity, that is, when it has just been removed from the atom to form a hydrogen ion, a doubly ionized helium ion, or a triply ionized lithium ion.

Insertion of the allowed values of r from Equation 42.4 yields allowed values of the energy, one corresponding to each allowed orbit. These are given by

$$E_n = -\frac{1}{n^2} \cdot \frac{mZ^2e^4}{8\epsilon_0^2 h^2} \qquad (42.5)$$

The difference between any two of these energies is

$$\Delta E = E_{n2} - E_{n1} = \frac{mZ^2e^4}{8\epsilon_0^2 h^2}\left(\frac{1}{n_1^2} - \frac{1}{n_2^2}\right) \qquad (42.6)$$

and if in a change between the orbits specified by the quantum numbers n_1 and n_2 this energy difference must be absorbed or emitted in a single package, one way to do this is to have the electron absorb or emit a train of electromagnetic waves whose frequency is given by the Planck formula

$$\Delta E = hf \qquad (42.7)$$

where ΔE is the energy difference between two allowed states (orbits) in joules, h is Planck's constant in joule-seconds, and f is the frequency of the radiation in hertz (seconds^{-1}). Substitution of Equation 42.7 in 42.6 yields

$$f = Z^2 \frac{me^4}{8\epsilon_0^2 h^3}\left(\frac{1}{n_1^2} - \frac{1}{n_2^2}\right)$$

which may be stated in terms of the radiation wavelength as

$$\frac{1}{\lambda} = Z^2 \frac{me^4}{8\epsilon_0^2 ch^3}\left(\frac{1}{n_1^2} - \frac{1}{n_2^2}\right) \qquad (42.8)$$

where c is the velocity of light and use has been made of the usual relation $f = c/\lambda$ between frequency and wavelength.

Equation 42.8 shows that if a hydrogen atom is excited to any high energy state, that is, if it has its electron placed in an orbit of relatively large radius specified by the value n_2 of the quantum number n, it may make transitions to any lower energy state (smaller radius orbit) specified by the smaller quantum number value n_1 by radiating electromagnetic waves whose wavelength is given by this equation. Various series in hydrogen are predicted by taking $Z = 1$ and setting $n_1 = 1$ while n_2 successively takes on the values 2, 3, 4, . . . for the first possible series; $n_1 = 2$ and $n_2 = 3, 4, 5, . . .$ for the next; $n_1 = 3$ and $n_2 = 4, 5, 6, . . .$ for the next; etc. The situation is illustrated in Fig. 42.1, which depicts the Bohr circular orbits schematically (not to scale) and shows various

possible transitions arranged according to series. Notice that the Balmer formula, Equation 42.1, is identical to Equation 42.8 when $Z = 1$, $n_1 = 2$, and the Rydberg constant is defined as

$$R_H = \frac{me^4}{8\epsilon_0^2 ch^3} \qquad (42.9)$$

Agreement between accurately measured values of the Rydberg constant and values calculated from this relation using the best available numerical data for the fundamental physical constants is remarkably close. This is particularly true if account is taken of the fact that the nucleus is not truly stationary but rotates along with the electron about the atom's common center of mass. To do this, a so-called reduced mass is calculated for the electron, which compensates for the assumption that the nucleus is a fixed point. The effect decreases for heavier nuclei, since the atom's center of mass then comes closer and closer to coinciding with the nucleus, but there is nevertheless an observable shift in the value of the Rydberg constant as applied to hydrogen, deuterium, singly ionized helium, etc., as shown in Table I at the end of the book. Notice that the change occurs in the fifth significant figure and thus affects only very precise measurements.

In addition to the excellent agreement between the theoretical and experimental values of the Rydberg constant, Bohr's prediction of other series besides that of Balmer has been verified. Thus the Lyman series ($n_1 = 1$, $n_2 = 2, 3, . . .$), the Paschen series ($n_1 = 3$, $n_2 = 4, 5, . . .$), the Brackett series ($n_1 = 4$, $n_2 = 5, 6, . . .$), and even the Pfund series ($n_1 = 5$, $n_2 = 6, 7, . . .$) have all been observed since Balmer's time and are found to follow Equation 42.8 with great precision. Special equipment is required

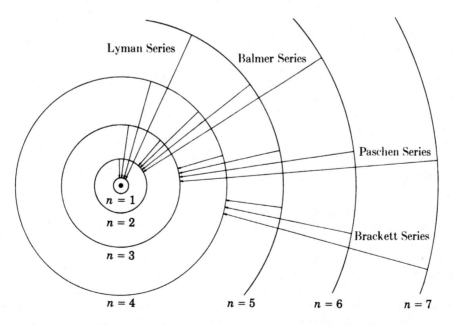

Figure 42.1 *Bohr circular orbits of hydrogen with transitions for different series of lines*

to observe these series, however, since the Lyman lines are all in the ultraviolet region of the spectrum and the other series are all in the infrared. Thus the Balmer series forms the only visible part of the hydrogen spectrum, and even its shorter wavelength lines fall in the ultraviolet. In the present experiment, only the four longest wavelength Balmer lines will be studied. In order of decreasing λ, these are often designated H_α, H_β, H_γ, and H_δ.

The presentation of energy states and the transitions between them can be made much more quantitative by using an *energy level diagram* than is possible with a diagram like Fig. 42.1. Fig. 42.2 shows the energy level diagram for hydrogen. The allowed energies are plotted on a vertical scale, a horizontal line being drawn at the appropriate level for each such energy and accordingly called an energy level. The levels are labeled both with the appropriate energies computed from Equation 42.5 and with the corresponding quantum numbers. When transitions are indicated by vertical lines drawn between the levels involved, the length of each line is proportional to the frequency of the emitted radiation and hence inversely proportional to the wavelength. Fig. 42.2 clearly indicates why the Lyman series should be in the ultraviolet and the Paschen and Brackett series in the in-frared. The energy level diagram greatly simplifies the description of the spectrum. It must be remembered that each series is an aggregate of infinitely many lines and that the complete spectrum contains an infinite number of series. An energy level diagram allows the whole spectrum to be simply represented by only one aggregate of infinitely many levels. From these levels, all the lines of all the series can be derived.

In the present experiment, a diffraction grating will be used to measure the hydrogen wavelengths. The theory and operation of such a grating is fully covered in Experiment 39, which should be reviewed in preparation for the present work. If you have not done that experiment, read the Theory and Procedure sections, noting Equation 39.1 in particular. You should also note, however, that the setup in Experiment 39 was an extremely elementary one that enabled you to concentrate your attention on the operation of the grating. In particular, the method used for determining the diffraction angle θ was not capable of great precision. We shall, therefore, now use a spectrometer similar to that studied in Experiment 37 with the transmission grating of Experiment 39 substituted for the prism, as shown in Fig. 42.3. Here again the collimator contains a slit through which light from the source enters

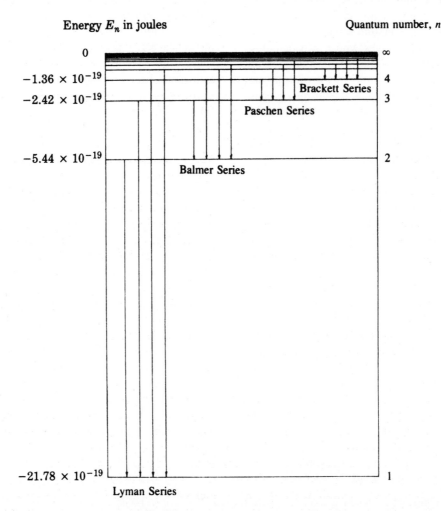

Figure 42.2 *Energy level diagram of hydrogen*

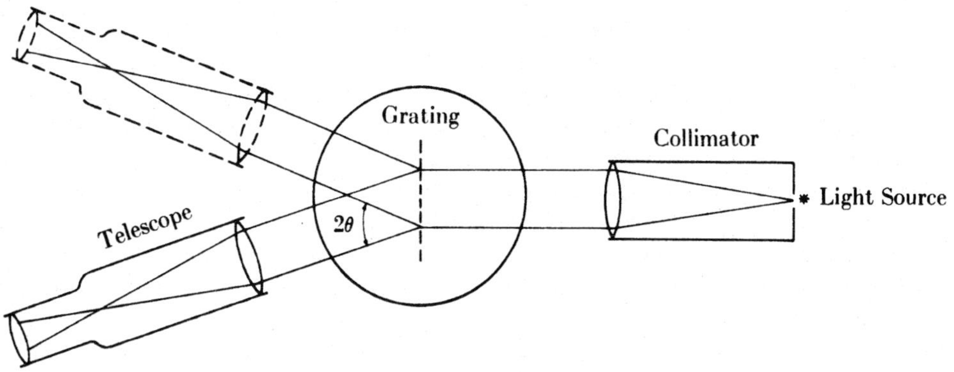

Figure 42.3 *The diffraction grating spectrometer*

the spectrometer and a lens whose function is to render the rays coming from the slit parallel. Compare this arrangement with that used in Experiment 39 to direct approximately parallel rays at the grating. The grating is mounted on the rotating table in the middle of the spectrometer, which is adjusted so that the plane of the grating is at right angles to the incoming rays. The diffracted image of the slit is then viewed with the telescope, which is rotated until the image falls on the cross hairs in the eyepiece. The angle θ can then be read directly from the spectrometer's divided circle rather than having to be calculated from length measurements as in Experiment 39. Note that with the grating, an absolute measure of the wavelength λ forming a particular image of the slit can be obtained if the grating constant is known. No calibration curve is required as it was with the prism in Experiment 37. A grating spectrometer is used in the present work both for this reason and to give the opportunity for experience with this important instrument.

APPARATUS

1. Spectrometer, for prism or grating
2. Diffraction grating
3. Grating holder
4. Atomic hydrogen discharge tube
5. Discharge-tube power supply
6. Tube holder (unless part of the power supply)

PROCEDURE

1. The spectrometer should be already adjusted by the instructor, with the collimator and telescope properly focused and aligned and the spectrometer table leveled so that its surface is perpendicular to its axis of rotation. CAUTION: Do not handle the spectrometer at all until the instructor has demonstrated its operation to you, as the adjustments are very easily upset. Then, be extremely careful in operating it.

2. Place the diffraction grating in its mounting at the center of the spectrometer table. The unruled side of the grating should face the collimator, and the plane of the grating should be made as nearly perpendicular to the collimator's axis as can be judged with the eye. The lines on the grating should be parallel to the collimator slit, and the center of the grating itself should be as nearly as possible in the center of the spectrometer table.

3. Place the stand containing the hydrogen discharge tube in front of the spectrometer, with the discharge tube just outside the collimator slit (see Fig. 42.4). Connect the leads from the tube holder terminals to the high-voltage power supply. Adjust the spectrometer so that the collimator points directly at the discharge tube's center. Close the power supply switch.

CAUTION: Do not touch the discharge tube or the high-voltage wires while the current is on.

4. Observe the central image produced by the diffraction grating. To do this, turn the telescope into line with the collimator and view this image through the eyepiece. It will be an image of the collimator slit and will have the same color as the hydrogen discharge. The slit should be made as narrow as possible and yet remain visible in the telescope. Make sure the telescope is adjusted so that when the eyepiece is focused on the cross hairs, the image of the slit is also in focus.

5. Move the telescope to the right so as to view the first-order hydrogen spectrum. Adjust the position of the telescope until the red hydrogen line falls on the intersection of the cross hairs of the telescope.

6. A clamping device that rotates with the telescope and permits it to be clamped by means of a radial screw is located directly under the telescope's supporting arm. A tangent screw on the arm allows a fine adjustment of the setting of the telescope; however, the tangent screw is used only after the telescope has been clamped with the radial screw. Clamp the telescope by means of the radial screw and locate the position of the red hydrogen line

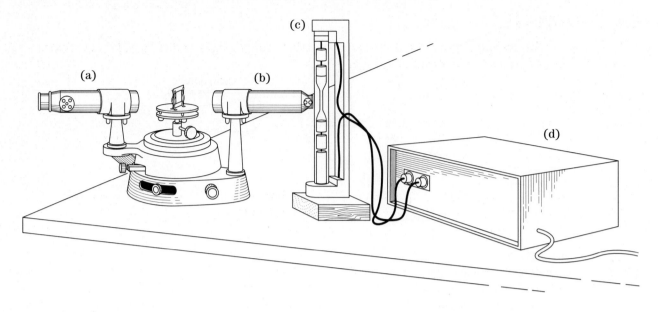

Figure 42.4 *Arrangement of apparatus for the diffraction grating spectrometer: (a) telescope, (b) collimator, (c) discharge tube, (d) power supply*

precisely by setting the intersection of the cross hairs exactly on that line by means of the tangent screw. Record the setting of the telescope to the nearest minute of arc, reading the divided circle with the help of the vernier.

7. Repeat Procedures 5 and 6 for the blue-green, the blue, and the violet hydrogen lines. Record the setting of the telescope to the nearest minute of arc for each of the lines measured.

8. Move the telescope to the left of the central image, so as to view the left-hand, first-order hydrogen

spectrum. Adjust the position of the telescope until the red hydrogen line falls exactly on the intersection of the cross hairs. Record the setting of the telescope to the nearest minute of arc.

9. Repeat Procedure 8 for the blue-green, the blue, and the violet hydrogen lines. Record the setting of the telescope to the nearest minute of arc for each of the lines measured.

DATA

Number of lines per inch on the grating _____   Grating constant in centimeters   _____

Hydrogen lines		Setting of telescope		Calculated values				
Wavelength, Å	Color	Right	Left	θ	$\sin \theta$	Computed wavelength	Percent deviation	Computed value of Rydberg constant
6562.8	red							
4861.3	blue-green							
4340.5	blue							
4101.7	violet							

Calculated value of Rydberg constant _____

Percent error _____

Accepted value of Rydberg constant for hydrogen _____

CALCULATIONS _____

1. Compute the grating constant, that is, the distance between the lines on the grating in centimeters, from the number of lines per inch stated on the grating.

2. Compute the angle of diffraction θ for each of the four lines measured. The angle of diffraction is one half of the angle between the two settings of the telescope, for each line.

3. Compute the value of sin θ for each of the four lines measured.

4. Compute the wavelength of each line, using the value of the grating constant previously calculated and applying Equation 39.1. The wavelength should be expressed in angstrom units, evaluated to four significant figures.

5. Calculate the percent deviation of the computed values of these wavelengths from the known values.

6. Calculate the value of the Rydberg constant from the computed wavelength of each line by applying the Balmer formula, Equation 42.1, in each case. Pay careful attention to your units.

7. Find the average value of the computed Rydberg constant and compare it with the known value from Table I at the end of the book by finding the percent error.

QUESTIONS _____

1. Calculate the angular separation between the D lines of sodium for a diffraction grating of 15,000 lines per inch in the first-order spectrum. Calculate the separation in the second-order spectrum. The wavelengths of the lines are 5890 and 5896 Å, respectively.

2. Calculate the frequencies in hertz and the wave numbers in centimeters^{-1} of the first four lines in the Balmer series.

3. Using the Balmer formula, calculate the wavelength of the fifth line in the Balmer series of hydrogen lines. What does your answer tell you about the possibility of seeing this line in your spectrometer?

4. Calculate the wavelength of the first line in the Balmer series in deuterium (heavy hydrogen), using the Balmer formula with the appropriate value R_D for the Rydberg constant. See Table I at the end of the book. Calculate the wavelength in angstrom units to five significant figures and find the wavelength difference in angstrom units between this line and the corresponding hydrogen line. It was the accurate measurement of these lines, and the interpretation given by the Bohr theory, that led to the discovery of heavy hydrogen by Urey, Brickwedde, and Murphy in 1932.

5. Explain the significance of the Fraunhofer lines in the spectrum of the sun.

6. A famous constant called the *fine structure constant* may be defined as the ratio of the speed of the electron in the lowest ($n = 1$) Bohr orbit to the speed of light. Derive an expression and calculate the numerical value of the fine structure constant. Also calculate the reciprocal of this value. It is a famous number, which you may already have heard of.

7. Carry out the derivation of Equation 42.5.

8. Suppose an electron makes a transition between two energy states in hydrogen, both of which have very high quantum numbers n_1 and n_2 but which are close together, that is, for which $n_2 - n_1$ is a small integer like 1, 2, or 3. Show that the frequency of the light radiated is approximately an integral multiple or harmonic of the fundamental frequency of revolution of the electron in the associated orbit. This is an example of Bohr's *correspondence principle,* which states that at high quantum numbers the quantum picture must become identical to the classical one.

9. (a) What is meant by an atom's *ionization energy*? (b) What is the ionization energy of hydrogen? (c) Show that the *series limit* of the Lyman series involves an energy identical to the hydrogen ionization energy.

10. Obtain a numerical value of the Rydberg constant from Equation 42.9 by inserting values of the quantities involved from Table I at the end of the book. Compare your result with the known value.

11. What lines in the spectrum of ionized helium lie in the visible range (4000 to 7000 Å)?

12. Which, if any, of the first four lines in the visible spectrum of hydrogen could you see in second order with your grating spectrometer? Explain your answer.

The Ratio *e/m* for Electrons 43

The physical nature of cathode rays is revealed by the deflections they undergo as they pass through electric and magnetic fields. The direction of these deflections shows that these rays consist of particles that have a negative electric charge and a definite mass. In 1897, the investigations of J. J. Thomson (1856–1940) led to a measurement of the ratio of charge to mass for these particles, which have been called electrons. The results of these measurements showed that the atom is not indivisible, that electrons make up a part of any atom, and that this particle is much smaller than the smallest atom. The object of the present experiment is to determine the value of the ratio of charge to mass of an electron. From the known value of the charge on the electron and the value of *e/m* so obtained, the mass of the electron can be computed.

THEORY

One of the simplest methods of measuring the ratio *e/m,* or the *ratio of charge to mass* of an electron, was first devised around 1900 by P. Lenard (1862–1947) and consists of measuring the deflection of a beam of electrons in a known magnetic field. The beam of electrons is bent into a circular path. The value of *e/m* can then be calculated from the known values of the accelerating potential, the strength of the magnetic field, and the radius of curvature of the path.

When an electron or any other charged particle moves in a magnetic field in a direction at right angles to the field, it is acted upon by a force perpendicular to the direction of the field and the direction of motion of the particle. The force on an electron traveling perpendicular to a magnetic field is given by

$$F = Bev$$

where B is the strength of the magnetic field, e is the charge on the electron, and v is the electron's velocity.

Since the force is always perpendicular to the electron's direction of motion, it acts as a centripetal force to make the electron move in a circular path whose plane is perpendicular to the field direction. The centripetal force required to keep a body moving in a circle is

$$F = m \frac{v^2}{r}$$

where F is the force in newtons, m is the body's mass in kilograms, v is the speed in meters per second, and r is the radius of the circle in meters. This force is exerted on the electron by the magnetic field so that we have mv^2/r = Bev or

$$\frac{mv}{r} = eB \tag{43.1}$$

Note that the *momentum mv* of the electron determines the radius of its path in a given magnetic field. Note also that in our SI system of units the field strength B must be measured in webers/meter^2, or (as they are now called) teslas. This is a large unit as far as normally encountered field strengths go, and so you will often find B expressed in *gauss*, a unit taken from the so-called *mixed system*, with which we shall have no contact and which is named after the famous German mathematician and physicist Karl Friedrich Gauss (1777–1855). However, it is often useful to remember that there are 10,000 gauss in a tesla.

When an electron is accelerated by an electric potential difference V, it gains a kinetic energy equal to the work done on it by the electric field. Because the potential difference is by definition the work done per unit charge, the work done on the electron is Ve. Therefore, the kinetic energy $\frac{1}{2}mv^2$ of an electron starting from rest and accelerated by voltage V is given by

$$\frac{mv^2}{2} = Ve$$

where m is the electron's mass in kilograms, v is its velocity in meters per second, V is the accelerating potential in volts, and e is the electron's charge in coulombs. Solving, we find that the velocity acquired by the electron is

$$v = \sqrt{2Ve/m}$$

Substitution in Equation 43.1 then yields

$$\frac{e}{m} = \frac{2V}{B^2 r^2} \tag{43.2}$$

This expression gives the ratio of charge to mass of an electron in terms of the accelerating potential, the strength of the magnetic field, and the radius of the circular path of the beam of electrons. When these quantities are known, the value of *e/m* can be computed.

373

The apparatus to be used for measuring the ratio of *e/m* is based on a design originally proposed by K. T. Bainbridge. It consists of a large vacuum tube mounted at the center of a pair of helmholtz coils that provide the uniform magnetic field needed for the operation of the tube. One arrangement is shown in Fig. 43.1. The beam of electrons is produced by an electron gun composed of a straight filament parallel to the axis of the coils and surrounded by a cylindrical anode containing a single slit parallel to the filament. Electrons emitted from the heated filament are accelerated by the potential difference applied between the filament and the anode. Some of the electrons come out as a narrow beam through the anode slit. The tube contains a trace of mercury vapor that serves to render the path of the electrons visible. When the elections collide with the mercury atoms, some of the atoms will be ionized; as these ions recombine with stray electrons, the characteristic mercury light is emitted. Since the recombination and emission of light occur very near the point where ionization took place, the path of the beam of electrons is visible as a bluish white streak.

A pair of helmholtz coils (named after the German physicist Hermann von Helmholtz, 1821–1894) is a particular arrangement of two coaxial, circular coils of radius *a* with their planes parallel and separated by a dis-tance *a* equal to their radius. This arrangement is very useful because it produces an almost uniform magnetic field over a fairly large region near the center of the coils. The magnetic field at the center is parallel to the coil axis and has a magnitude given by

$$B = \frac{8\mu_0 NI}{a\sqrt{125}} \qquad (43.3)$$

where B is the magnetic field strength in teslas, N is the number of turns of wire on each coil, I is the current through the coils in amperes, and a is the mean radius of each coil in meters.

Each coil of the pair used in the setup of Fig. 43.1 has 72 turns of copper wire with a resistance of approxi-mately 1 Ω. The mean radius a is 0.33 m. The coils are supported in a frame that can be adjusted with reference to the angle of dip of Earth's magnetic field so that the field produced by the coils will be parallel to that of Earth but in the opposite direction. A graduated scale in-dicates the angle of tilt.

The magnetic field of the helmholtz coils causes the beam of electrons to follow a circular path whose radius decreases as the magnetic field increases. The field can be adjusted until the sharp outer edge of the beam coin-cides with the outer edge of one of the five bars spaced at

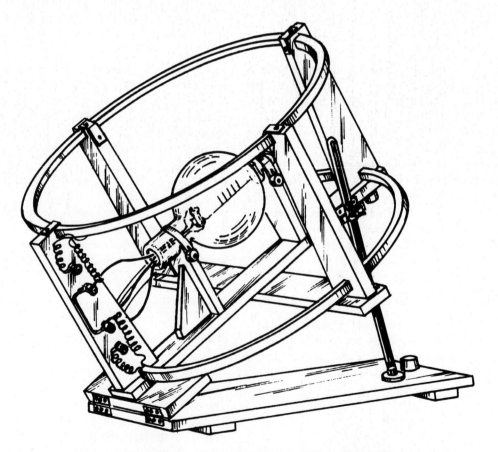

Figure 43.1   *e/m vacuum tube and helmholtz coils*

different distances from the filament. The vacuum tube shown here has been so designed that the radius of the circular path can be conveniently measured. In its manufacture, five crossbars have been attached to the staff wire, and this assembly is attached to the filament structure in such a way that the distance between each crossbar and the filament is accurately known. The distance in centimeters between the filament and the outer edge of each bar is given in the following table. These distances are the diameters of the circles that the electron beam will be made to describe. Note that these distances are *diameters* of the electron beam paths and must be divided by 2 to give the radii.

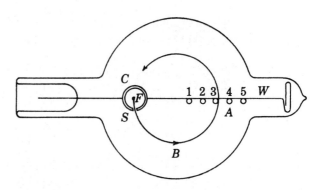

Figure 43.2 *Sectional view of e/m tube and filament*

Crossbar Number	Distance to Filament
1	6.48 cm
2	7.75 cm
3	9.02 cm
4	10.30 cm
5	11.54 cm

Fig. 43.2 is a sectional view of the tube and filament structure. *A* represents the five crossbars attached to staff wire *W*, *B* shows a typical path of the beam of electrons,

C represents the cylindrical anode with slit *S*, and *F* is an end view of the filament.

The circuit for this tube and its associated helmholtz coils is shown schematically in Fig. 43.3. The coils are powered from a variable dc power supply capable of delivering 5 A of well-filtered direct current continuously. Because the total resistance of the two coils in series is about 2 Ω, this supply will not be called upon to produce over 10 V, but it should have a smooth output control to enable you to set the coil current precisely. A similar supply is used to heat the tube's filament. Note that this filament must not be run at currents greater than 4.5 A, and a fuse is included in the apparatus to prevent operation at

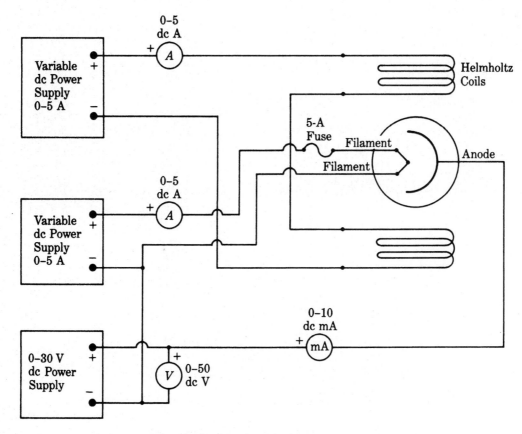

Figure 43.3 *Circuit for the setup of Fig. 43.1*

currents in excess of this value. A regulated, low-current 0–30-V dc supply is used to establish the accelerating voltage between the filament and the anode.

Fig. 43.4 shows an alternative arrangement of the vacuum tube and helmholtz coils which is now available and may be the one provided in your laboratory. Although the theory already discussed applies equally well in this case, this apparatus differs from that of Fig. 43.1 in the following respects:

a. The vacuum tube has a much more sophisticated electron gun than the one already discussed. It uses an indirectly heated cathode (that is, an electron-emitting surface heated by a separate heater rather than being itself the heater element) and has a first or focusing anode in addition to the second or main accelerating anode. It is thus remarkably like the electron gun in the cathode ray tube described in Experiment 33 and diagrammed in Fig. 33.1, even having a pair of electrostatic deflection plates, although these will not be used in our present work. There is, however, a focusing control that sets the voltage on the first anode to allow adjustment of the beam focus. Even if you have done Experiment 33 you should review the description of the cathode ray tube in the Theory section of that experiment. Remember, "cathode ray" is an old name for a stream of electrons (they come from the cathode or negative electrode in any beam tube and move in a line; hence the appellation "ray"), and that's what we're dealing with here.

b. The voltages and magnetic fields used in the apparatus of Fig. 43.4 are quite a bit higher than those in the apparatus of Fig. 43.1, so that Earth's magnetic field is negligible compared to the values of *B* to be encountered now. Consequently, no special procedure is needed to compensate for Earth's field.

c. The radius and spacing of the helmholtz coils, as well as the number of turns in them, are not given for this apparatus. Thus, instead of using Equation 43.3, we are told that $B = (7.80 \times 10^{-4})\,I$, a direct relationship between the magnetic field strength *B* in teslas and the coil current *I* in amperes for this particular helmholtz coil pair.

d. The apparatus of Fig. 43.4 uses a different method for determining the diameter of the circular electron beam path. An illuminated glass mirror is mounted behind the tube with a horizontal centimeter scale engraved on it. The experimenter sights across the luminous beam to the scale. Because the scale is on a mirror, a reflection of the beam can be seen there, and parallax in the observation is removed by moving the eye until the direct view of the beam and its reflection coincide. With care, a determination of the point at which the beam path crosses the scale can be made to 0.1 cm.

e. This apparatus is mounted on a base that contains a lot of the wiring, with connections to meters and power supplies brought out to terminals on a front panel. The complete circuit, the counterpart of that in Fig. 43.3, appears in Fig. 43.5, in which a dashed line encloses the parts and wiring located inside the base.

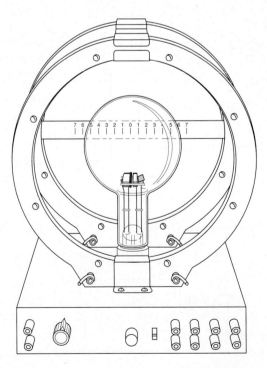

Figure 43.4 *Alternate version of the* e/m *vacuum tube and helmholtz coils*

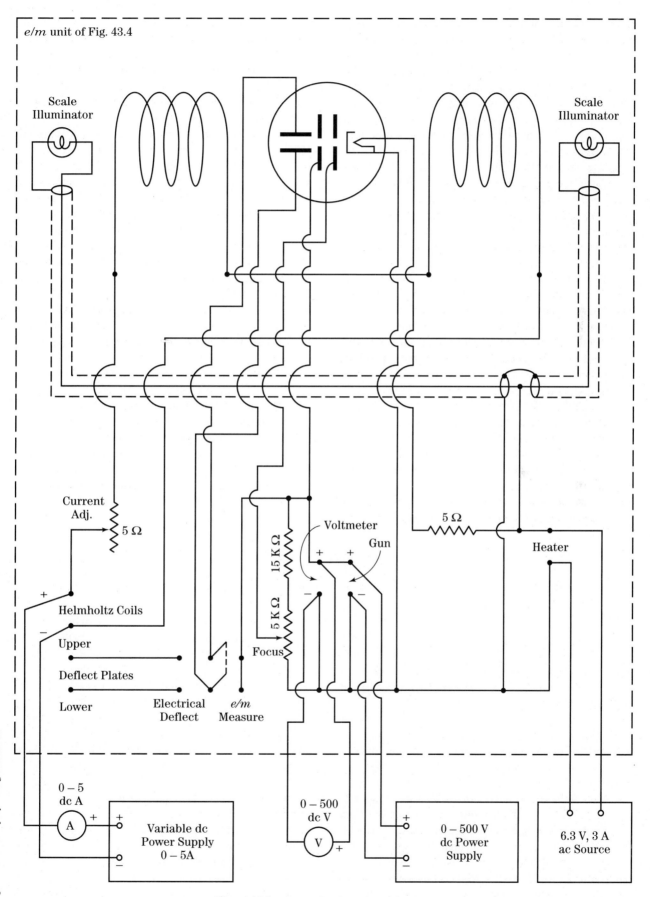

Figure 43.5 *Circuit for the setup of Fig. 43.4*

Either setup may be used equally well to measure *e/m*, but the approach with the setup of Figs. 43.4 and 43.5 is a little different; hence a separate procedure section and data table are provided for this system.

It is of interest to note that the apparatus used in this experiment demonstrates the principle of the *mass spectrometer*, an instrument that can determine the mass of individual atoms. Thus if ionized atoms could be projected from a source as the electrons are from the filament of your *e/m* tube, Equation 43.2 shows that, for a given accelerating voltage *V* and magnetic field *B*, all particles having a particular charge-to-mass ratio will follow a circular path of radius $(1/B)\sqrt{2mV/e}$. But if the ions in the beam all have charge *e* but different masses, they will follow separate paths, one for each mass, and can thus be separated. Such an arrangement is one form of mass spectrometer, but we should note that the determination of mass depends on our prior knowledge of the charge on the particle. It is most important to realize that *no deflection experiment with charged particles* (such as the present one) *can ever determine the charge or mass separately; it can only determine their ratio.* This follows from Newton's second law, $F = ma$, in which we note that a deflection experiment basically measures some sort of acceleration produced by a force due to an electric or magnetic field. Such a force is proportional to the particle's charge, so that we are always measuring $a = F/m \sim e/m$. Thus either *e* or *m* must be obtained from some independent measurement.

During the years 1909 to 1913, Robert A. Millikan (1868–1953) devised a method of measuring *e* by essentially determining the charge-to-mass ratio of oil drops small enough to carry only a few excess electrons. He could determine the masses of his drops by measuring their velocities when falling under gravity, and he knew that their charges had to be small integer numbers of electron charges. Thus, having found the charges on a large number of drops, he could conclude that the electronic charge would be the lowest common multiple, that is, a value that would divide evenly into all his results. Strange as it may seem, this method, although admittedly after much repetition and refinement, remains today as the basic source of our knowledge of *e*.

APPARATUS

1. *e/m* tube mounted between helmholtz coils
2. Power supply capable of delivering 5 A at 10 V dc (two required for the apparatus of Fig. 43.1)
3. Variable line transformer ("Variac" or "Powerstat") if not included in the above power supply
4. 0–30 V dc power supply (apparatus of Fig. 43.1) or 0–500 V dc power supply (apparatus of Fig. 43.4)
5. 6.3 V, 3 A filament transformer if not included in the 500 V supply (apparatus of Fig. 43.4 only)
6. dc voltmeter (0–50 V for the apparatus of Fig. 43.1; 0–500 V for the apparatus of Fig. 43.4)
7. dc ammeter (0–5 A, two required for the apparatus of Fig. 43.1)
8. dc milliammeter (0–10 mA, apparatus of Fig. 43.1 only)
9. Compass (apparatus of Fig. 43.1 only)
10. Dip needle (apparatus of Fig. 43.1 only)
11. Assortment of connecting leads

PROCEDURE

Apparatus of Fig. 43.1

1. Wire the tube and helmholtz coils in the circuit shown in Fig. 43.3. Make sure all power supply controls are turned all the way down and that their power switches are in the off position. Have your wiring approved by the instructor before plugging in any of the supplies.

2. The helmholtz coils have been designed to provide the uniform magnetic field at the location of the tube given by Equation 43.3. For quantitative measurements, the axis of the coils should be made parallel to Earth's magnetic field. This is done by orienting the apparatus so that the *e/m* tube will have its long axis in a magnetic north-south direction, as determined by a compass. Then measure the magnetic dip at this location by using a good dip needle. Finally, tip the coils up until the plane of the coils makes an angle with the horizontal equal to the complement of the angle of dip. The axis of the coils will now be parallel to Earth's magnetic field. For example, in New York State the magnetic dip is about 74° and thus the plane of the coils should make an angle of about 16° with the horizontal. A graduated scale connected to the apparatus indicates the angle of tilt of the plane of the coils. If the dip needle is now placed in a north-south direction on the inclined board supporting the coils, the needle should point along the axis of the coils.

3. Earth's magnetic field should be directed along the axis of the helmholtz coils in a direction opposite to that of the field you will apply. Because in the Northern Hemisphere the vertical component of Earth's field is down, the vertical component of the field due to the coils should be up when the apparatus is used in this hemisphere. The field of the coils will be so directed if the *e/m* tube is oriented so that the electrons travel in a counterclockwise direction, which means that the slit in the anode should face to your right when you are standing looking at the base end of the tube (the end where the leads to the filament and anode come out). Make sure the tube is mounted in the apparatus with this orientation.

4. Apply the accelerating potential to the tube's anode by plugging in and turning on the 30-V power supply. Turn up its output control to give an accelerating voltage of about 25 V as read on the dc voltmeter.

5. Make sure that the filament power supply's output control is turned all the way down; then plug it in and turn it on. Slowly increase the filament current by turning up the output control while watching both the milliammeter in the anode circuit and the light from the filament wire viewed through the slit in the anode. *Be very careful in this step.* You will not see any emission current until the filament current gets up to about 3 A, but further increases will cause the emission to rise rapidly. As soon as you see the filament lighting up, proceed with great caution, while continually monitoring the value of the anode current. Set it at 8 mA, after which the filament current should not be increased any further. This will avoid overheating the filament and possibly burning it out. Remember that under no circumstances is the filament to be run at a current higher than 4.5 A. Record the values of the anode and filament currents you have just set.

6. The thin ray of light produced by the electrons as they pass through the mercury vapor should now be clearly visible, especially in a darkened room. We will refer to this ray as the electron beam. If the tube has been oriented as described in Procedure 3, the crossbars will extend upward from the staff wire and the electron beam will emerge from the anode slit to your right as you stand facing the tube from the base end. But the beam will be deflected slightly toward the base by Earth's magnetic field. To balance out this effect, a small current will be sent through the helmholtz coils, but first you must determine whether these coils are properly connected. Plug in their power supply; first make sure that its output control is turned all the way down, then turn it on, and increase its output slightly from zero. If the magnetic field of the coils straightens the beam, and, upon further increase, bends it around toward the crossbars, the coils are properly connected. If the field increases the deflection toward the base, the field current is flowing in the wrong direction and the connections from the power supply and meter should be interchanged.

7. Adjust the value of the field current until the electron beam is perfectly straight and perpendicular to the long axis of the tube. This adjustment should be made very carefully and accurately. When the beam is straight, the magnetic field of the coils is just equal to Earth's magnetic field, so that the resultant field at the beam is zero. Record this value of the current as I_1. It will be constant throughout the experiment, since the accelerating voltage is kept constant.

8. Increase the field current slowly until the electron beam describes a circle. Adjust the value of this current carefully until the bright, sharp, outside edge of the beam just clears the outside edge of crossbar number 5, the one that is farthest away from the filament of the tube.

Record this value as the I_2 for crossbar number 5 in the table. Also record the accelerating voltage indicated by the voltmeter.

The outside edge of the beam is used because it contains the electrons emitted from the negative end of the filament. These electrons fall through the greatest potential difference between filament and anode and thus have the greatest velocity. This is the potential difference that the voltmeter measures, and it is the one to be used in the calculations. The electron beam spreads as it goes around the tube, but all of the spreading is toward the inside of the circle. It is caused by the fact that electrons coming from the positive end of the filament are accelerated by a potential less than the voltage applied to the anode by the amount of the voltage drop across the filament. This can be seen in Fig. 43.3, which shows the accelerating voltage supply connected between the negative end of the filament and the anode and the filament supply connected between the negative end of the filament and the positive end. Thus the potential difference between the positive end and the anode is the difference between the two supply voltages, the potential difference between the negative end and the anode is the full anode supply voltage, and intermediate points along the filament are at intermediate potentials.

9. Keep on increasing the field current slowly and adjust its value carefully until the bright, sharp, outside edge of the electron beam just clears the outside edge of crossbar number 4. Record this value of the current as the I_2 for crossbar number 4 in the table. Record the accelerating voltage to be sure there has been no change. Repeat the same procedure for crossbars 3, 2, and 1 and record the value of the current I_2 and the accelerating voltage in the boxes appropriate for each case.

Apparatus of Fig. 43.4

1. Wire the unit containing the tube and helmholtz coils to the power supplies and meters as shown in Fig. 43.5, noting that all wiring enclosed in the dashed line in that figure is contained inside the unit's base. Your wiring thus consists of simply connecting the power supplies and meters to the terminals on the unit's front panel. Note also that many 0–500 V power supplies incorporate a 6.3 V ac source because this is a very common vacuum-tube heater/filament requirement. You will be given a 6.3 V so-called filament transformer to use if your supply does not have this feature. Make sure all power-supply controls are turned all the way down and that their power switches are in the off position. Have your wiring approved by the instructor before plugging in any of the supplies.

2. When your setup has been approved, apply 6.3 V to the tube's heater by either turning on the 500 V supply (with its voltage control, which does not affect the heater supply, turned all the way down) or plugging in the filament transformer. You should be able to see the red glow of the heater by looking at the back end of the electron gun in the tube. Allow a minute or two for the heater to bring the cathode to operating temperature.

3. The apparatus of Fig. 43.4 uses considerably higher fields than that of Fig. 43.1; hence no special procedure to compensate for Earth's magnetic field is needed. You can therefore proceed at once to bring up the accelerating voltage *V* with the voltage control on the 500 V supply. Begin by setting it at 150 V. You should be able to see a bluish streak coming straight out of the electron gun and striking the glass envelope of the tube. You may need to darken the room or throw a dark cloth over the apparatus to make this streak (which will now be referred to as the electron beam) clearly visible.

4. Turn on the power supply for the helmholtz coils and gradually increase its output. You should observe the electron beam bending around into a circle in the upper part of the tube. *Note 1*: There is a rheostat (a variable resistor) built into the helmholtz coil circuit inside the *e/m* unit's base that can be used to vary the coil current. In general, however, power supplies of the sort used here have output controls that are much smoother than this rheostat. You may therefore find it convenient to set the rheostat's knob (on the unit's front panel) somewhere between halfway and all the way up and control the current entirely from the power supply. Do *not* set the rheostat all the way *down*, because this is an off position that opens the circuit. *Note 2*: If the positive lead from the supply has been connected (through the ammeter) to the *red* helmholtz-coil terminal, the coil polarity should be correct and the beam should curve upward into the empty upper part of the tube. If by any chance the beam bends *down* when the coil current is increased, the polarity is wrong and the leads to the coil terminals should be interchanged.

5. Adjust the coil current to give a good circle and note where the beam comes around and strikes the back of the electron gun. It should hit the middle of the gun's rear; that is, the beam path should be a circle, not a spiral. A spiral results from the axis of the gun not being exactly perpendicular to the magnetic field so that the electrons have a velocity component parallel to it. Such a component is unaffected by the field and displaces the path sideways, thus producing a spiral as the main velocity component takes the electrons around in a circle. It is possible to rotate the socket holding the tube to align the gun exactly perpendicular to the field direction, but this should be done before you come into the laboratory. Consult the instructor if this adjustment proves necessary.

6. Adjust the helmholtz coil current to deflect the beam in a circle 10 cm in diameter. This measurement is made using the illuminated reflecting scale mounted behind the apparatus. Sight across the right-hand side of the beam circle to its reflection in the scale mirror and remove parallax by moving your eye until the direct view of the beam coincides exactly with its reflection. Note that you should move your eye vertically so that the curve of the beam matches the curve of its reflection, thus insuring that your line of sight is at the right level, as well as horizontally to bring the beam and its image into coincidence. The coil current is adjusted so that this coincidence falls on the scale's 5-cm mark. Then repeat this procedure where the left-hand side of the beam circle crosses the scale, noting that the scale has its zero at the center and is graduated in increasing numbers of centimeters both to the right and to the left. If the zero mark is not exactly in line with the center of the circle, the beam will not line up on the 5-cm mark on the left when it has been lined up with the one on the right. In this case, adjust the coil current to give left and right scale readings that total 10 cm so as to make this your beam circle diameter. The radius must then be 5 cm.

7. To facilitate the determination of *r* in the last procedure, use the focus control to bring the beam to a focus—that is, make it as fine as possible—where it crosses in front of the scale on the right. This adjustment should also produce a narrow beam when it again crosses in front of the scale on the left, although scattering of the electron stream along the path will make the focusing on the left side much less precise. When you have optimized the beam size, repeat Procedure 6 to establish the beam circle diameter as accurately as possible at 10 cm. Then record the coil current that produced this diameter with your 150-V accelerating potential.

8. Repeat Procedures 6 and 7 with accelerating voltages of 200, 250, 300, 350, and 400 V, recording the current required to give a 10-cm diameter circle each time. Note that what you're doing is maintaining *r* constant as *V* and *B* are increased. When you have finished taking data, turn both the accelerating voltage and the coil current down to zero and switch off the respective power supplies. Do not run the *e/m* apparatus with the accelerating voltage at 400 V any longer than necessary, because the electron gun gets quite hot under these conditions.

DATA _____

Apparatus of Fig. 43.1

Number of turns on each coil   _____   *e/m* calculated value   _____

Mean radius of each coil   _____   *e/m* accepted value   _____

Filament current   _____   Percent error   _____

Anode current   _____

Crossbar number	Radius of circle	Anode voltage	I_1	I_2	I	B	*e/m*
1							
2							
3							
4							
5							

Apparatus of Fig. 43.4

$B = 7.80 \times 10^{-4}\ I$ $r = 5\ \text{cm} = 5 \times 10^{-2}\ \text{m}$

Accel. voltage V, volts	Coil current I, amperes	Coil current squared, I^2	Accel. voltage V, volts	Coil current I, amperes	Coil current squared, I^2
150			300		
200			350		
250			400		

Slope of graph   _____   *e/m* accepted value   _____

e/m from slope   _____   Percent error   _____

CALCULATIONS _____

Apparatus of Fig. 43.1

1. Calculate the *radius in meters* of the circle of the electron beam for each crossbar distance.

2. For each value of the field current I_2, subtract the current I_1 to obtain the value of I, the current needed to produce the magnetic field that deflects the electrons in the circle of the corresponding radius.

3. For each value of the field current I, calculate from Equation 43.3 the field strength B of the magnetic field produced within the helmholtz coils.

4. For each circular path described by the electron beam, compute the value of *e/m* by substituting in Equation 43.2 the appropriate value of the accelerating potential, the magnetic field strength obtained in Calculation 3, and the radius of the circle in question.

5. Find the average of the values of *e/m* obtained in Calculation 4 and record this as the calculated value of *e/m*.

6. Compare your experimental result with the accepted value given in Table I at the end of the book by finding the percent error.

Apparatus of Fig. 43.4

1. Calculate the square of each of your values of *I* and enter your results in the appropriate boxes in the data table.

2. Plot a graph of *V* against I^2 using linear graph paper. Choose scales that spread your points over as much of the paper as possible. According to Equation 43.2, your points should lie on a straight line that includes the origin. Draw the best straight line you can through these points and extrapolate it to the edge of the graph sheet.

3. Calculate the slope of the line you drew in Calculation 2 and enter your result in the data table.

4. Find your experimental value of *e/m* from the slope obtained in Calculation 3. Compare this result with the accepted value as instructed in Calculation 6 for the setup of Fig. 43.1.

QUESTIONS _____

1. Name two sources of error in your measurements that are most likely to affect the accuracy of your results.

2. Compute the velocity of an electron that has been accelerated through differences of potential of 100 V and 400 V. Express your results in meters per second. Do you think relativistic effects will have to be considered in either of these cases?

3. (a) Calculate the strength of the magnetic field required to bend electrons that have been accelerated through a potential difference of 100 V and 400 V into a circle 10 cm in diameter. Express your results in gauss, remembering that there are 10,000 gauss in a tesla. (b) What helmholtz-coil currents are needed to produce these fields in the two apparatuses considered in this experiment? (c) Could the coils take these currents? You should be able to answer this question easily for the apparatus you used and should be able to give an educated guess for the other one based on the descriptions given in the Theory and Procedure sections of this experiment.

4. Find the mass of the electron from your result for *e/m* and the value of the electronic charge *e* given in Table I at the end of the book.

5. (*Apparatus of Fig. 43.1*) Calculate the expected spread in the beam at crossbar number 5 due to the voltage drop across the filament. Do this by noting that the accelerating voltage between the anode and the negative end of the filament is the voltage read on the voltmeter, whereas the voltage between the anode and the positive end is the voltmeter reading minus the filament voltage. This latter may be taken to be 3.0 V unless your instructor gives you a better value. (*Apparatus of Fig. 43.4*) Briefly discuss the advantages and disadvantages of the indirectly heated cathode of the Fig. 43.4 apparatus as compared to the directly heated filamentary cathode of the Fig. 43.1 apparatus.

6. Show that if the magnetic field is held constant, the time *T* required for an electron to make a complete circle in your *e/m* tube and return to the anode is independent of the accelerating voltage by deriving an expression for this time. The reciprocal $f = 1/T$ is called the electron's *cyclotron frequency*.

7. (*Apparatus of Fig. 43.1*) Substitute Equation 43.3 in Equation 43.2 to obtain an expression for *e/m* in terms of known constants and the quantities you actually measured in this experiment. (*Apparatus of Fig. 43.4*) (a) Substitute $B = 7.80 \times 10^{-4} I$ in Equation 43.2 to obtain an expression for *e/m* in terms of *V*, *I*, and *r* in this case. (b) Rearrange this equation to reveal *V* as a linear function of I^2 and show how you extracted *e/m* from the slope of this function in Calculation 4.

8. (a) Answer Question 5 of Experiment 33 if you have not already done so. (b) Does the ratio *e/m* appear in the formula you obtained in Part (a) of that question? (c) If not, explain why not.

9. (*Apparatus of Fig. 43.4 only*) It was pointed out in Procedure 5 for this apparatus that if the electron gun was not aligned perpendicular to the magnetic field, the resulting beam would form a spiral because of the component of the electron's velocity parallel to the field direction. This phenomenon is used to produce *magnetic focusing* in some cathode ray tubes not set up for the electrostatic focusing method described in Experiment 33. Most large cathode ray tubes used as the picture tubes in modern television sets are focused magnetically. Here a uniform magnetic field is imposed *along* (that is, parallel to) the gun axis, so that an electron that comes straight out of the gun proceeds undeflected in a straight line. However, a practical electron gun never has all the electrons emerging in this ideal axial line, many of them coming out at small angles to the axis. Their velocities therefore have a component perpendicular to the field as well as the main component along it; hence a spiral results. Show that an electron emerging from the gun with velocity *v* making a small angle θ with the axis will be returned to the axis after making one turn of the spiral, and that in view of your result in Question 6, all electrons will be returned to the same point regardless of the value of θ (provided it's small), so that this point is a true focus. Then derive an expression for the "focal length" of this "magnetic lens," namely, the distance from the gun exit to the focal point just described. *Hint:* The gun axis is *not* the axis of the spiral described by the electron. Avoid making this common mistake by drawing diagrams of both the side view and the head-on view of an electron emerging from the gun into the axial magnetic field. Try to visualize what actually goes on, remembering that we are in three dimensions so that the lateral component of the electron's velocity may be in *any* direction perpendicular to the axis. Draw your head-on view for several electrons having different directions for their lateral velocity components and see what your picture begins to look like.

The Photoelectric Effect and Planck's Constant 44

This is one of the most important experiments in any basic physics curriculum, for it represents the vindication of the early quantum theory as it was proposed at the turn of the century, which at that time sparked considerable controversy. Indeed, when Max Planck (1858–1947) proposed the idea of quantization of radiant energy, he was widely derided, especially since his introduction of the constant h, which now bears his name, looked simply like the insertion of an adjustable parameter that would enable his theory to be brought into agreement with experiment. It was only later, when Albert Einstein (1879–1955) applied Planck's idea to a problem with the photoelectric effect that had confounded classical physicists and not only resolved the problem but got the same value for h in the process, that Planck's work was shown to re-

veal a fundamental property of our universe. Ironically, Planck's ideas, which had originally engendered so much criticism and even ridicule, won him the Nobel Prize in 1918. Also of interest is the fact that Einstein, who is usually known for his development of relativity theory, won *his* Nobel Prize (1921) not for relativity but for his application of Planck's work to the photoelectric effect (and hence suggesting the very experiment we shall perform here). In the course of this experiment we shall determine the value of Planck's constant, which, far from being a "finagle factor," has turned out to be a fundamental ingredient of the quantum theory that revolutionized our views on radiation, the structure of atoms and molecules, and the interaction between radiation and atomic particles.

THEORY

At the turn of the century, many physicists had a rather smug outlook on the status of their knowledge of the physical world, one of them even saying that he would not want to embark on a scientific career in the new century because all the major discoveries had been made and future work would be limited to measuring the parameters of that world with ever greater precision. At the same time, there were three little chinks in the beautiful edifice of classical physics that had been erected over the years—chinks that most people thought would soon be filled in but that were in fact destined to bring down the entire structure. One of these, the problem of atomic spectra, has already been considered in Experiment 42, where we saw how Niels Bohr used some of Planck's ideas to explain the existence of spectral series. But there were two others. One was the problem of so-called black-body radiation, which deals with the electromagnetic radiation from a hot body due to the thermal agitation of the charged particles in that body's atoms. It is called "black" because a black surface, as a result of its being the best absorber of radiation (such as light), must also, according to one of Kirchoff's laws, be the best radiator. Classical physics had no trouble predicting that a hot body should radiate electromagnetic waves and could even present a perfectly good derivation of the Stefan-Boltzmann law, which states that the intensity of this radiation is proportional to the fourth power of the absolute temperature, or

$$I = \sigma T^4 \qquad (44.1)$$

where I is the intensity in watts radiated from each square meter of surface, T is the *absolute* temperature in degrees *Kelvin*, and σ is the Stefan-Boltzmann constant, which, for a true black body, has a value of 5.67×10^{-8} watt/meter^2-^{0}K^4. The trouble comes when one investigates how much of this energy is radiated at each frequency, for a hot body does not radiate electromagnetic waves at a single frequency like a radio station but rather at all frequencies; hence, the question of how much at each one naturally arises. A detailed analysis based on strictly classical (that is, no quantum theory or relativity) ideas leads inexorably to the so-called Rayleigh-Jeans law, which predicts that the radiated power should increase without limit as the frequency gets higher. Such a result is clearly wrong and became known as the *ultraviolet catastrophe* because its prediction that the world should have blown up long ago indicated a serious flaw in the theory and because this catastrophe became apparent at high frequencies (the short wavelengths characterizing ultraviolet light). Then, on December 14, 1900, often called quantum theory's birthday, Max Planck proposed in a talk before the Berlin Physical Society that the classical theory needed to be modified only by putting in the idea that the radiant energy was not continuous but came in discrete packages he called quanta. Although mankind's imagination had never seemed to have any trouble in accepting the granularity of matter (the idea that an atom is the smallest possible piece of matter, or, nowadays, that the atom itself is made up of discrete particles) or the idea that there is a smallest possible electric

387

charge (namely that of the electron), we balk at the thought of energy not being continuous so that its addition to or removal from a body must take place in jumps. Objections to Planck's hypothesis were compounded by the fact that his quantum was not a fixed package but varied in size, the amount of energy therein being proportional to the frequency of the radiation regarded as a wave, an idea later taken up by Bohr, as illustrated in Equation 42.7 of Experiment 42. Many of Planck's colleagues thought he had devised his theory simply to suit the facts, for the actual spectral distribution of energy was known from experiment, and in particular that he had introduced his variable-size package both to satisfy a well-known thermodynamic requirement and to give himself a constant that he could choose at will to force his formula to agree with the experimental results. His work was thus not well received until it was found also to resolve the other flaw that had been bothering the classicists—the details of the photoelectric effect.

Like the emission of radiation by a hot body and its general description by the Stefan-Boltzmann law, photoelectric emission—that is, the emission of electrons from the surface of an illuminated conductor—was not a surprise to classical physicists. We have already seen in Experiment 26 that a metal functions as a conductor because its atomic electrons are not bound to their respective atoms but are free to move through the metal's lattice structure. There is an energy barrier at the surface that prevents the electrons from emerging into the surrounding space, just as the wall of a water pipe constrains the water to flow along the pipe without leaking out. But the metal's energy barrier is not infinite, so that an interior electron *can* acquire enough energy to overcome it and leave the surface. Any method of giving electrons this energy should work, and *thermionic emission*, which means giving it to them by heating a metal electrode so that some of the electrons in its surface get enough to break through the barrier, has been known since the days of Thomas Edison (1847–1931). Edison placed a hot cathode (so called because electrons came *from* it) in an evacuated bulb (so that the emitted electrons wouldn't hit gas molecules) along with a second metal electrode (not heated and called the *anode* because electrons move *to* it, or sometimes simply the *plate*) and observed that a current passed through the apparently empty space between cathode and anode. Moreover, he observed that this current increased both when he made the cathode hotter (thus causing more electrons to be emitted) and when he made the anode positive with respect to the cathode by connecting a battery between them with its positive lead going to the anode. He also observed that if he interchanged these connections so that the anode was negative, current flow ceased because, of course, the negative electrons were repelled back to the cathode and could not be replaced by emission from the unheated anode. This result was called "the Edison effect" and was regarded as something of a

curiosity, although in fact Edison had made the first rectifier (that is, conductor of electricity in one direction only) diode (two-element vacuum tube), the forerunner of the entire array of vacuum tubes that served the emerging electronic industry up to the introduction in the late 1950s of solid-state diodes and transistors.

It was thus not surprising that electromagnetic waves such as light, which were known to carry energy, could cause electron emission from a metal surface on which they impinged. The effect was first discovered by Heinrich Hertz in 1887 and was investigated a year later by Hallwachs, who found that a freshly polished zinc plate that was charged negatively would lose its charge when exposed to ultraviolet light but would not do so if charged positively. Ten years after that, Philipp Lenard, whose work on the charge-to-mass ratio of the electron was discussed in Experiment 43, studied photoemission in a special vacuum tube shown schematically in Fig. 44.1. It featured a cathode that was not heated but was instead arranged to have ultraviolet light fall on it from a source S through quartz window Q. Emitted electrons could be collected at anode A when the cathode was made negative (anode positive, but in Lenard's experiment the anode was taken to be at zero or ground potential), but a positive cathode stopped the electron flow, as in the Edison effect. Lenard noticed, as Edison had, that electrons were still collected at the anode with the cathode only slightly positive, showing that they were emitted with some initial velocity and could reach the anode unless stopped by a sufficiently large opposing voltage. Note that a small hole in anode A allowed a beam of electrons to proceed to anode P_1 unless deflected by a magnetic field, which Lenard applied with a pair of helmholtz coils (represented by the dashed circle in the figure), thus measuring these particles' charge-to-mass ratio by the method described in Experiment 43. He applied a field that bent the beam around so that it was collected at anode P_2, whose position he knew, and could so calculate the radius of the circular path. He was thus able to identify the "cathode rays" as a stream of electrons just like those observed by J. J. Thomson, as discussed in Experiment 43, and thereby established that any such rays, whether produced by photo or thermionic emission, were all streams of electrons, a fact that was not well understood at the time. For this, Philipp Lenard was awarded the Nobel Prize in 1905.

So what's wrong? The fundamental problem lies in the fact that if we consider the incident radiant energy to be carried by a wave, we have to understand that this energy is distributed over the wave front. We considered this aspect of electromagnetic radiation—in particular, light—in Experiments 36, 38, and 39, and must conclude that the energy put into these waves by the emitting source should be thought of as distributed over the outward propagating wave fronts. Classically speaking, then, we must assume that any particular electron can only receive energy from the portion of the incident wave

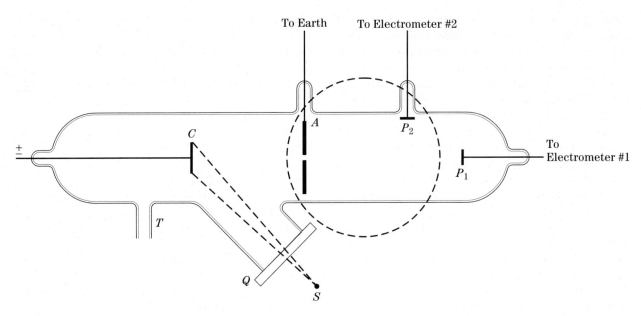

Figure 44.1 *Lenard's apparatus for determining the ratio
e/m for photoelectrons.* C *is the photocathode and* T *the tube to
the vacuum pump. (Reprinted by permission from F. K. Richtmyer
and E. H. Kennard,* Introduction to Modern Physics, *McGraw-Hill, New York, 1947, p. 86)*

front that it intercepts, and, given the fact that an electron is a very small particle, that can't be much. Calculation (of a strictly classical nature—see Question 8) shows that an electron in a metal surface would have to sit there for years acquiring energy from each successive wave front before accumulating enough to get out, whereas some electrons are emitted immediately, even with low-intensity light.

An analogy may be useful at this point. Consider a professor addressing a large class on the wonders of physics. Despite his impassioned rhetoric, he observes that one of his back-row students has fallen asleep. Wishing to awaken this delinquent so that he will not miss any of the intellectual gems being put before him, our lecturer notes that he can choose between two ways of doing this. The first, which we will call the wave picture, would be to yell "**WAKE UP!**" in a loud voice. The professor would then be considered to be the point source of an acoustical wave directed at the class. The trouble is that the energy he put into his yell would be distributed over the entire outward-propagating spherical wave front, so that everyone in the class, including all the wide-awake types who hadn't missed a word, would get some of it. Indeed, the cross-sectional area of the sleeping student's ears (especially when the component perpendicular to the direction of propagation of the yell is taken) would be quite small, so that the fraction of the originally transmitted energy he would receive would probably be insufficient to wake him, while each of the other students would be treated to an equal amount despite not needing it.

But our professor has another option, which we will call the particle picture. He can throw chalk. He thus picks up a piece of chalk and, putting the same amount of energy into his throw that he would otherwise have put into his yell, hurls it at the sleeper, catching him with unerring aim squarely between the eyes. As all of the energy is now concentrated on this one student (no other member of the class getting any part of it), our delinquent is awakened *en sursaut,* and, when his subsequent language has been cleaned up for public consumption, is heard to exclaim, "Good grief, I am the recipient of a large quantum of energy!"

Herein lies the crux of the situation. Although our professor would have put out the same amount of energy in either case, with the particle picture all of the energy was concentrated on the one sleeper whereas with the wave picture every member of the class got an equal share, no one of which was enough to do much. It was this idea that led Einstein to look at Planck's work and note that the proposed discontinuity in radiant energy suggested a return to the model (originally proposed by Newton) that light could be regarded as a stream of particles (later called photons) each having energy E, which would be related to the frequency f of the light (regarded as a wave) by the Planck relation $E = hf$. An electron struck by such a package would get all the energy therein without sharing it with other electrons, just as the sleeping student received the full impact of the thrown chalk. But is the energy in the package enough to get the electron over the surface barrier and out into the space beyond? Well, according to Planck, the energy in the package is proportional to frequency *and there should therefore be a threshold frequency below which no electrons can come through the surface regardless of the light intensity.* The utterly devastating fact is that such a threshold was exactly what was observed, this dependence on frequency (and thus wavelength) having frustrated classical physicists, who had no explanation for it, for years. Einstein's resolution of the problem, once Planck's hypothesis was adopted, was quite simple. If an incident photon carrying energy hf is completely absorbed by and thus gives all its energy to an electron in a metal, that electron can emerge through the surface provided hf is greater than the surface's energy barrier, which we will denote by E_b. If hf is indeed greater than E_b, then the emerging electron will come forth with kinetic energy E_k, so that

$$hf = E_b + E_k \qquad (44.2)$$

The electron's kinetic energy E_k is what Lenard and others measured with their stopping potential, that is, the negative voltage V_s they applied to the anode in tubes such as that of Fig. 44.1 and which was adjusted to just stop the electron current flow. We can easily see that in this case the electron, starting from the cathode with kinetic energy E_k and no potential energy, has traded all this energy for potential energy eV_s at the anode surface. Having adjusted V_s to establish these conditions, we can write $E_k = eV_s$ and, by substitution in Equation 44.2, obtain

$$hf = E_b + eV_s \qquad (44.3)$$

It is usual to divide this equation through by the electronic charge e, after which a rearrangement of terms yields

$$V_s = \left(\frac{h}{e}\right)f - \phi \qquad (44.4)$$

where ϕ is defined as E_b/e and is called the *work function* of the metallic surface in question. Note that because a volt is a joule per coulomb (that is, voltage is energy per unit charge), the work function is expressed in volts and is the energy required to get a coulomb's worth of electrons (that's 6.25×10^{18} electrons) out of the surface. It represents the energy barrier we've been talking about and is a characteristic of the particular metal involved. Table XII at the end of this book lists the work functions of some common metals.

Experimental investigations of the photoelectric effect were carried out in 1912 by A. L. Hughes and also by O. W. Richardson and K. T. Compton. Their work showed that the energy of the photoelectrons increased proportionately with the frequency of the light and that the constant of proportionality had about the same value Planck had found earlier, showing that his constant was no mere stopgap in the explanation of black-body radiation. This was the first experimental confirmation of Einstein's photoelectric equation. In 1916, R. A. Millikan, who was to receive the Nobel Prize for both this work and the measurement of the electronic charge e as mentioned in Experiment 43, carried out a series of experi-

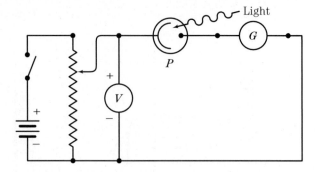

Figure 44.2 *Circuit for the study of the photoelectric effect*

ments that confirmed Einstein's explanation of the photoelectric effect completely and led to a very accurate determination of h. Millikan showed beyond a peradventure that the energy given an electron in a metal surface is proportional to the frequency of the impinging radiation; that whether an electron emerges or not, and how much kinetic energy it has if it does, depends solely on the radiation's frequency; that if electrons come out at all, a few do so immediately, even with very low radiation intensity; and that therefore this intensity controls how many electrons come out per second (as is easily understood if we think of intensity as measuring how many photons go by per second in the incident beam) but not how much energy they have.

Millikan's equipment is diagrammed in simplified form in Fig. 44.2. Light of a particular frequency selected by appropriate filters impinges on the cathode of photocell P, an evacuated tube containing two elements as shown. In this arrangement, variable voltage V is applied between the cathode and the anode, the anode being considered at ground potential and the cathode being made positive as in Lenard's setup. The stopping potential then appears as a positive voltage in both cases. Fig. 44.3 shows typical plots of the photoelectric current (the anode current measured by galvanometer G in Fig. 44.2) against the retarding voltage V for four different wavelengths of the incident light. In each case, the stopping potential V_s, that is, the voltage that just reduces the current to zero, is marked with a small arrow, and these values are plotted against frequency in Fig. 44.4. The plotted points are seen to lie on a straight line as predicted by Equation 44.4, and extrapolation of that line allows calculation of h from its slope and ϕ from its intercept with the zero-voltage axis.

This is the time-honored method of confirming Einstein's hypothesis and measuring Planck's constant, but it has a serious difficulty, namely the determination of the *exact* voltage at which the photoelectric current becomes zero for each frequency. Fig. 44.3 illustrates how imprecise the selection of V_s might be, for the curves become nearly horizontal near zero current, and if an extremely sensitive electrometer is used to measure this current, stray currents will mask the actual disappearance of that due to the photoelectrons. We shall thus use a

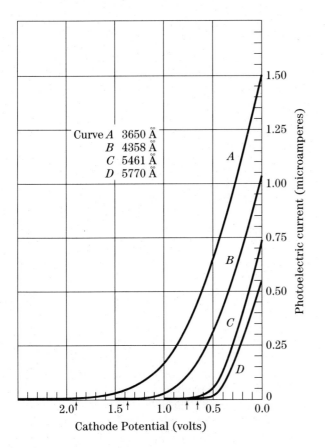

Figure 44.3 *Photoelectric current plotted against retarding potential for light of different wavelengths*

newly developed apparatus for measuring V_s, shown in highly schematic form in Fig. 44.5. A standard vacuum photodiode tube, type 1P39, is connected to the input of a 1:1 operational amplifier represented by the horizontal

triangle in that figure. The 1:1 means that this amplifier has a gain of unity, its output voltage being always identical with that at its input. In what sense is it therefore an amplifier? It is an amplifier in that its input circuit is of very high impedance ($\sim 10^{13}$ ohms), whereas its output has a much lower impedance suitable for driving an appropriate voltmeter. The unit may thus be considered a current amplifier, since $i = V/R$, or even a power amplifier, since $P = V^2/R$, but for our purposes we should consider it an impedance transformer that does not require ac for its operation (see Experiment 32). Because of the very high input impedance, this amplifier's input terminal is practically an open circuit, so that photoelectrons ejected from the cathode leave behind a positive charge that has no place to go and must therefore collect in the capacitance formed between the cathode (together with the wire connecting it to the amplifier input and this input itself) and ground. This capacitance thus fills up with positive charge, the voltage across it (and therefore the voltage measured at the amplifier's output) building up according to Equation 34.7 in Experiment 34. But this voltage is a retarding voltage opposing the movement of electrons to the anode, and when it reaches the cutoff level at which no further electrons can leave the cathode, it will cease increasing and assume a steady, constant value. This, then, will be V_s, and we have only to connect a voltmeter to the amplifier's output terminals to measure it. Very precise values of h and ϕ can be obtained from such an arrangement, especially if the wavelength (and hence the frequency) of the incident light is accurately selected by means of a special grating (see Experiments 39 and 42). In this way, we shall be able to repeat with considerable precision one of the most crucial experiments of modern physics.

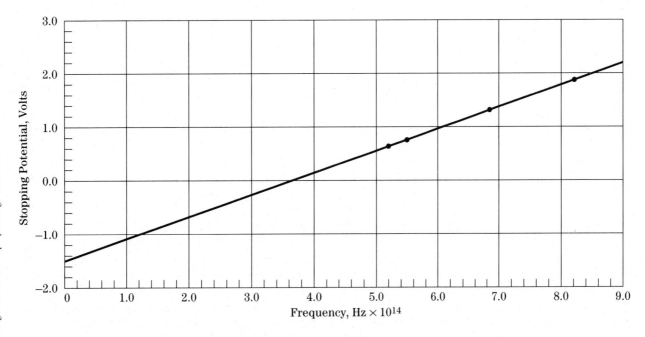

Figure 44.4 *Stopping potential* V_s *plotted against light frequency* f

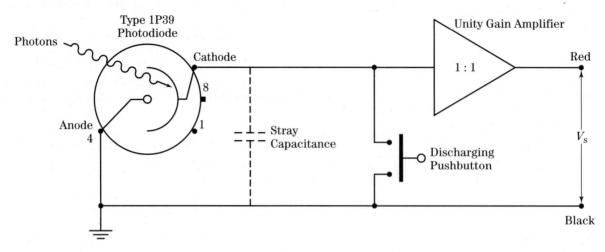

Figure 44.5 *Circuit of the photoelectric head*

APPARATUS

1. Photoelectric head, including a 1P39 photodiode and the associated operational amplifier, together with its circuit and supply batteries
2. High-intensity mercury arc lamp
3. Support base and spacing rod for the photoelectric head
4. Coupling bar assembly
5. Lens/grating assembly and mount
6. Voltmeter with low-voltage range
7. Stopwatch or stop clock

PROCEDURE

1. Arrange the apparatus as shown in Fig. 44.6 if this was not already done before you came into the laboratory. The photoelectric head is mounted on its support base and can be rotated to line the entrance slot up with the incoming beam. The coupling bar assembly slides into the lower mounting groove in the mercury arc lamp housing and is secured in place with a thumbscrew, which you should tighten against the housing front. A pin at the end of the coupling bar fits in a hole in the rod attached to the support base so that the head can be moved around the lamp to make light of successive colors fall on the entrance slot in the white reflective mask on the front of the head as shown in Fig. 44.6. The light from the mercury arc is separated into its various component colors by a special lens/grating that is held in front of the lamp on supporting rods fixed to a light aperture assembly. You should slide this assembly's mounting plate down the central groove on the front of the lamp housing until it hits a stop which positions the aperture directly in front of the lamp's light exit port. Secure the mounting plate in place by tightening the two thumbscrews against the front face of the lamp housing and slide the lens/grating onto its mounting rods as shown in the figure, with the lens (round aperture) side toward the lamp.

2. Study the controls and connection terminals on the side panel of the photoelectric head. The black terminal is common ground, the red is the unity-gain amplifier's output, and the blue ones are connected one to each of the batteries inside the head so that you can check their condition with the voltmeter. You might now make this test by connecting one voltmeter lead to the ground terminal and the other to each of the blue terminals in turn. Note that in one case you'll be reading a positive voltage and in the other a negative one (the two batteries are in series, the connection between them being grounded to produce both a positive and a negative supply for the amplifier), so that you must either shift the voltmeter's polarity switch (if it has one) or interchange its leads as you make these two measurements. Use an appropriate voltmeter scale such as 0–10 V and consider the batteries to be satisfactory if both readings are greater than 6 V. If either is 6 V or less, call the instructor, as the batteries will have to be replaced. Also check that the on-off switch on the side panel is in the "off" position to save battery drain until you are ready to run the experiment, and observe the push button that discharges the amplifier input between measurements. Refer to Fig. 44.5 to see how this push button is connected.

3. Turn on the mercury lamp and allow it to warm up for a few minutes until it reaches full intensity. Check that the light aperture in the assembly holding the lens/grating is centered so that light from the lamp falls

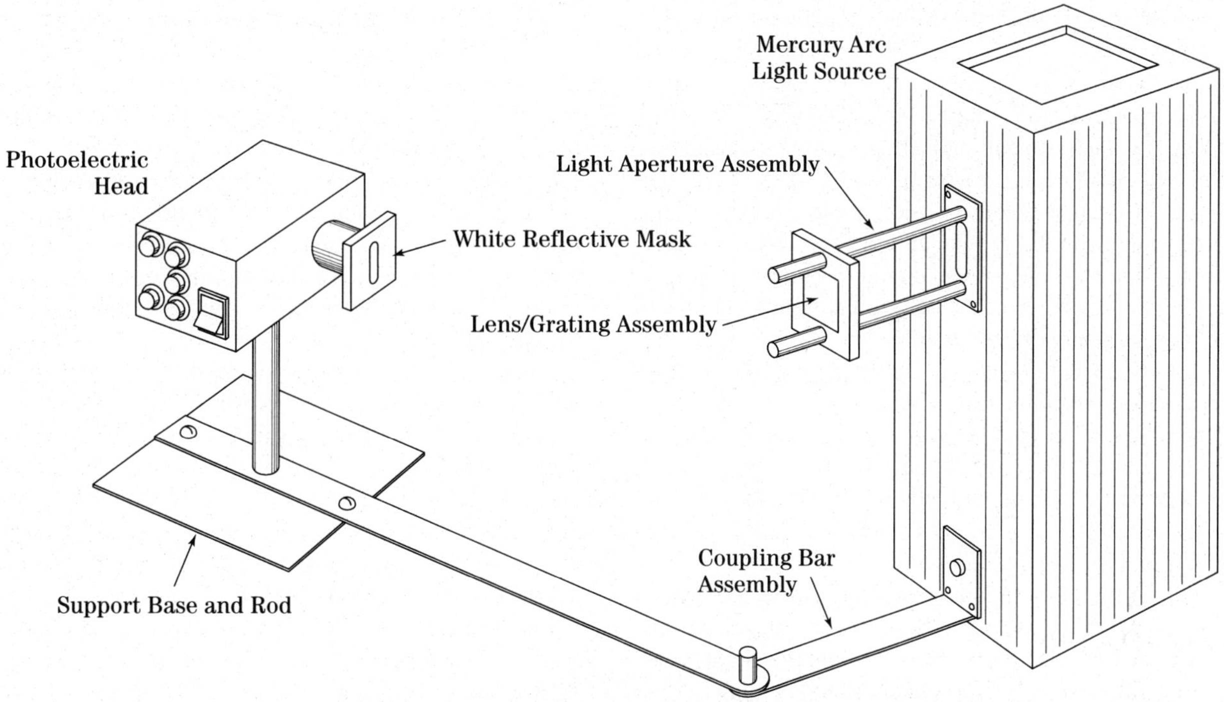

Figure 44.6 *Experimental setup for the study of the photo-electric effect. (Adapted from PASCO Scientific Instruction Manual, October, 1995. Courtesy of PASCO Scientific.)*

on the center of the lens. This should have already been done before you came to the laboratory, but if you find that the beam is not centered on the lens, call your instructor. He or she will have the Allen key needed to loosen the screws on the rear of the assembly's mounting plate and will help you remove it from the lamp housing (where it was attached in Procedure 1) and take off the lens/grating to protect it while this adjustment is being made. With its setscrews loosened, the aperture plate can be slid left or right and also rotated slightly to insure it will extend straight up and down. The mounting plate can then be replaced on the lamp housing and the lens/grating slid back onto its supports.

4. Move the photoelectric head in an arc about the pin at the end of the coupling bar assembly so that the white reflective mask receives the direct beam from the mercury lamp, and slide the lens/grating back and forth on its support bars to focus the light on this mask. Check that the vertical bar of light seen on the mask is indeed vertical. If it is not, call your instructor, as the light aperture will have to be readjusted as described in Procedure 3.

5. Loosen the thumbscrew fixing the photoelectric head to its support base so that it can be rotated and roll the cylindrical light shield out of the way so that you can see the white photodiode mask inside the head. Move the head about the coupling bar pin and rotate it on its support base until the image of the light aperture (which you

focused on the reflective mask in Procedure 4) is positioned right on the slot in that mask, passes through the center of the hole in the head housing, and falls on the photodiode mask inside. Tighten the thumbscrew to retain the head in this position.

6. Move the photoelectric head in its arc about the coupling-bar pin to bring the ultraviolet line at 3650 Å onto the slot in the white reflective mask. Note that this line in the mercury spectrum is not in the visible range but that the reflective mask has been coated with a special fluorescent material that glows blue when struck by ultraviolet light. You can thus see the line's position on the mask. Note also that the grating has been *blazed*; that is, the engraving tool has been canted at an angle during ruling so that the first-order spectrum is enhanced and the refracted light thrown mostly to one side. A brighter spectrum is thus obtained on one side of the central ray than on the other, unlike what you observed in Experiment 39, where the spectra were seen with equal intensity on either side of the centerline. If you faced the lens side of the lens/grating assembly toward the lamp as mentioned in Procedure 1, the proper direction in which to move the head is as shown in Fig. 44.6. Light entering the head through the slot in the white reflective mask must pass through the cylindrical light shield and the hole in the head housing as noted in Procedure 5 before falling on the white mask that shields the photodiode's anode from the incoming beam. The cylindrical shield keeps stray light out of the housing and must be rolled out of the way when you wish to see the photodiode mask. Do this now and observe a small, square window in the mask that admits light to the photodiode's cathode. Make slight adjustments of the head's orientation and position about the coupling-bar pin to insure both that the slot in the white reflective mask is centered on the ultraviolet line from the grating and that the window in the photodiode mask is centered on the resulting bar of light falling on it. Then roll the cylindrical light shield back into its normal operating position.

7. Connect the voltmeter to the red and black output terminals on the photoelectric head, noting that the stopping potential as read between the photodiode *cathode* (and hence the amplifier output) and ground is positive so that the red terminal is positive with respect to the black as usual. Use a 2- or 3-volt range on the voltmeter and then turn on the power switch on the head.

8. Press the push button on the photoelectric head's control panel to discharge the apparatus and release it when the voltmeter reads zero. Note that the output voltage builds up rapidly at first and then more slowly to a maximum value and stabilizes there. This value is V_s, the stopping potential for the electrons released from the photodiode's cathode by light having a wavelength of 3650Å. Practice pressing the push button and releasing it a few times to see that your measured voltage always comes back to the same value.

9. Place the variable transmission filter over the slot in the white reflective mask. Your filters have frames with magnetic strips that will hold them in place on the mask. The variable transmission filter controls the intensity of the light transmitted into the photoelectric head by having five sections of successively increasing obscuration. These sections are calibrated to transmit 100%, 80%, 60%, 40%, and 20% of the incident light, respectively. Begin by positioning the 100% section over the slot in the mask. Press the push button and (with your third hand!) start the stopwatch and release the button simultaneously. Stop the watch at the instant you see the voltage stabilize. You may want to practice this procedure a few times before recording both the observed value of the stopping potential V_s and the time required to reach it.

10. Repeat Procedure 9 four times using a different section of the variable transmission filter each time.

11. The brightest visible spectral lines of mercury are the yellow doublet at 5790 and 5770 Å, the famous mercury green line at 5461 Å, and the blue and violet lines at 4358 and 4047 Å, respectively (see Table 37.2 in Experiment 37). Set the variable transmission filter for 100% transmission and move the photoelectric head to bring the violet line onto the slot in the white reflective mask. Recheck the alignment and focus of the apparatus as described in Procedure 6 and then repeat Procedure 9 for this line.

12. Move the photoelectric head to receive the blue line at 4358 Å and repeat Procedure 11.

13. Move the photoelectric head to receive the green line at 5461 Å. As you get to the longer wavelength lines, you should use color filters in addition to the variable transmission filter to keep shorter-wavelength ambient room light and ultraviolet light from the possibly overlapping second-order mercury spectrum from entering the apparatus. The longer wavelengths correspond to lower frequencies and hence lower energies, so that stray light can easily cause the emission of electrons at higher energies and so produce a false stopping potential reading. Therefore, for the present procedure, remove the variable transmission filter, place the green filter directly on the white reflective mask, and then mount the variable transmission filter set for 100% transmission on top of it. Repeat Procedure 9 with this arrangement.

14. Move the photoelectric head to receive the yellow doublet and replace the green filter with the yellow one. Note that the grating cannot separate the members of this doublet enough to allow only one at a time to enter the apparatus, but the shorter wavelength (higher frequency) one will dominate in ejecting electrons from the photocathode, the corresponding photons having the higher energy. With this arrangement, repeat both Procedures 9 and 10 in order to get both the stopping potential for this wavelength and the times needed for the output to reach this voltage with each of the several light intensities.

15. Be sure to switch off the mercury arc lamp and the photoelectric head when you have finished the experiment.

DATA _____

Color	Wavelength, angstroms	Frequency, hertz × 10¹⁴	Percent transmission	Stopping potential, volts	Charge time, seconds
Ultraviolet	3650		100%		
			80		
			60		
			40		
			20		
Violet	4047		100%		
Blue	4358		100%		
Green	5461		100%		
Yellow	5770		100%		
			80		
			60		
			40		
			20		

Value of h/e from slope _____   Value of work function from intercept _____

Experimental value of Planck's constant _____

Accepted value of Planck's constant _____   Percent error _____

CALCULATIONS _____

1. Calculate the frequency of the light in hertz for each of your five different wavelengths and enter the results in the spaces provided in the data table. Be careful of your units and note that these frequencies will be in the 10^{14} Hz range.

2. Plot a graph of stopping potential V_s against the frequencies obtained in Calculation 1. Use linear graph paper and choose scales so that your plot will cover most of the sheet, but start your frequency scale from zero so that by extrapolating the straight line that you should be able to draw through your plotted points you can obtain a zero-frequency intercept. See Fig. 44.4 and Equation 44.4.

3. Find the slope of the line plotted in Calculation 2. According to Equation 44.4, this is *h/e*. Then find your experimental value of Planck's constant *h* by multiplying this result by the electronic charge listed in Table I at the end of the book.

4. Look up the accepted value of Planck's constant in Table I and find your percent error.

5. Read the work function of the 1P39's cathode surface as your graph's intercept on the zero-frequency axis. Refer to Equation 44.4 and show how the algebraic signs come out right.

QUESTIONS

1. How do your data on the charging times for the ultraviolet and yellow lines and the stopping potentials observed with all the lines confirm Einstein's hypothesis?

2. Briefly discuss the sources of error you think you encountered in this experiment.

3. The so-called *photoelectric threshold* f_0 is the frequency of incident light that will just get an electron out of the cathode surface, leaving it with no kinetic energy. Calculate f_0 for your 1P39 photodiode's cathode by two methods: (a) from the value of the work function found in Calculation 5 and the accepted values of h and e, and (b) from your graph's intercept with the $V_s = 0$ axis. In doing so, show that this intercept is indeed f_0 by modifying Equation 44.4 so that the threshold frequency appears in place of the work function. How do the two values of f_0 so obtained compare?

4. The work function of a metal to be used for the cathode of a photodiode (often called a photoelectric cell) is 1.5 volts. What is the wavelength of the light having the corresponding photoelectric threshold frequency? What color is this light? Could a photoelectric cell using this metal for its cathode give a signal when it "sees" ordinary daylight?

5. Can you make a reasonably good guess as to what the cathode of your 1P39 is made of based on your measured value of its work function?

6. For the yellow line in your experiment, did the stopping potential seem to be lower for the lower light intensities? If so, can you explain why? *Hint:* In your form of the experiment, how was the stopping potential developed?

7. A yellow sodium lamp can be thought of as sending out light of wavelength 5893 Å. (a) Calculate the amount of energy contained in a photon whose corresponding wavelength has this value. (b) How many quanta of radiation are emitted each second by a sodium lamp whose output is one watt? (c) A photon, considered as a particle, is a very peculiar one in that it has no rest mass; that is, if it's not going with the speed of light it ceases to exist. Its energy cannot be given by the usual expression $1/2\, mv^2$ for kinetic energy but can, according to Einstein's relativity theory, be written as mc^2 where m may be regarded as its apparent mass and c is the velocity of light. Find the apparent mass of the photons considered in this question, and compare it with the rest mass of an electron.

8. Suppose the classical rather than the quantum theory were true. How long would it take for an electron to get through the surface of the 1P39's photocathode if it were held 10 cm away from a 100-watt light bulb? Assume the bulb to be a point source radiating a total of 100 W uniformly in all directions.* Assume also that an electron in the cathode surface collects only the energy it intercepts from the passing wave fronts and that it is a tiny sphere 10^{-13} cm in diameter.

9. Because some experiments, such as Experiments 38 and 39 on interference, seem to establish that light is a wave whereas the present experiment says it's a stream of particles, the question, Which is it really? naturally arises. State briefly your own answer to this question. *Note:* Contrary to the usual practice in physics texts, this is an essay question. It is also a question that has intrigued the philosophers of science since Einstein's explanation of the photoelectric effect threw doubt on what had until then been the accepted idea that light was an electromagnetic wave. Think about this for a while before attempting an answer, and when you do, be brief and bear in mind that there is no simple, pat, definite answer to this question even though we're tempted to believe that we can always find one in physics.

---

*A poor assumption if the 100 watts is a commercial rating, for this would refer to the electrical input to the bulb and not to its output, which is more heat than light. Ordinary incandescent light bulbs are notoriously inefficient light producers.

Tables

Table 1: Physical Constants

Velocity of light in vacuum	$\begin{cases} = 3.00 \times 10^8 \text{ meters per second} \\ = 186{,}000 \text{ miles per second} \end{cases}$
Velocity of sound in air at 0°C	$\begin{cases} = 331.4 \text{ meters per second} \\ = 1087 \text{ feet per second} \end{cases}$
Acceleration of gravity	$\begin{cases} = 9.80 \text{ meters per second per second} \\ = 32.15 \text{ feet per second per second} \end{cases}$
1 standard atmosphere	$\begin{cases} = 76 \text{ centimeters of mercury} \\ = 1.013 \times 10^5 \text{ newtons per square meter (pascals)} \\ = 14.70 \text{ pounds per square inch} \end{cases}$
1 horsepower	$\begin{cases} = 550 \text{ foot-pounds per second} \\ = 746 \text{ watts} \end{cases}$
1 faraday	= 96,500 coulombs
Avogadro's number	$= 6.022 \times 10^{23}$ molecules per mole
Mechanical equivalent of heat	= 4.186 joules per calorie
Absolute zero	= −273.18°C
Electron charge	$= 1.602 \times 10^{-19}$ coulomb
Electron charge-to-mass ratio	$= 1.7588 \times 10^{11}$ coulombs per kilogram
Proton-electron mass ratio	= 1836
Stefan-Boltzmann constant	$= 5.6705 \times 10^{-8}$ watt per meter^2 − °K^4
Boltzmann's constant	$= 1.381 \times 10^{-23}$ joule per degree Kelvin
Planck's constant	$= 6.626 \times 10^{-34}$ joule-second
Rydberg's constant	$\begin{cases} \text{Hydrogen: } R_H = 1.09678 \times 10^7 \text{ per meter} \\ \text{Deuterium: } R_D = 1.09707 \times 10^7 \text{ per meter} \\ \text{Helium: } R_{He} = 1.09722 \times 10^7 \text{ per meter} \\ \text{Infinitely heavy nucleus: } R_\infty = 1.09737 \times 10^7 \text{ per meter} \end{cases}$
Permittivity of free space	$= (1/36\pi) \times 10^{-9}$ farad per meter
Permeability of free space	$= 4\pi \times 10^{-7}$ weber per ampere-meter
Heat of fusion of water at 0°C	= 79.6 calories per gram
Heat of vaporization of water at 100°C and 1 atmosphere pressure	= 539 calories per gram

Table II: Metric and English Equivalents

1 inch	= 2.540 centimeters (exactly)		1 centimeter	= 0.3937 inch
1 foot	= 30.48 centimeters (exactly)		1 meter	= 39.37 inches
1 mile	= 1.609 kilometers		1 kilometer	= 0.6214 mile
1 ounce	= 28.35 grams		1 gram	= 0.03527 ounce
1 pound	= 453.6 grams		1 kilogram	= 2.205 pounds
1 quart	= 0.9463 liter		1 liter	= 1.057 quarts

Table III: Metric Prefixes

Deka (da)	$= 10^1$		Deci (d)	$= 10^{-1}$
Hecto (h)	$= 10^2$		Centi (c)	$= 10^{-2}$
Kilo (k)	$= 10^3$		Milli (m)	$= 10^{-3}$
Mega (M)	$= 10^6$		Micro (μ)	$= 10^{-6}$
Giga (G)	$= 10^9$		Nano (n)	$= 10^{-9}$
Tera (T)	$= 10^{12}$		Pico (p)	$= 10^{-12}$
Peta (P)	$= 10^{15}$		Femto (f)	$= 10^{-15}$
Exa (E)	$= 10^{18}$		Atto (a)	$= 10^{-18}$

Table IV: Densities (in grams per cubic centimeter)

Solids				Liquids		Gases (at 0°C, 760 mm pressure)	
Aluminum	2.7	Glass, crown	2.5 –2.7	Alcohol, ethyl	0.81	Air	0.001293
Brass	8.47	Glass, flint	3.0 –3.6	Alcohol, methyl	0.79	Carbon dioxide	0.001977
Copper	8.96	Granite	2.7	Carbon tetrachloride	1.60	Helium	0.0001785
Gold	19.3	Marble	2.6 –2.8	Ether	0.74	Hydrogen	0.00008988
Iron, cast	7.20	Wood		Gasoline	0.68	Nitrogen	0.001251
Iron, wrought	7.80	Balsa	0.11–0.13	Mercury	13.6	Oxygen	0.001429
Lead	11.3	Cork	0.22–0.26	Olive oil	0.92		
Nickel	8.90	Ebony	1.11–1.33	Turpentine	0.87		
Platinum	21.5	Maple	0.62–0.75	Water, pure	1.00		
Silver	10.5	Oak	0.60–0.90	Water, sca	1.025		
Steel	7.86	Pine, white	0.35–0.50				
Zinc	7.14						

Table V: Young's Modulus (in newtons per square meter)

Aluminum	7.0×10^{10}
Brass	9.2×10^{10}
Copper	12×10^{10}
Phosphor bronze	12×10^{10}
Steel	19.2×10^{10}

Table VI: Coefficient of Linear Expansion (per degree Celsius)

Aluminum	0.000023	Lead	0.000029
Brass	0.000018	Nickel	0.000014
Copper	0.000016	Platinum	0.000009
German silver	0.000018	Silver	0.000019
Glass	0.000009	Steel	0.000011
Gold	0.000014	Tin	0.000027
Iron	0.000012	Zinc	0.000026

Table VII: Specific Heat (in calories per gram, per degree Celsius)

Aluminum	0.22	Mercury	0.033
Brass	0.092	Nickel	0.11
Copper	0.093	Steel	0.12
Glass	0.16	Tin	0.054
Iron	0.11	Water	1.00
Lead	0.031	Zinc	0.093

Table VIII: Resistivity of Metals (in ohm-meters)

Aluminum	2.6×10^{-8}	German silver	33.0×10^{-8}
Brass	$6.4–8.4 \times 10^{-8}$	Iron	10×10^{-8}
Constantan	44.1×10^{-8}	Manganin	44.0×10^{-8}
Copper	1.72×10^{-8}	Nichrome	100×10^{-8}
		Silver	1.63×10^{-8}

Table IX: Electrochemical Data

Element	Atomic mass	Valence	Electrochemical equivalent, g/C
Aluminum	27.1	3	0.0000936
Copper	63.6	2	0.0003294
Copper	63.6	1	0.0006588
Gold	197.2	3	0.0006812
Hydrogen	1.008	1	0.00001046
Iron	55.8	3	0.0001929
Iron	55.8	2	0.0002894
Lead	207.2	2	0.0010736
Nickel	58.7	2	0.0003040
Oxygen	16.0	2	0.00008291
Silver	107.9	1	0.0011180

Table X: Vapor Pressure of Water

Temperature, °C	Vapor pressure, mm of Hg	Temperature, °C	Vapor pressure, mm of Hg
0	4.58	55	118.0
5	6.54	60	149.4
10	9.21	65	187.5
15	12.79	70	233.7
20	17.54	75	289.1
25	23.76	80	355.1
30	31.82	85	433.6
35	42.18	90	525.8
40	55.32	95	633.9
45	71.88	100	760.0
50	92.51		

Table XI: Dielectric Constant and Strength

Material	Dielectric constant	Dielectric strength*, kV/mm
Vacuum	1.000 00	∞
Air	1.000 54	0.8
Paper	3.5	14
Ruby mica	5.4	160
Amber	2.7	90
Porcelain	6.5	4
Fused quartz	3.8	8
Pyrex glass	4.5	13
Polyethylene	2.3	50
Polystyrene	2.6	25
Teflon	2.1	60
Neoprene	6.9	12
Pyranol oil	4.5	12
Titanium dioxide	100	6

*This is the maximum potential gradient that may exist in the dielectric without the occurrence of electrical breakdown. Table from D. Halliday and R. Resnick, *Physics*, Third Edition, John Wiley & Sons, Inc., New York, 1978, p. 659.

Table XII: Work Functions (in volts)

Aluminum	4.28
Copper	4.65
Gold	5.10
Iron	4.50
Lead	4.25
Molybdenum	4.60
Nickel	5.15
Platinum	5.65
Silver	4.63
Tantalum	4.25
Tungsten	4.55
Zinc	4.90

Table XIII: Index of Refraction for $\lambda = 5890$ Å

Water	1.33	Fused quartz	1.46
Ethyl alcohol	1.36	Glass, crown	1.52
Carbon disulfide	1.63	Glass, dense flint	1.66
Air (1 atm and 20°C)	1.0003	Sodium chloride	1.53
Methylene iodide	1.74		

Table from D. Halliday and R. Resnick, *Physics*, Third Edition, John Wiley & Sons, Inc., New York, 1978, p. 939.

Table XIV: Standard Wire Sizes (resistance for standard copper wire)

A.W.G. No.	Diameter, in. $\times 10^{-3}$	Resistance of 1000 f, Ω at 25°C
12	80.8	1.619
14	64.1	2.575
16	50.8	4.094
18	40.3	6.510
20	32.0	10.35
22	25.3	16.46
24	20.1	26.17
26	15.9	41.62
28	12.6	66.17
30	10.0	105.2

Table XV: Resistor Color Code

Black	= 0	Blue	= 6
Brown	= 1	Violet	= 7
Red	= 2	Gray	= 8
Orange	= 3	White	= 9
Yellow	= 4	Silver	= 10% tolerance
Green	= 5	Gold	= 5% tolerance